Urban Regeneration in Australia

Drawing together leading urban academics, this book provides the first detailed and cohesive exploration of contemporary urban regeneration in Australian cities. It explores the multiple aspects and processes of regeneration, including planning policy (strategic and regulatory), development financing, sustainability, remediation and transport.

The book puts forward a unique and innovative 'scaled' analysis of urban regeneration, which positions urban regeneration as more than just large-scale redevelopment projects. It examines the processes of urban change which occur outside inner suburbs, which contribute to regenerating the city as a whole. The book moves beyond the planning and economic considerations of the regeneration process to describe the social and cultural aspects of regeneration. In doing so, it focuses on the management of higher-density environments, culture as a trigger for regeneration, and community opposition to the regeneration process.

Urban Regeneration in Australia would benefit academics, students and professionals of urban geography and planning, as well as those with a particular interest in Australian urbanism.

Kristian Ruming is Associate Professor of Urban Geography at the Department of Geography and Planning, Macquarie University, Australia. His main research interests include urban regeneration and renewal, affordable and social housing provision, urban governance, strategic planning, and planning system reform. Before joining Macquarie University, he was a Research Fellow at the City Future Research Centre, University of New South Wales, Australia.

Urban Regeneration in Australia

Policies, Processes and Projects of Contemporary Urban Change

Edited by Kristian Ruming

LONDON AND NEW YORK

First published 2018
by Routledge
2 Park Square, Milton Park, Abingdon, Oxon OX14 4RN

and by Routledge
605 Third Avenue, New York, NY 10017

First issued in paperback 2022

Routledge is an imprint of the Taylor & Francis Group, an informa business

Publisher's Note
The publisher has gone to great lengths to ensure the quality of this reprint but points out that some imperfections in the original copies may be apparent.

British Library Cataloguing-in-Publication Data
A catalogue record for this book is available from the British Library

Library of Congress Cataloging-in-Publication Data
Names: Ruming, Kristian, editor.
Title: Urban regeneration in Australia : policies, processes and projects of contemporary urban change / edited by Kristian Ruming.
Description: New York : Routledge, 2018. | Includes bibliographical references and index.
Identifiers: LCCN 2017043380| ISBN 9781472471635 (hardback) | ISBN 9781315548722 (ebook)
Subjects: LCSH: Urban renewal--Australia. | Sociology, Urban--Australia.
Classification: LCC HT178.A8 U69 2018 | DDC 307.3/4160994--dc23
LC record available at https://lccn.loc.gov/2017043380

ISBN 13: 978-1-03-240199-7 (pbk)
ISBN 13: 978-1-4724-7163-5 (hbk)
ISBN 13: 978-1-315-54872-2 (ebk)

DOI: 10.4324/9781315548722

Typeset in Sabon
by HWA Text and Data Management, London

Contents

Figures

Tables

Contributors

Hazel Easthope is a UNSW Scientia Fellow at the City Futures Research Centre, UNSW Sydney, Australia. With a background in sociology and human geography, she has built up a strong research track record in urban studies. She has a particular interest in residential decision-making, and the politics and practice of apartment living.

Robert Freestone is Professor of Planning in the Faculty of Built Environment at UNSW Sydney, Australia. His research interests include planning history, urban development, metropolitan planning, and planning education. His books include *Urban Nation* (2010), *Designing Australia's Cities* (2007) and *Model Communities* (1989). His recent co-edited books include *Planning Metropolitan Australia* (2018), *Dialogues in Urban and Regional Planning* 6 (2017), *Place and Placelessness Revisited* (2016), and *Exhibitions and the Development of Modern Planning Culture* (2014).

Nathalie Gentle graduated from the University of Newcastle, Australia, with Bachelor of Development Studies, Honours and University Medal in 2013. She is currently employed by the Benevolent Society, working in community development and engagement in Logan City, Queensland. Nathalie is a member of the Logan Financial Literacy Action Group, and is actively involved in a collaborative research project focusing on identifying and mapping the enablers and barriers to financial wellbeing in Logan City. Previously, she worked as a research assistant at the Centre for Urban and Regional Studies, UON and the Institute for Social Science Research, UQ.

Stephen Glackin is a Senior Research Fellow at Swinburne University, Australia, where his focus is currently on geographical information systems, complex data and optimising the regeneration of urban environments.

Nicole Gurran is Professor of Urban and Regional Planning at the University of Sydney, Australia, where she leads the university's Urban Housing Lab research incubator and Australian Housing and Urban Research Institute node. She has authored numerous publications on land-use planning, housing and the environment in Australia, including *Australian Urban Land Use Planning; Policy, Principles and Practice* (2011, in its second edition). Nicole is also a professionally qualified planner and Fellow of the Planning Institute of Australia and serves as Practice Editor of the journal *Urban Policy and Research,* published by Taylor and Francis.

Mike Harris is a landscape architect and urban design researcher, teacher and practitioner. He is a lecturer in landscape architecture at UNSW, Australia, and is currently undertaking a PhD on mixed-use megaprojects in Australia and Denmark through the University of Sydney, Australia and the Architecture School, KADK in Copenhagen, Denmark. His PhD research investigates how the aims mixed-use megaprojects with an explicit narrative of global economic competitiveness are reconciled with the delivery of strategic infrastructure, liveability goals, local identity and social equity.

Phil Heywood has been associated with planning in Queensland for more than thirty years, both as head of the Department of Planning and Landscape Architecture, QUT, Australia, and subsequently as a community planner, advocate and activist. A Life Fellow and previous President of the Queensland PIA, he is author of *Community Planning: Integrating Social and Physical Environments* (2011); *The Emerging Social Metropolis (*1997*);* and *Planning and Human Need* (1974) as well as numerous articles on social planning, community governance, public participation and planning for truly independent societies. He remains a Visiting Associate with QUT, supervising research projects and teaching planning processes and practice.

Crystal Legacy is a Senior Lecturer in Urban Planning at the University of Melbourne, Australia. Crystal's research focuses on questions of urban conflict, the post-political city, citizen engagement, urban politics and infrastructure implementation. Her current research examines the politics of urban transportation planning with a specific focus on the role of the citizen in contested transport processes in Australian and Canadian cities. She is the co-editor of *Building Inclusive Cities: Women's Safety and the Right to the City* (2013) and of *Instruments of Planning: Tensions and Challenge for more Equitable and Sustainable Cities* (2015).

Paul J. Maginn is Associate Professor and Programme Co-ordinator of the Masters in Urban and Regional Planning at the University of Western Australia. He is co-editor of *(Sub)Urban Sexscapes: Geographies and Regulation of the Sex Industry* (2014) and *Planning Australia: An Overview of Urban and Regional Planning* (2007).

Pauline McGuirk joined UOW, Australia, as Professor of Human Geography in 2016. Previously, she was Director of the Centre for Urban and Regional Studies at UON, Australia. As an urban political geographer, her research focuses on geographies of urban governance and unpacking the emergence of new practices, alliances, sites and scales of governance and their politics. Her three current projects investigate diverse practices of urban regeneration; the material means and social practices through which energy transition governance is playing out through urban spaces; and the governance processes and practices involved in making cities smart in Australia. She is an editor of *Progress in Human Geography.*

Kathy Mee is an Associate Professor in the Discipline of Geography and Environmental Studies and the Centre for Urban and Regional Studies at the University of Newcastle, Australia. She is a cultural geographer with expertise in the areas of housing studies, urban and regional transition, everyday practices, belonging, and stigmatisation. Her research explores three major themes (i) the changing nature of social vulnerability in urban and regional areas, (ii) housing and transition for socially vulnerable groups, and (iii) the diverse workings and practices of urban regeneration.

Peter Newton, FASSA, is a Research Professor in Sustainable Urbanism at Swinburne University of Technology in Melbourne, Australia, where his research interests focus on sustainable built environment planning and design, low-carbon living and urban transition. He is involved in the CRCs for Low-Carbon Living, Spatial Information and Water Sensitive Cities and is on the Board of AURIN. Prior to joining Swinburne he was a chief research scientist at the CSIRO. His most recent books include: *Resilient Sustainable Cities* (2014), *Urban Consumption* (2011), *Technology, Design and Process Innovation in the Built Environment* (2009) and *Transitions. Pathways Towards More Sustainable Urban Development* (2008).

Hal Pawson joined UNSW, Australia, in 2011 as a Professor of Housing Research and Policy at the City Futures Research Centre. His key interests include the governance and management of social housing, private rental markets and urban renewal. Formerly a researcher at Edinburgh's Heriot-Watt University, UK (1995–2011), Hal's prolific record of academic publication includes numerous journal papers and three co-authored books – most recently *After Council Housing: Britain's New Social Landlords* (2010). He is a member of the Australasian Housing Institute, a Fellow of Chartered Institute of Housing, and Managing Editor (Australasia) for the leading international journal *Housing Studies.*

Peter Phibbs is the Chair of Urban Planning and Policy at the University of Sydney, Australia, and Director of the Henry Halloran Research Trust at the same university.

Simon Pinnegar is Associate Professor of City Planning and Associate Director of the City Futures Research Centre in the Faculty of Built Environment at UNSW Sydney, Australia. His main research interests include strategic spatial planning, urban renewal and regeneration, and questions of fairness and justness in cities. From 2013 to 2017, he served as Program Director of the UNSW Planning Program. Prior to joining UNSW in 2005, he was a senior research analyst at the Office of the Deputy Prime Minister in the UK.

Bill Randolph is Professor and Director of the City Futures Research Centre at UNSW, Australia. Bill has 35 years' experience as a researcher on housing and urban policy issues in the academic, government, non-government and private sectors in both the UK and Australia. His current research focuses on housing markets and policy, urban inequality, urban renewal, high density housing and affordable housing.

Kristian Ruming is Associate Professor of Urban Geography at the Department of Geography and Planning, Macquarie University, Australia. His main research interests include urban regeneration and renewal, affordable and social housing provision, urban governance, strategic planning, and planning system reform. Before joining Macquarie University, he was a research fellow at the City Future Research Centre, UNSW, Australia.

Justine Saint Hilaire is an engineering student engaged in a dual master degree in both Denmark and France. Her major is in industrial engineering management. Justine is particularly interested in the sustainability of cities and supply chains. In 2016 Justine joined the Institute for Sustainable Futures at the University of Technology Sydney, Australia, for an internship. During this time she assisted in a number of research projects on urban regeneration and water efficiency with most of her research concentrating on interviews and research associated with One Central Park.

Glen Searle is an Adjunct Associate Professor of Urban Planning at the University of Sydney, Australia, and at the University of Queensland, Australia. He was previously Director of the Planning Programs at UQ and the University of Technology, Sydney, Australia. Before that he had research and policy positions with the NSW Department of Decentralisation and Development, the NSW Treasury and the UK Department of the Environment. He was manager (policy) in the NSW Department of Planning, where he was in the team that produced the 1988 Sydney Metropolitan Strategy. He is the author of *Sydney as a Global City* (1996).

Kate Shaw is a critical urban geographer at the University of Melbourne, Australia. She is interested in the cultures of cities and the political-economic and social processes that shape them, in particular, urban planning and policy practices and their capacity to deliver social equity and cultural diversity. She is co-editor with Libby Porter of *Whose Urban Renaissance? An International Comparison of Urban Regeneration Policies* (2009) and author of the forthcoming *Squander and Salvage of Global Urban Waterfronts*, which looks at the redevelopment of the deindustrialised docks and harbours of London, Melbourne, Sydney, Toronto, Vancouver, Berlin, Hamburg and Copenhagen.

Jill Sweeney is a Lecturer in the Discipline of Geography and Environmental Studies at the University of Newcastle, Australia. She has previous research experience in the fields of post-human and marine geographies and is currently involved in research into processes of urban regeneration, media studies, and placemaking.

Elizabeth Taylor is a Vice Chancellor's Post-Doctoral Research Fellow in the Centre for Urban Research (CUR). She was previously a McKenzie Fellow in the Faculty of Architecture Building and Planning at the University of Melbourne, Australia. Her interests are in policy-focused research across urban planning, housing markets, property rights and locational conflict and her research often makes use of geographical information systems (GIS). An increasing research focus is car-parking policy.

Andrea Turner is a chartered civil engineer with a postgraduate degree in environmental engineering. She has over 25 years' engineering and research experience having worked in the UK, HK and Australia. Since joining the Institute for Sustainable Futures at the University of Technology Sydney, Australia, Andrea has managed and been the key researcher on a wide range of water, energy and solid waste research projects. These projects have concentrated on resource efficiency and best practice decision-making for a broad range of clients in Australia and overseas. Andrea has presented the findings of her research and provided workshops on best practice water management nationally and internationally. Andrea is currently a committee member on both the Australian and International Water Associations.

Stuart White is Professor and Director of the Institute for Sustainable Futures at the University of Technology Sydney, Australia, and has over 30 years' experience in sustainability research. His work focuses on achieving sustainability outcomes for a range of government, industry and community clients in Australia and internationally including both the design and assessment of programs for improving decision-making and resource use. Stuart has written and presented widely on sustainable futures and is a regular commentator on sustainability issues. He is a member of the Board of the Australian Alliance for Energy Productivity, a member of the Strategic Council of the International Water Association and in 2012 he was awarded the Australian Museum Eureka Prize for Environmental Research.

Ilan Wiesel is an urban geographer at the School of Geography, University of Melbourne, Australia. His main research interest is in social inequality in cities, with a particular focus on the housing and neighbourhood experiences of diverse social groups, including social-housing tenants, people with disability and the super-rich. Before joining the University of Melbourne in May 2016, Ilan was a senior research fellow at the University of New South Wales, Australia (2009–2016). He has a PhD in urban planning (Melbourne University, Australia, 2009) and an MA in geography (Tel Aviv University, Israel, 2005).

Acknowledgements

For the past 15 years I have been conducting research into issues surrounding urban regeneration. Most of this research has been conducted with friends and colleagues from universities across Australia. This book emerges from this collaborative endeavour. Many of the ideas and topics explored in this book are the product of various research projects and conversations held throughout the years. Special thanks go to Ray Bunker, Nicole Gurran, Donna Houston, Paul Maginn, Pauline McGuirk, Kathy Mee, Simon Pinnegar, and Bill Randolph. I look forward to many collaborative research adventures in the future.

All chapters have been double peer-reviewed. My thanks go to the referees who so generously gave their time and expertise to review and improve the chapters: Paul Burton, Sebastien Darchen, Michael Darcy, Hazel Easthope, Sharon Fingland, Robert Freestone, Robin Goodman, Shane Greive, Phil Heywood, Donna Houston, Louise Johnson, Crystal Legacy, Heather MacDonald, Paul Maginn, Tony Matthews, Pauline McGuirk, Phil McManus, Kathy Mee, Trivess Moore, Clare Mouat, Peter Newton, Phillip O'Neill, Peter Phibbs, Simon Pinnegar, Dallas Rogers, Stephen Rowley, Glen Searle, and Ilan Wiesel.

I am indebted to Liz Shoostovian for her excellent copy editing.

Finally, my thanks go to Julie, Darcy and Fraser for their continued love and support.

1 Urban regeneration and the Australian city

Kristian Ruming

Introduction

In the face of mounting pressures as diverse as population growth, economic restructuring, housing supply and affordability concerns, climate change, resource depletion and environmental crises, social polarisation, and shortfalls in transport and other infrastructure, urban regeneration has emerged as a central urban policy across Australian cities (Newton and Thompson 2017). It is likely that this emphasis will remain for decades to come. Yet, urban regeneration is far from a simple process. While the state might view regeneration as a process which delivers economic growth and performance, concerns around social equity emerge as it reconfigures the city, potentially creating winners and losers (Shaw and Porter 2009). The reliance on the private sector to deliver urban regeneration gives it a significant amount of power in determining the form and function of the city. As a result, social polarisation and inequality can be exacerbated as profitable locations across cities are regenerated, often forcing low- and moderate-income citizens out, while other parts remain largely ignored. Yet, urban regeneration also offers a set of opportunities and innovations, as new development addresses concerns around housing supply, infrastructure delivery and environmental sustainability. However, the capacity to realise these opportunities are often constrained by policy, governance, and funding arrangements, as well as wider market conditions. The purpose of the book is to examine the complex processes, tensions and challenges which surround the planning and delivery of urban regeneration. In doing so, the book provides analysis of the multiple scales and locations of regeneration which collectively contribute to the regeneration of Australian cities.

In examining urban regeneration across Australia, contributions to this collection explore two core themes. The first theme centres on the complex planning, governance and funding arrangements which frame urban regeneration. Contemporary urban regeneration projects are enacted within the context of neoliberal governance and financial arrangements which have dominated Australian cities for at least the last 30 years (Haughton and Mcmanus 2011; McGuirk and Dowling 2009; Cook and Ruming 2008;

Gleeson and Low 2000). Here the delivery of urban regeneration by the private sector emerges as a cost-effective solution for financially constrained governments, where principles of market efficiencies, entrepreneurialism, private property, and profit emerge as paramount (Evans and Jones 2008; Forster 2006). Thus, urban regeneration takes place within the context of market parameters and needs to meet the objectives of market actors who are increasingly tasked with delivering urban change (Pinnegar and Randolph 2012). Mirroring international trends, urban regeneration is no longer the task of the state alone, but is "contingent upon the outcomes of a much wider array of processes and actors" (Karadimitriou et al. 2013: 1). It is not solely driven by planning frameworks and decisions, but a product of wider economic processes, such as land ownership, project funding and finance, and taxation regimes (Bunker 2007).

Despite a continued reliance on the private sector, rather than the complete "hollowing out" of the state, the state has been reconfigured as a facilitator of urban change (Jones and Evans 2013). The emphasis is increasingly on how the state can direct or enable market and non-market stakeholders to address wider social and economic challenges. Facilitating development via funding arrangements or the planning system, is a core responsibility of the state. Thus, urban regeneration emerges as a "precarious mix of market, state planning and local communities" (Ruming et al. 2007: 10). It is here where the notion of the partnership has emerged as the dominant governance arrangement for large-scale redevelopment projects (de Magalhaes 2015; Karadimitriou et al. 2013; Ruming 2009).

The second theme is the re-scaling of urban regeneration. Typically, international literature and policy on urban regeneration has tended to focus on individual sites, primarily large brownfield sites or social housing estates (de Magalhaes 2015; Karadimitriou et al. 2013). While these sites are important, they represent just part of the urban regeneration processes occurring across cities. As such, contributions to this collection provide a wider analysis of urban regeneration, recognising the diverse forms, locations and scales of regeneration occurring across cities. The focus is on the regeneration of cities, not just individual sites. Here the regeneration processes occurring across middle and outer regions are equally important to the ongoing evolution of Australian cities as regeneration sites located in the inner city.

Defining urban regeneration

Urban regeneration is difficult to define. For example, Smith (2012) argues it is often difficult to separate urban regeneration from wider urban policy and processes, while Tallon (2013: 4) characterises urban regeneration as a subset of urban policy, which "at the most general level ... has come to be associated with any development that is taking place in cities and towns". Along a similar theme, Colantonio and Dixon (2010: 55) suggest:

> urban regeneration conjures up different meanings for different people and can range from large-scale activities promoting economic growth through to neighbourhood interventions to improve the quality of life.

For de Magalhaes (2015), urban regeneration represents a "wicked issue", given the multiplicity of state and non-state actors involved, as well as the long planning and development timelines, which often span multiple electoral cycles. What is common to many definitions is a recognition of the multidimensional, complex and multiscalar processes it involves.

The difficulty surrounding definition is compounded by the multiplicity of terms which have been mobilised to explain and promote regeneration. In particular, renewal, revitalisation and renaissance have emerged as ideas which cover similar urban processes (de Magalhaes 2015; Leary and McCarthy 2013). These terms are often used interchangeably; however, each has a slightly different emphasis. For example, Lees (2003) suggests renewal reflects a form of urban intervention more likely to be led by the public sector, with an emphasis on addressing sites of social disadvantage. In contrast, regeneration is seen to emphasise economic growth and private development in response to limited public funding (Tallon 2013). For Leary and McCarthy (2013: 2) "renewal *tends* to mean physical approaches and regeneration more holistic responses" (emphasis in original). Shaw and Porter (2009: 3) illustrate how the notion of renaissance has emerged as a central objective of neoliberal urban policy advocating for "well-being, creativity, vitality and wealth" across cities. The use of multiple terms has led to some confusion. However, at the foundation of all of these terms is a desire to reconfigure the form and operation of cities in response to a series of social, economic and environmental challenges.

Internationally, urban regeneration is often mobilised as a policy intervention designed to address the needs of blighted communities and places (Bunce 2004; Furbey 1999). Evans (2005: 967) defines urban regeneration as:

> the transformation of a place – residential, commercial or open space – that has displayed the symptoms of physical, social and/or economic decline.

While for Karadimitriou et al. (2013: 9) urban regeneration is:

> concerned with dealing with the physical, social and economic consequences of urban change in the context of a retreating welfare state.

The focus of such interventions is often on poor people and poor places, where "inward looking" regeneration projects can address local problems (Matthews 2010). This form of urban regeneration typically adopts an area-based approach as a means of focusing limited resources and addressing complex issues facing disadvantaged communities. Historically, urban regeneration projects have been property-led, where the physical transformation is the principal goal. This interpretation of urban regeneration often advocates for a top-down, project-based model of development, most easily enacted via new developments on inner city brownfield sites and large social housing estates (Tallon 2013). Thus, urban regeneration projects which "entail[ed] higher density, mixed-use developments on brownfield sites'" emerged as the dominant form of development (Winston 2010: 1787).

Evans and Jones (2008: 1418) argue that the regeneration agenda has been adopted by many cities as a "magic bullet for delivering 'better' cities". In this context, regeneration is positioned as a process capable of "fixing" urban problems, be they abandoned urban land or disadvantaged communities. Urban regeneration has emerged as a normative concept which promotes a vision of what should happen to the city, or locations within it, and how the vision should be achieved (Smith 2012). However, this understanding downplays the diversity of objectives expressed by the public sector, private sector and urban communities. It is unlikely that any urban regeneration vision will deliver the outcomes desired by all stakeholders, resulting in tensions surrounding what a "better city" might look like (de Magalhaes 2015). Thus, some have argued that urban regeneration has emerged as a lever of neoliberal and entrepreneurial urban governance where the priorities of private sector/markets are given precedence over other concerns (Leary and McCarthy 2013). Further, place-based, property-led forms of urban regeneration have been criticised due to their failure to address the wider societal and economic processes which led to the need for urban regeneration in the first place (de Magalhaes 2015).

In response to some of these critiques, there has been somewhat of a retreat from market-led/property-led definitions and interventions, which are seen to place too much emphasis on the configuration of the built environment. Increasingly regeneration schemes are accompanied by place-based social interventions able to "rehabilitate" environments and communities (Charmes and Keil 2015). Thus, the definition provided by Roberts and Sykes (2000: 17) recognises the diverse processes and impacts of urban change, where urban regeneration is seen as a:

> comprehensive and integrated vision and action that leads to the resolution of urban problems and which seeks to bring about a lasting improvement in the economic, physical, social and environmental condition of an area that has been, or is subject to change.

For Roberts and Sykes (2000), the movement away from property-led regeneration in the UK has seen the emergence of three new pillars, represented by the concepts of sustainability (economic, social and environmental), strategic visions and partnership. This interpretation is more holistic, with notions such as partnership, integration, competition, empowerment and sustainability becoming increasingly important. Many of the chapters in this book map how these ideas are enacted across Australian cities.

For the purpose of this book, a broad definition of urban regeneration (and renewal in some chapters) is adopted. Here urban regeneration is considered any process which reconfigures the city and contributes to its ongoing evolution. This definition recognises urban regeneration as a process occurring simultaneously at the local, project scale, and at a wider urban scale. This approach recognises that the regeneration of an individual site changes the nature of the city. Regeneration projects (no matter what their size) should not be viewed in isolation, but collectively as drivers of city-wide change. With some exceptions (Baxter and Lees 2008), most urban regeneration research fails to grasp the connections between official, large-scale regeneration and the regenerative capacity of less-recognised sites, such as middle-ring infill development and greenfield suburban developments. Thus, in this book, urban regeneration is not confined to large-scale brownfield projects; rather it is a multiscalar process which acknowledges the vital role that small and medium-sized projects, such as knock-down and rebuild or shopping centre-led regeneration, play in reconfiguring the city. The city emerges as a mosaic of regeneration projects and processes, be they initiated by the state or private sector, or a partnership between them. Each chapter in this book offers a detailed exploration of one or more forms of urban regeneration. Yet collectively they present an overarching picture of the complex, multiscale, contradictory and controversial processes of regeneration occurring across Australian cities.

Australian urban regeneration: an urban mosaic

In contrast to urban regeneration in the UK or the USA, Australian urban regeneration projects tend to be driven by capitalising on the commercial or strategic value of land rather than issues of poverty and area-base disadvantage. Typically, policy, public and academic debate around urban regeneration has coalesced around two primary types: high-value brownfield sites and large social housing estates. However, focusing only on these sites ignores the diversity of regeneration processes occurring across the city. Beyond large post-industrial sites and social housing estates, a new wave of urban regeneration has emerged across the middle and outer parts of Australian cities. This new form of regeneration comes in response to the ageing of Australian suburbia, as suburbs emerge as viable redevelopment opportunities. Across these diverse geographies of regeneration, complex planning, governance and funding arrangements emerge to shape and direct regeneration. In order to examine the complex urban regeneration process occurring across Australian cities, this book is divided into three parts.

Part I: Planning and funding urban regeneration

Beyond the occasional foray into urban policy by the Commonwealth Government – such as the Department of Urban and Regional Development under the Whitlam government (Oakley 2004), the Better Cities Program under the Hawke/Keating government (Neilson, 2008) or the City Deals program recently introduced by the Turnbull government (AHURI 2017) – it has been largely left to state and local governments to manage urban regeneration. The planning system has emerged as a central mechanism for governments to promote and direct urban regeneration across our cities. In Chapter 2, Ruming reviews contemporary metropolitan strategies across all state capital cities. Across the strategic plans, ideals such as urban consolidation or the compact, polycentric or 30-minute city emerge as central ambitions. The emphasis on infill development, densification, activity centres, nodes, transport corridors and transit oriented development (TOD) promotes a new form of Australian urbanism which comes in direct opposition to historical dominance of fringe development (Newton and Thompson 2017; Dodson 2010). Concentrating development within designated centres emerges as a panacea to many of the social, economic and environmental challenges facing cities. These strategies aim to set the framework for urban regeneration across Australian cities, identifying a hierarchy of regeneration sites, ranging from large-scale brownfield regeneration sites to small local centres.

However, urban scholars have highlighted the inability of state governments to implement these visions on the ground, with the link between metropolitan strategic planning and local development processes historically weak (Bunker et al. 2005). As such, it is through a series of regulatory measures that state and local governments seek to manage urban regeneration. At the highest level, large regeneration projects are often assessed under special-purpose planning legislation which removes planning authority from local government. The planning conditions (e.g. development density and height) are often in excess of those placed on surrounding development activity. These developments can emerge as sites of conflict as opponents challenge the degree to which private sector interests are supported at the expense of wider public benefit (Vigneswaran et al. 2017) or concerns over the assessment process itself and claims of undue process (Ruming and Gurran 2014). Other critiques emerge around the size and design of developments and their interrelationship with the existing urban form. State governments have also introduced a series of legislative arrangements which enable the operation of state land development authorities. This legislation enables these agencies to take a lead role in planning and developing major regeneration projects, a process examined in depth by Gurran and Phibbs (Chapter 3) and Searle (Chapter 4).

Beyond the use of special-purpose legislation for large regeneration sites, state governments have promoted regeneration by replacing local government planning policies. The most common is the introduction of policies which override local zoning in efforts to increase allowable housing densities and trigger private-sector development. Even in cases where state governments have not intervened, the process of "up-zoning" emerges as a regulatory mechanism for local governments seeking to promote regeneration. Up-zoning involves changing land-use zoning in order to increase the permissible size of development. Increases in zoning allow landowners/developers to construct more units on their property, thereby improving the financial viability of projects. However, as a tool for promoting regeneration, changing zoning is a crude lever at best, with no guarantee that development of the desired size or style will be delivered within the context of a wider housing market. The process of re-zoning to promote regeneration has also been opposed by local residents and community advocates. The re-zoning decision is often seen to come in response to private-sector pressures to maximise yield and profit, while limiting public benefit and reconfiguring the local built environment (van den Nouwelant et al. 2015; Ruming et al. 2012). Both Gurran and Phibbs (Chapter 3) and Searle (Chapter 4) provide a detailed examination of the re-zoning process as a means of facilitating and funding regeneration.

Although regulatory frameworks have been used to promote regeneration, the capacity of the planning system to deliver wider public benefits from regeneration has been questioned. One example is the limited ability of the state to capture some of the value of the re-zoning and development-assessment process. Funds collected via such a mechanism could be reinvested into the city in the form of infrastructure, open/public space or affordable housing provision. While value capture has been championed as a means of addressing the infrastructure backlog facing some parts of Australian cities, it has not been widely implemented and represents an opportunity lost to provide wider benefits from urban regeneration (Gurran and Phibbs 2013). Likewise, the reluctance of governments to enforce inclusionary zoning or mandated affordable housing targets within urban regeneration projects remains an issue. Australian cities have experienced significant house price inflation over the past decade, with large parts of our cities becoming unaffordable for low- and moderate-income households (Gurran and Whitehead 2011). Inclusionary zoning – where new developments are required to include a proportion of dwellings available for sale or rent at prices affordable for low- and moderate-income households – represents one way of addressing inequality issues which arise as part of the regeneration process. Importantly, a number of large regeneration projects have begun to include affordable housing (Chapter 10). Likewise, state development corporations have begun to include affordable and social housing dwelling targets, usually set between five and ten per cent of dwellings. On redeveloped social housing estates, the number of affordable and social dwellings is likely to be larger, however, the final number is often balanced against concerns around financial viability and project marketing (Chapter 15). Finally, some local councils have set affordable housing targets for developments in their area. While these are important steps in ensuring low- and middle-income households have a place in the regenerated city, they fail to go far enough, falling well below targets set in the UK or the USA, while implementation remains a challenge as targets are often negotiated down by the private sector in order to maintain the economic viability of projects (Gurran and Whitehead 2011).

It is vital to recognise the diverse funding arrangements used to facilitate regeneration projects. At one end of the spectrum are large privately funded brownfield regeneration projects with budgets into the hundreds of millions of dollars. At the other end are small privately funded regeneration projects of the knockdown-rebuild variety reliant on personal savings or mortgage financing. In between, lies a complex array of public–private partnerships and co-funded projects. The partnerships often draw together various state government agencies, private developers, builders and planning consultancies, and private financiers (Chapters 3, 4 and 15; Davison 2011). In the case of social housing estates, community housing providers, as future tenancy managers, are also included. For social housing estates, public–private partnerships are often formalised following a public tender process, where competing development consortia design and bid for a regeneration project announced by the state (Chapter 15; Rogers 2013). While development parameters are often set by the state, such as the percentage of social housing to be provided, the exact design and configuration of the regeneration project is left to the private sector. For projects led by state development agencies, the planning and management of the regeneration is often undertaken by the development agency itself, while development occurs in partnership with the private sector (Davison and Legacy 2014; Chapter 3). However, challenges arise over the capacity of state development corporations to negotiate development conditions with the private sector, especially when it comes to the delivery of wider public benefits (Davison 2011). The requirement that development corporations provide a return/profit to the government also acts as a potential inhibitor to innovation and the inclusion of socially just elements. Thus, development corporations are constrained by the dominant neoliberal governance model that requires state agencies to align with market principles, limit public expenditure and generate a profit (Davison and Legacy 2014; Davison 2011). Concerns of conflict of interest can also emerge around the role of state development agencies in delivering urban change given that the state is simultaneously the development proponent (the development agency) and the assessment authority (such as the Department of Planning, or minister). In an attempt to overcome this challenge, “independent” planning panels or commissions have been introduced by state governments to assess projects (Ruming and Gurran 2014; Chapter 9).

One of the most significant barriers to regeneration is lot amalgamation. The inability to combine land holdings to take advantage of zoning conditions which allow for higher densities or new land uses is a major impediment to achieving the objectives set out in strategic planning documents. This is particularly the case for strata buildings – buildings with multiple private ownership managed under strata legislation (legislation allowing individual ownership of part of a property) – where the inability to get buildings into single ownership or get owners to agree to redevelop is a significant challenge for regeneration (Pinnegar and Randolph 2012). In Chapter 5, Easthope and Randolph examine in detail the challenges strata legislation/ownership pose for urban regeneration. For Easthope and Randolph, not only does strata legislation act as a barrier to lot consolidation, so too does strata ownership play a significant role in mediating the way residents experience the city. Strata ownership raises concerns of build quality, defects and maintenance issues. Strata ownership promotes a new form of micro governance as owners are required to engage with body corporates responsible for managing shared assets. While lot amalgamation remains a challenge for regeneration, recent years have seen an increase in landowners voluntarily coming together to sell large parcels of lands to developers, taking advantage of new zoning and favourable housing market conditions (e.g. Creer 2015). While such projects might result in regeneration in some parts of the city, the timing and location of such development is likely to be sporadic, while concerns around the provision of necessary infrastructure also emerge.

The planning, funding and governance issues outlined in Part I play out across Australian cities in uneven and diverse ways. Parts II and III of the book examine how these challenges manifest in particular types and locations of urban regeneration.

Part II: Inner-city regeneration

Part II examines the regeneration of the inner city. The focus is on some of Australia's largest brownfield projects. These sites are viewed as unique opportunities to revitalise inner city regions, promote new mixed-use developments and trigger cultural and tourism industries. Beyond these large brownfield sites, the inner city is also characterised by a diverse set of regeneration processes, including informal regeneration initiatives, smaller-scale precinct projects and large social housing estate regeneration projects.

Brownfields

Brownfield urban regeneration sites are redundant industrial or port lands typically located in the inner and middle regions of cities. Large-scale urban regeneration projects on these post-industrial sites have emerged as a response to economic blight, increased land values, demand for high-value residential and commercial property, and the desire for world-class entertainment and leisure precincts (Smith and Garcia Ferrari 2012). Brownfield regeneration sites are becoming key post-industrial consumption spaces, characterised by a mix of (pseudo) public space, retail and lifestyle elements, and tourist and entertainment attractions (Oakley and Johnson 2013; Oakley 2009; Dovey 2005; McGuirk et al. 1996). Brownfield regeneration projects are now central to the vision for Australian cities as outlined in metropolitan strategic plans (Chapter 2), which position these sites as central to promoting the compact city (Bunker and Searle 2009) and core to global city ambitions (Newton and Thompson 2017; Baker and Ruming 2015; McGuirk et al. 1996).

These sites are often in state ownership and are high-value real estate given their waterfront location or proximity to the central business distruct (CBD). This, in theory at least, gives the state the capacity to leverage state assets to deliver innovative and equitable development outcomes. Yet doubts have emerged over the extent to which brownfield regeneration sites provide wider public benefits, implying that the form of brownfield regeneration has tended to disproportionately benefit domestic and international urban elites at the expense of local communities (Smith and Garcia Ferrari 2012; Dovey 2005). Indeed, concerns around gentrification have been raised as projects displace local residents, while the provision of new expensive residential units and high-end leisure facilities represent an opportunity lost in terms of providing an equitable and accessible city (Shaw and Porter 2009).

Brownfield regeneration sites emerge as highly politicised projects, central to electoral strategies and mobilised by state governments to demonstrate their capacity to "deliver" urban change (Chapter 7). As such, these sites are likely to "win" internal government competition for resources and funding, raising concerns about the equitable distribution of resources across the city. Examples include the tendency for state governments to supply new transport infrastructure to support the viability of large regeneration projects. In instances where such support is not forthcoming, inner city regeneration is likely to stall (Ruming et al. 2016).

An important characteristic of brownfield regeneration is the role of key individuals in directing regeneration. These individuals often move between state government agencies and between the state and private sector over the life of a project (Chapters 6 and 7). On one hand, such movement could be viewed as a process of policy transfer (McCann and Ward 2011) as ideas, approaches and innovations circulate between projects. Such movement is not necessarily a bad thing as it potentially allows the circulation of "best practice". The movement of individuals between the public and private sectors is viewed by some as beneficial to project delivery as it allows state officials to appreciate the market realities of regeneration. However, on the other hand, this movement raises significant questions about the planning and development process with concerns around due process and claims of corruption emerging as individuals shift between public- and private-sector roles.

From a planning perspective, brownfield regeneration sites tend to be defined as "state significant", reflecting their perceived role in promoting economic growth and global status. This categorisation shifts planning and approval responsibility to the state government. These sites are increasingly planned and managed by new development agencies, or transferred to the private sector (Davison and Legacy 2014; Davison 2011). These development arrangements often result in a "death by a thousand cuts" as projects are repeatedly renegotiated over the planning and implementation process, to the point where the final redevelopment often bears little resemblance to the plan originally approved (Chapters 6 and 7). This process can exclude local councils and communities who feel sidelined and might oppose the size or scale of development. These processes, issues and conflicts emerge across the brownfield case studies explored in Part II.

In Chapter 6, Harris provides a detailed analysis of the planning and delivery of Baragaroo in Sydney. Harris unpacks the conflicts between key players at the core of the project which, over time, led to a radical reconfiguration of the site. Barangaroo emerges as an example of project creep (where the design of a development is constantly renegotiated), where the size and design of the project was progressively reconfigured to the benefit of commercial interests. Examples include increasing development heights, an increase in the amount of residential floor space and the introduction of new and controversial land uses, such as a casino. It is claimed that Barangaroo fails to maximise public benefit, as public space is restricted and opportunities for affordable housing are dismissed.

In Chapter 7, Shaw examines the planning and development of brownfield regeneration sites along the Yarra River in Melbourne. In her detailed historical analysis, Shaw reveals the highly political nature of brownfield regeneration projects as they emerge as central elements of a neoliberal state government's effort to facilitate economic growth. The Melbourne case study reveals how a lack of coherent and overarching vision results in a disjointed, inefficient and visually unappealing urban environment. The redevelopment of brownfield sites along the Yarra is the product of pragmatic and reactionary planning and development decisions made by state planning departments, ministers and development agencies. Like Barangaroo, the regeneration sites in Melbourne are characterised by project creep, where the planning vision shifts, progressively allowing more development. Shaw argues that the resultant developments fail to deliver wider public benefit with planning decisions allowing landowners and developers to reap largely unearned financial windfalls.

The politics of brownfield regeneration also emerge as central in Brisbane, as examined by Heywood in Chapter 8. The regeneration of a number of key brownfield sites along the Brisbane River are identified by the state government as essential in achieving a "New World City". As with the case of Sydney and Melbourne, urban regeneration emerges as a site of political contest, as state and local governments and politicians battle to reconfigure the inner city. Heywood examines two brownfield regeneration sites: North Bank/Queens Wharf and the Kurilpa Riverfront Renewal. These projects became sites of conflict as residents and professional bodies questioned project design and bulk; how they fit into the existing built form of the city; and, how they responded to environmental constraints, such as flooding. Again, the development of a casino acted as a point of opposition. While the Kurilpa Riverfront project has been dropped in response to public opposition, the Queens Wharf development continues to pursue the dominant form of global brownfield regeneration by pursuing tourism, gaming and high-value employment.

In Chapter 9, Maginn examines the push for large-scale urban regeneration projects across Perth. As with other cities, strategic planning for Perth promotes a compact urban form, where large-scale urban regeneration projects in the city centre are positioned as vital in creating a global city. In investigating a number of large regeneration projects across Perth (East Perth, Subi Centro, Perth Foreshore/Elizabeth Quay), Maginn provides examples of the central role of redevelopment authorities in planning, funding and delivering large regeneration projects. Despite a heavy reliance on the private sector to fund and develop projects, he argues that regeneration projects across Perth demonstrate an interventionist policy approach, especially compared to other states. Nevertheless, regeneration sites across Perth have been criticised on the grounds of size, design, and limited public consultation.

A more optimistic view emerges in Chapter 10 where White, Turner and Saint Hilaire, examine the potential sustainability benefits of large-scale brownfield regeneration projects. The Central Park project in Sydney is the focus. Here brownfield regeneration is positioned as a form of development which provides a series of opportunities and innovations for reconfiguring our cities in the face of growing environmental challenge. The Central Park project has emerged as the prototype for new environmentally sustainable brownfield regeneration. Beyond the development's signature green walls, the Central Park project offers valuable lessons in terms of heritage preservation, onsite water collection and electricity generation, and urban design. Yet the Central Park project also highlights a series of challenges and obstacles which need to be addressed in order to deliver large-scale sustainable projects, including new governance and funding mechanisms.

Gentle and McGuirk explore one of the key underlying assumptions of inner-city regeneration – that regeneration facilitates a form of urban creativity. Chapter 15 provides an analysis of culture-led urban regeneration and the complex assemblages which shape the inner city. Focusing on the regeneration of inner-city Newcastle, Gentle and McGuirk highlight the importance of community-based, temporary creative use of vacant inner-city properties as a means of reinvigorating a rundown urban centre. The focus is the *Renew Newcastle* movement which has emerged as a model for culture-led regeneration. The form of culture-led regeneration occurring in inner-city Newcastle emerges as a localised, informal regeneration activity which differs from the prevailing narrative around global creative capital which dominates urban regeneration policy and literature (Tallon 2013).

While it is clear that brownfield sites in the inner city play a significant role in the charging mosaic of Australian cities, the chapters in Part II present a cautionary tale highlighting the challenges and tensions which surround planning and implementation. Inner-city regeneration projects offer the possibility to deliver a more sustainable, efficient, economically comparative and equitable city. Yet, the case in Australia suggests that, to date, this opportunity has not been fully realised, as many inner-city regeneration projects have been compromised as wider public benefit has been downplayed in an attempt to meet the needs of private sector interests increasingly required to fund and deliver these projects.

Part III: Middle-ring and suburban regeneration

Part III examines regeneration processes occurring across the middle and outer rings of Australian cities. Contemporary middle-ring suburbs were developed as part of the first wave of Australian outward suburban expansion – they were our cities' first outer suburbs. Driven by the post-war ideal of the "quarter acre block" and private home ownership, many of these properties are now coming to the end of their lifespan and are ripe for redevelopment (Randolph and Freestone 2008). A number of authors argue that there is a need to acknowledge and support more incremental, ongoing processes of urban regeneration occurring across the middle regions of our cities (Pinnegar and Randolph 2012; Pinnegar et al. 2010). While some middle- and outer-ring locations are identified within metropolitan strategic plans as "centres", particularly those located within transport corridors, Pinnegar and Randolph (2012) argue that one of the major absences in urban regeneration policy relates to the suburban, private-sector housing renewal occurring outside designated centres. Despite the opportunities and, indeed, necessity for the regeneration of middle and outer suburban regions, a number of barriers remain, including the existing built form, complex patterns of ownership, complicated and disjointed planning legislation, and community expectations (Pinnegar and Randolph 2012). Part III examines five types of regeneration occurring in middle and outer regions: shopping centre-led regeneration, greyfield regeneration, knockdown and rebuild; social housing estate regeneration; and, new fringe development. Concerns around community opposition to regeneration are also examined.

Shopping centre-led regeneration

Town centre style regeneration, typically based on TOD principles, now represents the dominant planning and development model for middle-ring suburbs. Within the context of metropolitan strategic plans which promote a polycentric city (Chapter 2), local and regional shopping centres have emerged as key strategic centres. Internationally, research on retail-led regeneration has tended to focus on "main street" style regeneration, typical of the UK (Imrie and Dolton 2014). This form of regeneration has attempted to reinvigorate the commercial success of centres by drawing in new retail and commercial entities, improving the public domain and streetscapes, and redeveloping land for new residential purposes. However, less attention has been paid to the role of large-floor-space shopping centres/malls, more characteristic of the US or Australia. In the case of Australia, large-floor-space shopping centres represent the main form of retail activity in middle- and outer-ring locations. Shopping centres evolved as a product of urban sprawl as complexes were built alongside car-dependent new suburbs as the city expanded. These shopping centres now emerge as opportunities for middle-ring suburban regeneration (Goodman and Coiacetto 2012).

In Chapter 13, Ruming, Mee, McGuirk and Sweeney explore the role of shopping centre-led regeneration by examining Charlestown in the greater Newcastle region. As a middle-ring "regional" centre, Charlestown reveals simultaneously the opportunities and tensions of shopping centres as catalysts for regeneration. While shopping centres are often located within designated strategic centres, it is often the redevelopment of the shopping centre itself which emerges as the trigger for wider precinct master planning and regeneration. This allows local governments to negotiate development contributions to facilitate improvements to the public domain and local services. Nevertheless, the size and economic impact of large shopping centres places their owners in a relatively powerful negotiating position, often resulting in favourable contributions and infrastructure provision as part of the planning process, potentially having a negative impact on commercial premises located outside the shopping centre itself. Tensions also arise between local and state government agencies required to coordinate change, with the failure to deliver appropriate and timely public transport emerging as a central consideration. The redevelopment of shopping centres can reconfigure local planning strategies. This offers opportunities for increased allowable densities and introduces new mixed use zonings. This has seen shopping centre owners include residential components as part of shopping centre redevelopments, and has led to an increased housing supply surrounding shopping centres as other landowners take advantage of zoning uplift.

Greyfields

The notion of greyfield regeneration is a relatively new one in Australia (Newton 2010). Greyfield development refers to the regeneration of middle or established suburbs in large cities (Newton and Glackin 2014). Underlying the notion of greyfield regeneration is the notion of obsolescence and "economically outdated, failing or undercapitalised real estate assets" (Newton 2010: 81). The underlying objective of greyfield regeneration is to coordinate middle-ring regeneration at a larger spatial scale. More traditional forms of middle-ring regeneration have been characterised as often "poorly integrated in terms of street design, with low-quality visual amenity and open space provision" (Randolph and Freestone 2012: 2561). In response, it is claimed that there is a need to draw together fragmented forms of development to deliver coordinated, precinct-level regeneration (Newton 2010). The rationale behind greyfield regeneration is that increased density and coordinated precinct-wide development allows for better integration of new urban sustainability technologies, new forms of open and community space, and innovative urban design possibilities (Newton 2010).

Despite these claimed benefits, greyfield-style regeneration has been limited across Australian cities. A major barrier to this form of regeneration is lot amalgamation, with owners required to come together to sell or redevelop their properties. This form often delivers increased urban density, but the increase in yield may be insufficient to enable private, market-led greyfield regeneration. A second barrier to delivering greyfield regeneration are existing planning frameworks which potentially work to stymie precinct-style regeneration. In particular, concerns are raised about the ability of existing planning frameworks to facilitate mixed-use, higher-density developments in traditional suburban locations. The capacity of the planning system to assess a precinct of non-contiguous lots is also limited. In order to facilitate this change, new forms of urban governance, funding and planning models, as well as construction techniques are required. This position is advanced in Chapter 12 where Newton and Glackin make a case for the need for increased greyfield regeneration as a means of integrating new medium-density development across Australian suburbs. Focusing on Victoria, the central focus of this chapter is how the planning system can be reconfigured to identify potential sites of greyfield regeneration, and to support the design and implementation of such precincts.

Knockdown and rebuild

In contrast to a precinct-style form of regeneration advocated by greyfield regeneration, knockdown and rebuild (KDR) development has emerged as a form of ad hoc urban regeneration occurring across middle and outer suburbs. KDR regeneration, as the name suggests, is the process whereby one existing detached dwelling is demolished and replaced by a new detached dwelling. This form of regeneration often occurs outside designated centres and is largely driven by the housing and investment decisions of individual homeowners and purchasers. KDR offers an opportunity for existing homeowners to upgrade their home, while remaining in locations which are often well serviced or where owners have strong community connections (Wiesel et al. 2013a). For purchasers, KDR-style redevelopment offers a means of securing a property in a desirable location at a lower price, with the full intention of replacing the existing dwelling (Wiesel et al. 2013a). The KDR process regenerates suburbs by replacing aged and worn-out housing stock which no longer meet the needs of the population. The new stock has the potential to address the housing needs of new household types (e.g. multigenerational households) and provide an adaptable form of housing which might be easier to regenerate in the future. However, one of the emergent challenges of the KDR process is the loss of affordable housing stock as it is replaced by more expensive stock (Chapter 14), leading to a new form of suburban gentrification as low- and middle-income households are displaced.

KDR emerges as a form of regeneration which seeks to maintain the dominant form of Australian suburbia, the single detached dwelling. Yet, the form of suburbia is reconfigured as new dwellings are typically much larger than the homes replaced, prompting some to question the introduction of "McMansions" in existing suburbs (Wiesel et al. 2013b). While some communities might lament the evolution of their neighbourhood towards a new form of detached suburbanism, KDR is unlikely to experience the extent of community opposition associated with new higher-density developments (Legacy et al. 2013). KDR-style regeneration represents a challenge for wider urban policy on a number of fronts. First, KDR fails to significantly add to housing stock as regeneration typically results in a one-to-one replacement of existing properties. There is some evidence to suggest that some KDR redevelopment does add to supply via the construction of duplexes or granny flats, however, it is unlikely that this form of development will contribute a significant net increase in housing (Wiesel et al. 2013b; Chapter 14). Second, KDR represents a form of largely uncoordinated urban regeneration. The lack of coordination stems from the fact that redevelopment is initiated by individual home owners/purchasers. There are few levers available to state or local governments to influence the location and timing of this redevelopment.

Building on their extensive work on the KDR phenomenon, in Chapter 14 Pinnegar, Freestone and Wiesel position KDR as a form of incremental urbanism, progressively reconfiguring the suburban landscape. While strategic planning is often silent on KDR, focusing instead on the densification of designated centres, the authors identify a growing local government interest in KDR regeneration. The empirical focus of the chapter is the Hills Shire Council in northwest Sydney which has recently introduced a Knockdown Rebuild Landscape Contributions Scheme as a means of promoting KDR regeneration across the council area. KDR emerges as a form of regeneration supported by local government due to its potential to reinvigorate ageing housing stock across middle and outer suburbs.

Social housing estate regeneration

Of the varieties of urban regeneration occurring across Australia, the redevelopment of social housing estates most closely resembles those initiated in the UK and the USA that target place-based social and economic disadvantage. Historically, intervention in social housing estates has been in the form of property-led regeneration, where the redevelopment of a degraded housing stock is positioned as a solution to local disadvantage (Judd and Randolph 2006). From the early 1980s, state housing authorities allocated significant funds to estate regeneration programs (Hughes 2004). More recent social housing estate regeneration programs have tended to emphasise "community regeneration" which seeks to simultaneously address physical challenges and social and economic issues via education, employment and community initiatives. As noted above, within the context of severe financial constraints placed on social housing authorities, public–private partnerships have emerged as the principal arrangement for facilitating regeneration of social housing estates (Pinnegar et al. 2011; Chapter 15). Yet as a place-based response to concentrations of disadvantage, regeneration schemes fail to address the wider systemic causes of concentrated disadvantage, funding shortfalls or the impacts of displacement of the existing tenants.

A key element of the regeneration of Australian social housing estates, as it has been internationally, is the reconfiguration of social/tenure mix (Arthurson 2012). This strategy centres on the assumption that altering the concentration of social housing tenants, by increasing the number of private tenants will address the issues facing disadvantaged communities, such as high unemployment rates, low educational attainment and rundown physical environment. The assumption has been challenged by authors who argue social mix re-scales oppression (Ruming et al. 2004) or is socially unjust given the tendency to relocate disadvantaged communities (Arthurson 2012). Some have claimed that estate regeneration, particularly in inner and middle-ring suburbs, represents a form of gentrification which results in the displacement of disadvantaged communities (Shaw and Porter 2009). Recent regeneration projects have sought to overcome these criticisms by offering tenants the option to return to the regenerated estate. However, it remains largely untested as to whether tenants actually return, while the long-term impacts of displacement are unknown.

In Chapter 15, Pawson and Pinnegar provide a detailed exploration of the regeneration of social housing estates in Australia. The chapter begins with a detailed review of social housing policy, before examining the principles which underpin social housing estate regeneration. The chapter concludes with an in-depth analysis of the regeneration of the Bonnyrigg social housing estate in Sydney. The Bonnyrigg example reveals many of the challenges associated with the delivery of estate regeneration, in particular the governance and financial arrangements framing redevelopment.

New suburban development

Urban development on the fringe of Australian cities is rarely considered as part of wider academic and policy debates around urban regeneration. Indeed much strategic planning and academic research positions fringe development as the antithesis of good urban planning (Chapter 2). Fringe development is criticised on the grounds of long commuting times to employment, unhealthy lifestyles leading to a rise in obesity, environmental impacts of suburban developments and social isolation (Dodson 2016). Yet, the fringe is expected to house the majority of our growing population, with the bulk of new development still occurring on the edge of Australian cities. Fringe development is still viewed by some planners and politicians as a solution to contemporary housing affordability crises (Burton 2015). Thus, fringe development emerges as a central consideration when analysing urban regeneration across the city; it is a vital element of the ongoing regeneration of the city.

The form of fringe/suburban development has evolved. Increasingly new development is adopting many of the planning and design logics of major regeneration precincts. Underlying much of this new development are the principles of New Urbanism. While New Urbanist design principles are far from new, this form of development is becoming more common as developers appeal to a market wanting access to infrastructure, services and a sense of community seen to be delivered by town centres (McGuirk and Dowling 2009). This new form of fringe suburban development might also be termed "post-suburban" (Keil 2015). The post-suburb emerges as a reaction to the more traditional form of suburban sprawl and recognises the tendency of new development across the urban fringe to concentrate around a series of nodes or town centres.

Like brownfield and greyfield sites, new fringe development offers a series of opportunities for innovation around the planning and development of the city. The process of planning and delivering new fringe development is explored in detail by Ruming, Mee, McGuirk and Sweeney in Chapter 17. The focus of this chapter is the Huntlee development, located in the greater Newcastle region. Four areas of innovation are examined: nurturing community; diversity of housing and built form; infrastructure and service provision; and, sustainability and environmental protection. Across these four areas, Huntlee emerges as a development which strives for, and reinforces the perceived benefits of, a consolidated urban form. New technologies are seen to address some of the environmental critique levelled at fringe development. Likewise, housing diversity and the early provision of social services and infrastructure are efforts to overcome concerns around isolation and lack of access. These innovations aside, concerns around connection with the wider city, particularly access to public transport networks, remain.

Consultation and opposition

A constant critique of urban regeneration projects has been the failure to engage communities living in locations targeted for change (Cook 2016; Ruming et al. 2012). It is common for questions about the timing, scale and form of community consultation to be raised surrounding the planning and development of regeneration projects of all sizes. This often results in significant community opposition. Concerns mobilised by opponents against urban regeneration are manifold. At the local scale, existing residents often oppose projects which increase density and transform the built form of their local area on the basis of loss of amenity, impacts on local services and infrastructure or potential impacts on house prices. Such local opposition has been characterised as both a selfish form of NIMBY (not in my backyard) action as residents seek to maintain their own standard of living, and as a form of urban democracy, as citizens rally against planning processes that limit participation (van den Nouwelant et al. 2015; Ruming et al. 2012). Such opposition is likely to increase for projects where planning and assessment authority is removed from local government (Cook et al. 2012). Opposition to large regeneration projects tend to centre on concerns around the extent of public benefit, the planning and approval process and the level of public funding provided.

In Chapter 16, Legacy and Taylor provide a detailed examination of community opposition to transport-led regeneration in Melbourne. In particular, they examine the introduction of a new tram line as a catalyst for town centre regeneration initiatives. The case study reveals challenges around the timing and process of formal consultation processes. While public transport is a vital element of successful urban regeneration, in this instance significant resistance from local retailers emerge over plans to remove car parking. As is the case for the Charlestown regeneration (Chapter 13), maintaining car parking is viewed as essential in ensuring the financial viability of centres. Issues surrounding the privatisation of public space, value uplift and the provision of affordable housing also emerge as sites of conflict.

Collectively the contributions to this book provide a wide-ranging analysis of the diverse forms and locations of urban regeneration which are shaping Australian cities. Emergent throughout the chapters is a series of challenges and obstacles which need to be addressed to ensure that our future cities are equitable, social and environmentally sustainable, and economically productive. Yet, the chapters also reveal a series of opportunities and innovations which urban regeneration sites offer in reconfiguring our cities. The broad definition of urban regeneration mobilised throughout the book, allows academics, practitioners, politicians and communities to simultaneously recognise the importance of urban regeneration in changing the nature of individual sites, together with its contribution to the ongoing evolution of the city as a whole.

References

Arthurson, K. (2012) *Social Mix and the City*, Melbourne: CSIRO Publishing.

Australian Housing and Urban Research Institute (AHURI) (2017) 'City deals: Coordinating the growth of Australian cities', https://www.ahuri.edu.au/policy/

ahuri-briefs/city-deals-cooordinating-the-growth-of-australian-cities (accessed 22 June).

Baker, T. and Ruming, K. (2015) Making 'global Sydney': Spatial imaginaries, worlding and strategic plans, *International Journal of Urban and Regional Research*, 39(1): 62–78.

Baxter, R. and Lees, L. (2008) 'The rebirth of high-rise living in London: towards a sustainable, inclusive and liveable urban form', in R. Imrie, L. Lees and M. Raco (eds), *Regenerating London*, London: Routledge.

Bunce, S. (2004) The emergence of 'smart growth' intensification in Toronto: Environment and economy in the new official plan, *Local Environment*, 9(2): 177–191.

Bunker, R. (2007) *Progress and Prospect with 'City of Cities: A Plan for Sydney's Future'*, Sydney: City Futures Research Centre.

Bunker, R. and Searle, G. (2009) 'Theory and practice in metropolitan strategy: Situating recent Australian planning', *Urban Policy and Research*, 27(2): 101–116.

Bunker, R., Holloway, D. and Randolph, B. (2005) 'Building the connection between housing needs and metropolitan planning in Sydney, Australia', *Housing Studies*, 20(5): 771–794.

Burton, P. (2015) 'The Australian good life: The fraying of a suburban template', *Built Environment*, 41(4): 504–518.

Charmes, E. and Keil, R. (2015) 'The politics of post-suburban densification in Canada and France', *International Journal of Urban and Regional Research*, 39(3): 581–602.

Colantonio, A. and Dixon, T. (2010) *Urban Regeneration and Social Sustainability: Best Practice from European Cities*, Oxford: Wiley and Blackwell.

Cook, N. (2016) 'Performing housing affordability: The case of Sydney's green bans', in N. Cook, A. Davison and L. Crabtree (eds), *Housing and Home Unbound: Intersections in Economics, Environment and Politics in Australia*, London: Routledge.

Cook, N. and Ruming, K. (2008) 'On the fringe of Neoliberalism: Residential development in outer suburban Sydney', *Australian Geographer*, 39(2): 211–228.

Cook, N., Taylor, E., Hurley, J. and Colic-Peisker, V. (2012) *Resident Third Party Objections and Appeals against Planning Applications: Implications for Higher Density and Social Housing*, Melbourne: AHURI.

Creer, K. (2015) 'Residents band together to sell 30-40 homes to tempt developers close to the NorthWest Rail Link', *Hills Shire Times*, 3 August, http://www.dailytelegraph.com.au/newslocal/the-hills/hills-sellers-band-together-to-sell-3040-homes-together/news-story/a3758af31ae787174b3f8c371e249b84 (accessed 5 July 2017).

Davison, G. (2011) 'The role and potential of government land agencies in facilitating and delivering urban renewal', *Proceedings of State of Australian Cities Conference 2011*, Melbourne: State of Australian Cities Research Network.

Davison, G. and Legacy, C. (2014) 'Positive planning and sustainable brownfield regeneration: The role and potential of government land development agencies', *International Planning studies*, 19(2): 154–172.

de Magalhaes, C. (2015) 'Urban regeneration', *International Encyclopaedia of the Social and Behavioural Sciences*, London: Elsevier.

Dodson, J. (2010) 'In the wrong place and the wrong time? Accessing some planning, transport and housing market limits to urban consolidation policies', *Urban Policy and Research*, 28(4): 487–504.

Dodson, J. (2016) 'Suburbia in Australian urban policy', *Built Environment*, 42(1): 7–20.

Dovey, K. (2005) *Fluid City: Transforming Melbourne's Urban Waterfront*, Sydney: UNSW Press.

Evans, G. (2005) 'Measure for measure: Evaluating the evidence of culture's contribution to regeneration', *Urban Studies*, 42(5–6): 959–983.

Evans, J. and Jones, P. (2008) 'Rethinking sustainable urban regeneration: ambiguity, creativity, and the shared territory', *Environment and Planning A*, 40: 1416–1434.

Forster, C. (2006) 'The challenge of change: Australian cities and urban planning in the new millennium', *Geographical Research*, 44(2): 173–182.

Furbey, R. (1999) 'Urban "regeneration": reflections on a metaphor', *Critical Social Policy*, 19: 419–445.

Gleeson, B. and Low, N. (2000) '"Unfinished business": Neoliberal planning reform in Australia', *Urban Policy and Research*, 18(1): 7–28.

Goodman, R. and Coiacetto, E. (2012) 'Shopping streets or malls: Changes in retail form in Melbourne and Brisbane', *Urban Policy and Research*, 30(3): 251–273.

Gurran, N. and Phibbs, P. (2013) 'Housing supply and urban planning reform: The recent Australian experience, 2003–2012', *International Journal of Housing Policy* 13(4): 381–407.

Gurran, N. and Whitehead, C (2011) 'Planning and affordable housing in Australia and the UK: A comparative perspective, *Housing Studies*, 26: 1193–1214.

Haughton, G. and McManus, P. (2011) 'Neoliberal experiments with urban infrastructure, in Sydney, Australia', *International Journal of Urban and Regional Research*, 36: 91–105.

Hughes, M. (2004) *Community Economic Development and Public Housing Estates*, Sydney: ShelterNSW.

Imrie, R. and Dolton, M. (2014) 'From supermarkets to community building: Tesco plc, sustainable place making and urban regeneration', in R. Imrie and L. Lees (eds), *Sustainable London? The Future of a Global City*, Bristol: Policy Press.

Jones, P. and Evans, J. (2013) *Urban Regeneration in the UK*, London: SAGE Publications.

Judd, B. and Randolph, B. (2006) 'Qualitative methods and the evaluation of community renewal programs in Australia', *Urban Policy and Research*, 24(1): 97–114.

Karadimitriou, N., de Magalhaes, C. and Verhage, R. (2013) *Planning, Risk and Property Development: Urban regeneration in England, France and the Netherlands*, Abingdon: Routledge.

Keil, R. (2015) 'Towers in the park, bungalows in the garden: peripheral densities, metropolitan scales and political cultures of post-suburbia, *Built Environment*, 41(4): 579–596.

Leary, M.E. and McCarthy, J. (2013) 'Introduction: Urban regeneration, a global phenomenon', in M.E. Leary and J. McCarthy (eds), *The Routledge Companion to Urban Regeneration*, Abingdon: Routledge.

Lees, L. (2003) 'Policy (re)turns: gentrification research and urban policy – urban policy and gentrification research', *Environment and Planning A*, 35: 571–574.

Legacy, C., Pinnegar, S. and Wiesel, I. (2013) 'Under the strategic radar and outside planning's "spaces of interest": knockdown rebuild and the changing suburban form of Australia's cities', *Australian Planner*, 50(2): 117–122.

Matthews, P. (2010) 'Mind the gap? The persistence of pathological discourses in urban regeneration policy', *Housing, Theory and Society*, 27(3): 221–240.

McCann, E. and Ward, K. (eds) (2011) *Mobile Urbanism: Cities and Policymaking in the Global Age*, Minneapolis, MN: University of Minnesota Press.

McGuirk, P. and Dowling, R. (2009) 'Neoliberal privatisation? Remapping the public and the private in Sydney's masterplanned residential estates', *Political Geography*, 28(3): 174–185.

McGuirk, P., Winchester, H. and Dunn, K. (1996) 'Entrepreneurial approaches to urban decline', *Environment and Planning A*, 28: 1815–1841.

Neilson, L. (2008) 'The Building Better Cities program 1991–96: a nation-building initiative of the Commonwealth Government', in J. Butcher (ed), *Australia under Construction: Nation-building: Past, Present and Future*, Canberra: ANU press.

Newton, P. (2010) 'Beyond greenfield and brownfield: The challenge of regenerating Australia's greyfield suburbs', *Built Environment*, 36(1): 81–104.

Newton, P. and Glackin, S. (2014) 'Understanding infill: Towards new policy and practice for urban regeneration in the established suburbs of Australia's cities', *Urban Policy and Research*, 32(2): 121–143.

Newton, P. and Thompson, G. (2017) 'Urban regeneration in Australia', in P. Robert, H. Sykes and R. Granger (eds), *Urban Regeneration*, London: SAGE Publications.

Oakley, S. (2004) 'Politics of recollection: Examining the rise and fall of DURD and better cities through narrative', *Urban Policy and Research*, 22(3): 299–314.

Oakley, S. (2009) 'Governing urban waterfront renewal: The politics, opportunities and challenges for the inner harbour of Port Adelaide, Australia', *Australian Geographer*, 40(3): 297–317.

Oakley, S and Johnson, L. (2013) 'Place-taking and place-making in waterfront Australia', *Urban Studies*, 50(2): 341–355.

Pinnegar, S. and Randolph, B. (2012) *Renewing the West: Prospects for Urban Regeneration in Sydney's Western Suburbs*, Sydney: City Futures Research Centre.

Pinnegar, S., Freestone, R. and Randolph, B. (2010) 'Suburban reinvestment through "knockdown rebuild" in Sydney', in M. Clapson and R. Hutchison (eds), *Suburbanisation in Urban Sociology*, 10: 205–229.

Pinnegar, S., Wiesel, I., Liu, E., Gilmour, T., Loosemore, M. and Judd, B. (2011) *Partnership Working in the Design and Delivery of Housing Policy and Programs*, Melbourne: AHURI.

Randolph, B. and Freestone, R. (2008) *Problems and Prospects for Suburban Renewal: An Australian Perspective*, Sydney: City Futures Research Centre.

Randolph, B. and Freestone, R. (2012) 'Housing differentiation and renewal in middle-ring suburbs: The experience of Sydney, Australia', *Urban Studies*, 49(12): 2557–2575.

Roberts, P. and Sykes, H. (2000) *Urban Regeneration: A Handbook*, London: SAGE Publications.

Rogers, D. (2013) 'Urban and social planning through public-private partnerships: the case of the Bonnyrigg Living Communities Project, Sydney Australia', In J. Colman and C. Gossop (eds.),*ISOCARP Review 09: Frontiers of Planning: Visionary Futures for Human Settlements*, Brisbane: International Society of City and Regional Planners.

Ruming, K. (2009) 'The complexity of comprehensive planning partnerships: The case of the Warnervale Town Centre', *Urban Policy and Research*, 27(1): 25–42.

Ruming, K. and Gurran, N. (2014) 'Planning system reform in Australia', *Australian Planner*, 52(2): 102–107.

Ruming, K., Mee, K.J. and McGuirk, P.M. (2004) 'Questioning the rhetoric of social mix: courteous community or hidden hostility?, *Australian Geographical Studies*, 42(2): 234–248.

Ruming, K., Randolph, B., Pinnegar, S. and Judd, B. (2007) 'Delivering sustainable renewal in Australia's middle and outer suburbs: Council reflections', *Proceedings of the State of Australian Cities Conference 2007*, Adelaide: State of Australian Cities Research Network.

Ruming, K., Amati, M. and Houston, D. (2012) 'Multiple suburban publics: Rethinking community opposition to consolidation in Sydney', *Geographical Research*, 50(4): 421–435.

Ruming, K., Mee, K. and McGuirk, P. (2016) 'Planned derailment for New Urban futures? An actant network analysis of the "Great [Light] Rail Debate" in Newcastle, Australia', in Y. Rydin, and L. Tate (eds), *Actor Networks of Planning: Exploring the Influence of Actor Network Theory*, London: Routledge.

Shaw, K. and Porter, L. (2009) 'Introduction' in K. Shaw and L. Porter (eds), *Whose Urban Renaissance? An International Comparison of Urban Regeneration Strategies*, London: Routledge.

Smith, A. (2012) *Events and Urban Regeneration: The Strategic Use of Events to Revitalise Cities*, London: Routledge.

Smith, H., and Garcia Ferrari, M. (eds) (2012) *Waterfront Regeneration: Experiences in City-Building*, London: Routledge.

Tallon, A. (2013) *Urban Regeneration in the UK*, London: Routledge.

van den Nouwelant, R., Davison, G., Gurran, N., Pinnegar, S. and Randolph, B. (2015) 'Delivering affordable housing through the planning system in urban renewal contexts: Converging government roles in Queensland, South Australia and New South Wales', *Australian Planner*, 52(2): 77–89.

Vigneswaran, D., Iveson, K. and Low, S. (2017) 'Problems, publicity and public space: A resurgent debate', *Environment and Planning A*, 49(3): 496–502.

Wiesel, I., Freestone, R. and Randolph, B. (2013a) 'Owner-driven suburban renewal: Motivations, risks and strategies in "knockdown and rebuild" processes in Sydney, Australia', *Housing Studies*, 28(5): 701–719.

Wiesel, I., Pinnegar, S. and Freestone, R. (2013b) 'Supersized Australian dream: Investment, lifestyle and neighbourhood perceptions among "knockdown-rebuild" owners in Sydney', *Housing, Theory and Society*, 30(3): 312–329.

Winston, N. (2010) 'Regeneration for sustainable communities? Barriers to implementing sustainable housing in urban areas', *Sustainable Development*, 18(6): 319–330.

Part I
Planning and funding urban regeneration

2 Metropolitan strategic plans

Establishing and delivering a vision for urban regeneration and renewal

Kristian Ruming

Introduction

The challenges facing Australian cities are diverse, including population growth, economic restructuring, access to services and employment, concerns around housing affordability and dwelling construction, infrastructure funding and provision, socio-economic polarisation and social justice, environmental protection and climate change. The policy responses to these challenges are equally diverse. At an urban scale, Australian cities have mobilised Metropolitan Strategic Plans (MSPs) as a central policy tool for framing urban change. MSPs outline the state-sponsored vision for our future cities, by prioritising key sites of change, establishing frameworks and conditions of development, and advocating the benefits of a consolidated urban form. The underlying goal of urban regeneration is often translated via a complex, diffuse and, sometimes, confusing set of terminology. Terms mobilised in strategic planning policy include infill, densification, compact city, centres and nodes, transit oriented development (TOD) and urban consolidation. All imply a change to the existing urban structure. All seek to regenerate the city in a particular way. In his review of MSPs, Forster (2006) identifies a general consensus within planning policy built on the principles of *containment*, *consolidation* and *centres*. For Bunker (2014), urban consolidation transitioned into the notion of the compact city as the planning orthodoxy driving MSPs from the early 2000s.

Urban consolidation or the compact city is a form of urban development which increases residential densities in existing urban areas, particularly around designated transport nodes. Urban consolidation comes in stark contrast to the historical form of Australian urbanism – fringe development. The consolidation narrative is inherently "anti-sprawl". The concept is far from new, first appearing in the late 1970s as planning authorities sought to simultaneously reduce the fringe expansion and address issues of inner-city population loss. Early incarnations of urban consolidation policy mobilised by state governments came in response to increasingly tight budget positions, a need to reduce state infrastructure expenditure on an expanding fringe, an ambition to deliver a more diverse housing stock, and a desire protect the viability of central cities (Hamnett 2000; Lennon 2000). Bunker (2012) identifies the early 1990s as a period transition in MSPs where urban consolidation and density emerged as the dominant spatial vision for Australian cities, replacing previous discourses which centred on infrastructure development and coordination, and whole-of-government urban growth management. Pinnegar and Randolph (2012) argue that urban consolidation has emerged as an illusionary "win-win" arrangement which simultaneously meets the profit objectives the development sector and the strategic planning objectives of the state. For some, the strategic policy push for the compact city represents an extension of neoliberal urbanism, a "new urbanology", where urban density is seen to promote efficient and cost effective delivery of services (Davidson and Gleeson 2013).

This chapter reviews contemporary capital city metropolitan plans (Table 2.1) and maps the role that urban regeneration plays in defining the future of Australia's cities. It unpacks the logic of a denser Australian urban form, primarily delivered by the redevelopment of land in the urban footprint. While MSPs originate from a variety of geographic, economic, demographic, political and environmental contexts, a common set of claims, ambitions and instruments emerge. In particular, this chapter focuses on the urban regeneration ambitions embedded on the strategic planning principles of centres and corridors, targets, development hierarchies, and TODs. These strategies seek to define and frame the multiple geographies of regeneration, as explored in detail in following chapters.

Table 2.1 Current metropolitan plans

City	*Plan*	*Year*	*Plan timeline*
Sydney	A Plan for Growing Sydney[1]	2014	2014–2031
Melbourne	Plan Melbourne[2]	2014	2014–2050
South East Queensland	South East Queensland Regional Plan 2009–2031	2009	2009–2031
Perth	Directions 2031 and Beyond	2010	2010–2031
Adelaide	The 30-Year Plan for Greater Adelaide	2010	2010–2040
Hobart	Draft H.30 Hobart Capital City Plan	2011	2011–2040
Canberra	ACT Planning Strategy	2012	2012–2040

Source: Various metropolitan planning strategies

Notes

1 Towards our Greater Sydney 2056 – an amendment to update A Plan for Growing Sydney – was released late 2016 (see postscript to this chapter).

2 A revised Plan Melbourne 2017–2050 was released early 2017 (see postscript to this chapter).

Australian metropolitan planning: promoting urban regeneration and consolidation

In contrast to strategic planning elsewhere, Australian strategic urban planning is characterised as inflexible, driven by long-term projections and targets, dependent on state infrastructure provision as the primary driver of spatial change and reliant on market forces to implement the plan (Bunker 2012; Dodson 2009). Bunker and Searle (2009) characterise this as a "blueprint" quality of metropolitan planning and advocate for a shift in the Australian strategic planning paradigm, closer to the relational planning systems which characterise Europe. In terms of urban regeneration, this shift might move from a strategic planning process which defines sites and forms of regeneration to one which acknowledges and responds to nascent, sporadic and potentially innovative forms of regeneration which might emerge across our cities. For Bunker and Searle (2009) Australian metropolitan planning is the product of an entrenched path dependency which limits the scope of strategic visions, delivery mechanisms and regulatory frameworks, as the plans tend to do little more than update the underlying logics and spatial structures of previous strategies. The principles of urban regeneration are firmly entrenched as part of this path dependency, as urban consolidation, densification and compact city development are repeated and often strengthened in subsequent MSPs (Ruming et al. 2013).

Another defining characteristic of Australian MSPs identified by Bunker and Searle (2009) is a long-range planning horizon, often seeking to define a city for the next generation. Contemporary MSPs have timelines of around 30 years into the future. In an effort to overcome critiques of demographic forecasts, MSPs have moved towards using demographic ranges and implemented rolling (typically five-year) updates. The updates might incorporate revised figures/data/targets, but rarely do they reconfigure the spatial vision of the regenerated city. The designated centres remain the same; it is their size and function which might alter. Further, the relatively rapid process of plan revision in Australia has correlated with political transition which has seen new MSPs implemented surrounding state government elections (Maginn et al. 2016). In this context, MSPs are mobilised by governments in an attempt to win broader (electoral) support. One striking example is the case of the East–West Link in Melbourne which was included in *Plan Melbourne* and approved by the Napthine Coalition government in 2014, only to be scrapped by the Andrews Labor government on winning power later that same year (Whitzman 2015). This, along with other issues, has triggered a revision of *Plan Melbourne* (*Plan Melbourne Refresh*, Victoria State Government 2015), with a new strategic plan for Melbourne due for release in 2016.[1] The expression of urban regeneration outlined in MSPs can come to represent a site of urban conflict as some residents oppose key elements of the future city vision

(Chapter 16). Despite constant revision, a series of overriding discourses supporting a compact, consolidated and regenerated city emerge across the strategies.

The compact city is routinely defined and justified in sustainability terms. The relationship between increased density and environmental sustainability remains largely "non-disputable" and "beyond political debate" (Charmes and Keil 2015: 588). In their recent review of sustainability principles in Australian MSPs, Davidson and Arman (2014) identify the compact city model as the underlying spatial planning strategy to deliver sustainable urban futures characterised by higher densities integrated with public transport. In the case of Victoria, *Plan Melbourne* claims to promote the transition to a more sustainable city via a series of

> innovative metropolitan planning approaches, including:
> - creating more compact cities
> - making better use of transport infrastructure
> - greening metropolitan areas
> - creating more open space
> - reforming energy and water supply and use
> - conserving biodiversity
> - improving building design
> - encouraging active forms of transport, such as walking and cycling.
>
> (VicDTPLI 2014: 11)

A second emergent discourse in contemporary MSPs justifying a more consolidated, regenerated urban form is that of liveability. Underlying the creation of liveable cities are ideas about accessibility, walkability and vibrant public spaces. At its most simplistic, the liveability claim centres on the capacity of a more dense city to allow people to "live where they want – close to jobs, services and transport" (NSWDPE 2014: 83). The redevelopment of sites as mixed-use centres is seen to offer opportunities for the integration community and social service facilities (ACTDESD 2012). According to current MSPs, urban consolidation promotes liveability via: reducing travel times which increases the amount of time spent with friends and family in one's local area; promoting mix within communities; ensuring spaces are active and safe; responding to heritage issues; and providing access to high quality open space and public infrastructure.

Increasingly urban consolidation promotes the principles of healthy built environments. The regeneration of sites within the urban footprint is seen to offer a series of health benefits to individuals and communities, including: increased physical activity and active transport; greater access to health food via markets and community gardens; and greater social cohesion and connection:

> Urban intensification and improving Canberra's cycling and walking networks will clearly help establish an economically and environmentally healthier city.
>
> (ACTDESD 2012: 35)

> Directions 2031 ... promotes walkable neighbourhoods, compatible mixed use development, promotion of a sense of place and support for a variety of housing types.
>
> (WADOP 2010: 43)

Nevertheless, claims used to support urban consolidation have been critiqued. A number of authors highlight the absence of significant debate about the potential negative impacts of increased density as part of the regeneration of Australian cities (Grosvenor and O'Neill 2014; Randolph 2006; Searle 2004). Further, the willingness and capacity of the state to intervene in urban land markets, and direct development to achieve strategic goals is also a challenge. In an era of neoliberal urban governance, where market-based actors (private owners, developers, builders, financiers) are largely responsible for urban change, questions remain about the capacity of the state to identify and deliver development opportunities. A major challenge for MSPs is the growing recognition that regeneration is a dynamic and fluid process enacted on a day-to-day basis. As illustrated in the chapters of this collection, regeneration and consolidation is contested and negotiated. Despite this "on-the-ground" diversity, complexity and fluidity, a set of standardised visions, objectives and tools has emerged across MSPs (Maginn et al. 2016). It is to these common themes and tools around urban regeneration and consolidation this chapter now turns.

Planning logics and tools: delivering regeneration, consolidation and the compact city

Balance and targets

When it comes to urban development, growth and planning, the emergent discourse across metropolitan plans is one of *balance*:

> In locating this additional housing we need to consider the balance we want between greenfield development (new residential estates) and urban intensification in existing metropolitan areas.
>
> (ACTDESD 2012: 24)

> We need to use land and infrastructure in an efficient manner and manage the expansion of the urban zone in balance with urban infill projects.
>
> (WADOP 2010: v)

Ultimately this balance is achieved via advocating for a significant proposition of new development to occur within the existing urban footprint. Urban consolidation/regeneration/compact city, therefore, represents the mechanism by which the "unbalanced'" Australian urban form (i.e. where fringe development historically dominates) can be addressed. According to these strategies, achieving a balance requires a shift in the form and spatial distribution of development. All MSPs seek to increase the amount and proportion of higher-density development within the existing urban footprint. Nevertheless, all MSPs acknowledge that fringe development will continue to provide a significant proportion of new housing supply and, thereby, continue to play a major role in the overall evolution of our cities (Chapter 17).

The push for spatial realignment of dwelling, population and employment growth, away from greenfield development to a more compact urban form, is most frequently expressed in Australian MSPs in the form of dwelling and density targets. Dwelling targets have come to represent, what Bunker (2014: 452) calls, "landmark performance indicators" to define how much development should occur in established areas compared to fringe greenfield development – they quantify the ratio of development required to achieve the ideal population and development balance.

MSPs have mobilised two types of targets to pursue densification, consolidation and regeneration: first, density targets often represented as dwellings per hectare which promote a broader densification across the urban footprint, and second, areal based targets which define either the total number of dwellings to be constructed in a defined area or the proportion of dwellings constructed in that area to be higher density. Historically, the most common form of target mobilised within MSPs has been per hectare residential densities. For example, McGuirk and O'Neill (2002) trace evolution of density targets in Sydney, noting the increase in density targets from 8 to 10 dwellings per hectare in the 1988 Sydney Metropolitan strategy (*Sydney into its Third Century*). This increased to 15 dwellings per hectare in the 1995 strategy (*Cities for the 21st century*). Per hectare dwelling targets have been used less frequently in more recent MSPs. In the case of Sydney, none of the four MSPs released since 2005[2] included per hectare density targets. Likewise, a per hectare density target is not included in *Plan Melbourne*. However, they have not disappeared from all MSPs. In Perth, *Directions 2031* increased the density target by 50 per cent in new zoned fringe residential areas to 15 dwellings per hectare (WADOP 2010: 4). *The 30-Year Plan for Greater Adelaide* seeks to increase the gross per hectare density across Greater Adelaide from 8 to 11 dwellings, and increase per hectare density in designated corridors and TODs to an average of 25–35 dwellings (SADPLG 2010: 18). Likewise the *SEQ Regional Plan* advocates a compact form of urban settlement to "achieve a minimum dwelling yield of 15 dwellings per hectare net for new residential development in Development Areas" (QLDDIP 2009: 91). One could infer that per hectare density targets remain important planning instruments for cities with lower overall densities, where increasing density itself remains a key strategic objective. In our larger, denser cities, simply increased density is no longer enough.

Thus, MSPs have progressively moved towards incorporating infill targets, either replacing or accompanying per hectare targets. Infill targets represent a more explicit urban regeneration tendency of MSPs by more clearly articulating the strategic intention of the location (designated centres) and form (higher-density) of housing development. However, no universal target for urban infill exists across Australian cities (Table 2.2). Nevertheless, two important points emerge. First, the target for infill development now accounts for around half of all projected new development, in some cities considerably more. Second, for most cities the targets for infill development have fluctuated through time. In the case of Sydney, targets for multi-unit/infill construction were set at 65 per cent in *Cities for the 21st Century* (NSWDoP 1995). By the 2005 strategy, *City of Cities*, the overall fringe development target remained consistent at between 30–40 per cent, but the infill target became increasingly focussed, with 82 per cent to be delivered in defined "centres" via redevelopment and renewal processes (Bunker 2012). The target for infill development in *Metropolitan Plan for Sydney 2036* (NSWDoP 2010) was 70 per cent concentrated in designated centres; however an explicit target was removed from the release of the *Draft Metropolitan Strategy for Sydney to 2031* (NSWDoPI 2013) with the new Minister for Planning advocating for a 50:50 development mix, citing growing concerns around housing affordability as justification for increasing fringe development (McKenny 2014). Likewise, for Melbourne the infill targets have fluctuated, with existing urban areas expected to accommodate 78 per cent of new development to 2030 in *Melbourne 2030* (VicDoI 2002), 53 per cent of new development to 2030 in *Melbourne@5 million* (VicPCD 2008) and 61 per cent of new development to 2051 under *Plan Melbourne* (VicDTPLI 2014).

The movement away from wider density-based targets towards infill targets has led to a more prescriptive vision for urban development, which is seen to reinforce the centres and TOD logic of Australian urban regeneration. Thus, MSPs have not only sought to redefine broad balance of development (fringe versus infill) but progressively sought to define the type and location of infill development occurring across our cities. The regeneration of Australian cities is not to be delivered in a form where density is increased across the whole urban area; rather, redevelopment is confined to a hierarchy of strategic centres/nodes and corridors.

The use of targets exemplifies the highly prescriptive detail of Australian metropolitan plans (Bunker and Searle 2009). However, dwelling targets have been critiqued for lack of critical engagement with the broader issue of housing affordability and social justice (Forster 2006; Pinnegar and Randolph 2013). Bunker (2012) argues targets lack credibility and reliability

Table 2.2 Dwelling targets

Metropolitan planning strategy	*Infill target*
A Plan for Growing Sydney	664,000 new dwellings to 2031. No overall target. Subregional strategies (due early 2017) will "set five year local housing targets that maximise the opportunities to growing housing supply" (p. 65). Targets will be established for local councils.
Plan Melbourne	960,000 (61.1%) of 1,570,000 new dwellings to 2051 (p. 62). Infill targets allocated to planning sub-regions.
SEQ Regional Plan	374,000 (49.6%) of 754,000 new dwellings to 2031. Infill targets allocated to local government areas (p. 91).
Directions 2031 and Beyond	154,000 (47%) of 328,000 new dwellings to 2031 (p. 11). Infill targets allocated to planning sub-regions.
The 30-Year Plan for Greater Adelaide	258,000 new dwellings. Ratio between infill and fringe growth to be 50:50 in the short to mid-term, shifting to 70:30 by 2038 (p.60). Infill targets allocated to planning sub-regions (Map D9, p. 97).
Draft H.30 Hobart Capital City Plan	No targets
ACT Planning Strategy	"50% or more" (p.5) of estimated 55,000 new dwellings to 2030. Targets at metropolitan scale.

Source: Various metropolitan planning strategies

due to their inability to respond to changing contexts, including market performance or demographic forecasts. For Searle (2013) the use of targets actively works against a relational planning approach. In his analysis of Sydney's *City of Cities* strategy, Bunker (2012) notes the semantic transition of "possible" metropolitan targets to "subregional housing capacity target" at a lower level. Thus the headline or aspirational targets of MSPs, can become a disciplinary mechanism that defines and constrains local planning initiatives. In other words, local planning instruments need to facilitate the delivery of targets established in MSPs. Finally, and perhaps most importantly, the targets for infill development are far from being achieved in most cities (Bunker 2014). This raises important questions about the validity of such targets as a planning mechanism designed to promote consolidation, densification and regeneration. Implementation remains an issue.

In recognising the delivery function of the private sector, dwelling targets are an attempt by state planning agencies to better understand and incorporate the housing market, via significant housing market analysis across capital cities. However, the pervasiveness of housing market dynamics underlying MSPs raises questions about their strategic intent. Questions arise as to whether dwelling targets facilitate an urban form that is most desirable in terms of equity, sustainability, efficiency and the preferences of urban citizens or if the targets simply respond to the profit/development ambitions of private market actors. If the latter is the case, urban regeneration and consolidation will be confined to the most profitable sub-markets of our cities, leaving much of the city untouched, despite potentially being identified in MSPs (Pinnegar and Randolph 2013).

Centres, urban hierarchy and transit oriented development

In attempting to "balance" the spatial distribution of growth, the "activity centre" has become the dominant discourse of urban form. The idea of centres as a means of structuring Australian cities is not new, with the *County of Cumberland Planning Scheme* for Sydney defining a clear hierarchy of centres (Cumberland County Council 1948), while strategies from the 1970s increasingly promoted "urban villages" (Orchard 1995). The activity centre model is based on the redevelopment/regeneration of buildings (commercial, residential and industrial) in defined locations, usually serviced by major transport links (Bunker 2012). These activity centres or nodes are to attract and concentrate higher-value employment, provide diverse housing options and provide transport via public transport systems (Dodson 2016). Contemporary MSPs have extended centres logic to advocate for the regeneration of designated corridors which link centres along key transport infrastructure. The most recent strategies in Sydney and Melbourne call for the regeneration of blighted urban transport corridors in inner and middle suburbs (Searle 2013). In Sydney, examples include the Parramatta Road corridor and the Central to Eveleigh corridor, while in Melbourne, E-Gate, the Dynon corridor and the Flinders Street to Richmond corridor are identified.

The push for growth in designated centres has led to the emergence of the polycentric city as the ideal urban form in Australia. The polycentric city seeks to respond to the suburbanisation of employment, the need for specialised industrial and hi-tech clusters, new service centres, and access to transport infrastructure. Most recently the idea of the polycentric city was advocated by Prime Minister Turnbull and his call for the "30-minute city", which advocated for high speed rail, greater urban densities and a form of value capture as mechanisms for reconfiguring Australian cities, where "The idea is to plan for cities where residents can access employment, schools, shopping, services and recreational facilities within 30-minutes of home" (DPMC 2016: 11). Both *A Plan for Growing Sydney* and *Plan Melbourne* adopt the idea (indeed, *Plan Melbourne* calls for a 20-minute city). The idea received criticism, with commentators highlighting the lack of infrastructure spending, poorly coordinated state–federal planning and infrastructure delivery and limited capacity to capture the economic benefits of value uplift which arises from infrastructure improvements (Newman 2016).

The underlying principle of centres-based regeneration is transit oriented development (TOD). TOD has emerged as a central planning logic and is now common across all MSPs (Burke 2016). Originating from ideas of the ideal European city where dense mixed-use urban centres surround transport infrastructure, the idea soon emerged in US planning as a possible solution to sprawling metropolises and soon became one of the key platforms of the New Urbanism planning paradigm (Newman 2009). The logic of this spatial strategy lies in the belief that large populations living around and using public transport is a cost-effective solution for state governments. According to *A Plan for Growing Sydney:*

> The most suitable areas for significant urban renewal are those areas best connected to employment and include…in and around centres that are close to jobs and are serviced by public transport services that are frequent and capable of moving large numbers of people.
>
> (NSWDPE 2014: 65)

The rationale behind centres-based TOD has shifted from initial goals of reducing infrastructure and transport expenditure, through concerns around environmental and sustainability claims, to claims around liveability and affordability. The apparent virtues of a consolidated form of urban regeneration of our cities appear to be ever-increasing. The logic for urban regeneration/densification of designated strategic/TOD centres rests on at least three broad underlying assumptions. First, concentrating higher density development in centres is seen to generate efficiencies in terms of access to and use of public transport infrastructure, thereby reducing private motor vehicle use. Increasing public transport use in lieu of private motor vehicles is seen to deliver efficiency, sustainability and liveability benefits:

> Locating new housing in the existing urban area allows efficient use of the existing infrastructure.
>
> (NSWDPE 2014: 33)

Second, consolidating work and employment within defined centres is seen to have efficiency and economic benefits by drawing together housing and employment spaces, thereby limiting the need for long travel to work:

> The focus on promoting mixed use activity centres throughout the region that provide both housing and employment opportunities will reduce the reliance on the capital city centre and potentially reduce the number of commuter trips.
>
> (WADOP 2010: 57)

> Future employment growth will be accommodated within urban areas through a combination of activity centres.
>
> (QLDDIP 2009: 11)

Third, the regeneration of centres at higher densities is seen to offer a cost effective way of providing urban services to a growing population by utilising existing infrastructure and reducing the need for new infrastructure. Access to existing road, rail, sewerage, electricity, gas and telecommunication infrastructure are all identified within MSPs as being a benefit of increasing density in designated centres. However, the most recent round of MSPs have begun to acknowledge that upgrades to infrastructure may be necessary, and that existing networks may be unable to accommodate the planned increase in service demand. In NSW, while there is recognition that large-scale redevelopment projects are often accompanied by coordinated infrastructure planning and delivery, this is often missing for smaller-scale infill development:

> While significant programs are in place for large-scale urban renewal projects, small-scale urban infill development can also have an impact on the demand for infrastructure services. Urban infill will be most successful where development is coordinated with social infrastructure delivery.
>
> (NSWDPE 2014: 70)

Behind the push for urban regeneration in centres is the "new emphasis on urban infrastructure planning as the primary mode of intervention in these cities" (Dodson 2009: 109). Investment in infrastructure provision has emerged as one of the few levers available to state governments looking to implement MSPs. This position is re-enforced in the current federal government *Smart Cities Plan*, which prioritises "investment in public transport projects that improve accessibility to job centres and promote urban renewal" (DPMC 2016: 14). This is what Dodson (2009) terms the "infrastructure turn". In the case of Sydney, Searle (2013) identifies the *Metropolitan Transport Strategy* (2010) as essential in delivering the outcomes outline in the *Metropolitan Plan for Sydney 2036* (NSWDoP 2010). Similar transport and infrastructure plans have been vital in assisting with the delivery of the current round of MSPs (Burke 2016). In addition, a number of state governments have invested in significant new public transport systems, designed to support growth in mixed-use urban centres. Examples include the Sydney Metro, Melbourne Metro, the Gold Coast light rail (G:link) and the Fremantle rail line. In Canberra, the Northbourne Avenue rapid transit corridor is identified as a key site in need of detailed plans for urban intensification (ACTDESD 2012: 6). Despite these major projects, the capacity of state governments to fund major transport infrastructure remains a challenge (Burke 2016). This limited funding often leads to complex public–private funding arrangements as a central aspect of delivery (Chapter 4).

One mechanism mobilised across MSPs promoting a regenerated urban form is the use of a centres hierarchy. The approach seeks to deliver the polycentric city by clearly differentiating the form and function of centres across the urban area:

> Together, these centres form a network of transport-connected hubs that help to make Sydney a networked and multi-centred city.
>
> (NSWDPE 2014: 46)

Centres are defined by population size and densities, dwelling types, employment and economic performance and transport infrastructure. The centre hierarchy included in current MSPs is outlined in Table 2.3. While the exact number of tiers and their definition differs between states, all MSPs tend to promote a three-tiered hierarchy of centres: central/CBD; regional; and local.

Table 2.3 Centres hierarchy[1]

Metropolitan plan	*CBD/Central city centres*	*Regional/primary centres*	*Local centres*
A Plan for Growing Sydney	CBD (Central Sydney/North Sydney and Parramatta)	Regional city centres Strategic centre	
Plan Melbourne	Expanded central city (state significant)	Metropolitan activity centres (state significant)	Activity centres (locally significant) Neighbourhoods centres (locally significant)
SEQ Regional Plan	Primary activity centre	Principal regional activity centres Major regional activity centres	
Directions 2031 and Beyond	Capital city	Primary centres Strategic metropolitan centres Secondary centres	District centres Neighbourhood centres Local centres
The 30-Year Plan for Greater Adelaide	Transit oriented developments	Transit oriented developments[2]	

Source: Various metropolitan planning strategies

Notes

1 Excluding employment, industrial, health and rural centres/precincts.
2 In Adelaide, transit oriented developments in both the CBD and in key regional locations across the urban area. The objectives and built form of these centres are the same.

CBD or *central city centres* focus on the historical core of the city, although *A Plan for Growing Sydney* defines Parramatta as Sydney's second CBD centre. The objectives of CBD centres are explicitly mixed-use. However, the goal is primarily around maintaining and expanding the economic performance, with employment objectives often taking precedence over other land uses. These centres are the location of high-skilled employment, government services, higher-order retail, entertainment and tourist services. The catchment for these services is the entire city. The CBD centres are the focus of gaining and maintaining "global city status" with planning and development frameworks alignment with this goal (Baker and Ruming 2015). Increasing housing and, in particular, architecturally innovative developments are core to this strategy. It is within these centres where large-scale urban regeneration projects are most likely to be located.

Regional or *primary centres* are the secondary urban centres. These centres have a sub-regional catchment and perform major service delivery functions in terms of government, health, justice and education services. These centres also play key commercial and retail functions. These centres, via increased high density development are targeted with accommodating a significant proportion of new development. As established urban centres, the form of urban regeneration enacted in these sites is often in the form of main street or shopping centre-led regeneration (Chapter 13). Unlike urban regeneration processes in *central city* centres where development tends to take the form of either brownfield precinct redevelopment or the construction of single, large high-rise developments, regeneration in *regional centres* is likely to be more piecemeal, sporadic and require greater coordination across state and market actors to deliver the urban vision. Nevertheless, these are the centres which are essential in delivering a regenerated, polycentric city (Randolph and Freestone 2012; Pinnegar and Randolph 2012).

Local centres have a much smaller catchment and tend to provide retail services capable of meeting weekly or day-to-day needs. Generally housing densities and targets are lower than other centres further up the centre hierarchy. Local centres are seen to play a vital role in community development and urban liveability through the delivery of walkable neighbourhoods, infrastructure such as parks and open space, and services such as local libraries. Importantly, *Plan Melbourne* recognises local centres as essential in the delivery of the polycentric, 20-minute city, where *neighbourhood centres* provide:

> access to local goods, services and employment opportunities. Planning in these locations will help to deliver 20-minute neighbourhoods across Melbourne.
>
> (VicDTPLI 2014: 31)

In addition, metropolitan plans have progressively identified a series of specialised centres, usually associated with a particular type of employment or economic activity. These centres are resposible for significant trip generation, but are not necessarily targeted for housing or population growth.

Importantly, recent strategies have also begun to identify specific renewal or regeneration sites. Across MSPs, these urban renewal centres, sites or precincts tend to align with the objectives of higher-order centres. Urban renewal centres are recognised as vital in meeting targets around housing and employment. They are also positioned as essential in delivering flagship global urban renewal projects (Chapters 6, 7, 8, 9 and 10). *A Plan for Growing Sydney* identifies a series of "priority revitalisation precincts" (priority precincts) as sites suitable for urban regeneration, capable of increased housing and able to coordinate "planning and investment to revitalise local centres, services and infrastructure" (NSWDPE 2014: 67). The priority precinct projects seek to: facilitate state, regional or local strategies relating to housing, employment and urban renewal; maximise infrastructure (especially transport); coordinate developments which impact upon more than one local council; ensure environmental and social sustainability; and, maintain financial viability within the context of market fluctuations (NSWDPE 2014: 67). In *Plan Melbourne* two scales of renewal precincts are defined – state significant and locally significant – which seek to:

> take advantage of underutilised land close to jobs, services and public transport infrastructure, to provide new housing, jobs and services. Renewal projects in defined precincts and sites will play an important role in accommodating future housing and employment growth and making better use of existing infrastructure.
>
> (VicDTPLI 2014: 31)

It is within the higher-order centres, and increasingly within designated renewal centres, where often large, quite specific, development projects are identified. These projects might be termed large-scale, official or flagship regeneration projects as they are positioned as vital to achieving the urban vision set out in the MSP. Contemporary MSPs are replete with locations targeted for large-scale urban regeneration projects (Table 2.4). The majority of these sites are located on brownfield urban sites. As Bunker and Searle (2009) note, the process of large brownfield regeneration projects is a relatively recent occurrence in Australia, compared to the UK, the USA and Europe. These sites offer a number of opportunities for the regeneration of Australian cities as they:

- tend to be located in inner- and middle-ring locations;
- are often in single ownership and, thereby, overcome the challenge of lot consolidation;
- often have fewer environmental constraints (although site remediation is often a challenge);

- are potentially subject to less local opposition given the absence of existing residents on the sites;
- are able to access existing infrastructure and services, limiting the need for significant expenditure;
- are often owned by the state government, allowing the government to leverage under-utilised land to facilitate development; and,
- offer large and profitable development opportunities to the private sector who might be interested in purchasing the site or developing it in partnership with the state.

Increasingly these large-scale regeneration projects are identified within MSPs (and urban and economic policy more widely) as key elements and exemplars of the global status of Australian cities. These large-scale regeneraton sites are seen to: provide high-quality office space for national and international corporations; appeal to both mobile urban elite looking to move to our cities from overseas and Australian professionals looking to live in fashionable (yet extremely expensive) inner-city areas; provide examples of innovative architecture; provide high-end leisure facilities (such as casinos); and, act as tourist destinations:

> Future development of the Fishermans Bend Urban Renewal Area, which could accommodate up to 40,000 new jobs and 80,000 residents, provides an important opportunity to expand the central city and to consolidate Melbourne's position as one of the world's most liveable cities, with a highly creative and competitive economy.
>
> (VicDTPLI 2014: 39)

Table 2.4 Large-scale regeneration sites

City	*Major regeneration sites*
Sydney	Barangaroo, Darling Harbour, Walsh Bay, The Bays precinct, Parramatta North, Central to Eveleigh corridor
Melbourne	CBD, Docklands, Southbank, Fishermans Bend Urban Renewal Area, E-Gate, East Werribee
SEQ	None listed although North Bank and Queens Wharf Development and Kurilpa Riverfront Renewal have subsequently been identified. The regeneration of the showgrounds at Bowen Hills has also been initiated, despite being absent from the regional plan (Searle, 2010)
Perth	Perth Waterfront project (Elizabeth Quay)
Adelaide	Bowden

Source: Various metropolitan planning strategies

> Barangaroo is Sydney's newest precinct. It will provide a hub for Sydney's financial and professional services and will further enhance the city's appeal for international investment and skilled workers.
>
> (NSWDPE 2014: 24)

> Perth waterfront currently contains a small entertainment precinct. However, state government plans are progressing to create a vibrant, exciting waterfront precinct which will be connected to the city and the Swan River. The plan contains office space, apartment living, hotels and retail space.
>
> (WADoP 2010: 79)

Despite a central position in MSPs, issues of equity and social justice are often raised around such large-scale regeneration projects. Key criticisms relate to the tendency of these developments to: limit (reduce) affordable and social housing in inner-city areas, thereby limiting access to the city by low- and middle-income populations; the removal or reduction of public space, infrastructure and facilities, thereby privatising urban space for elites able to purchase dwellings or high-end services in the area; and, concerns around developer influence of the planning process, leading to claims of undue influence which shift the economic and social benefits away from the public to the private sector. These issues are explored in more detail in Chapters 6, 7, and 8.

A further critique of the centres model has been the absence of suburban spaces in MSPs (Dodson, 2016). In the case of Sydney, a number of previous metropolitan plans have been largely silent on Western Sydney (Pinnegar and Randolph 2012). Importantly, the most recent strategy has moved to address this silence, and position middle and outer suburbs as important sites of regeneration. *A Plan for Growing Sydney* places a greater emphasis on securing employment in the region, while also seeking to build new housing and facilitate "urban renewal around centres in Western Sydney" (NSWDPE 2014: 16). While the focus in the outer suburbs is welcomed, the spatial planning strategy for these regions amounts to little more than transplanting the TOD/centres based model to the suburbs. MSPs remain largely silent on alternative forms of regeneration such as knockdown and rebuild (Chapter 14) and greyfield precinct regeneration (Chapter 12).

The ability of the state to implement this strategic vision and deliver effective activity centres rests upon funding, planning and assessment regulatory mechanisms, the state of the housing market and the development objectives of the private-market actors. Bunker (2014) argues that while population and economic growth has been achieved in the central city core and some inner suburbs, growth in major activity centres has been mixed. Newton and Glackin (2014: 133) claim that less than 14 per cent of the net increase of infill housing in Melbourne between 2002 and 2010 was due to the activity centres policy (most of which was confined to large strategic brownfield development). Further, Burke (2016) argues that there is only limited up-zoning and TOD around rail stations, with most centres surrounded by relatively low density land uses. Development in activity centres has been restricted due to "difficulties of redevelopment to higher densities" (Bunker 2014, p. 453), such as issues of lot consolidation, retrictive regulatory framework (floor space ratios,[3] zoning) which limit development options, and community opposition. Newton and Glackin (2014) argue that government routinely "underestimate the level of innovation, strategic thinking and resourcing" needed to facilitate regeneration. There is a disjuncture between the vision outlined in MSPs and the regulatory tools and development mechanisms available to the state.

Implementing the urban regeneration vision

The tools available to state governments to deliver their strategic urban visions are rather limited and tend to lie with reformed (streamlined) regulation and assessment process and changes in zoning, all of which lie outside the remit of MSPs. The lack of statutory authority in many states represents a significant challenge, with state environmental and planning legislation used to translate metropolitan strategic visions to local planning frameworks. Queensland represents an unusual case in Australia with the South East Queensland Regional Plan always having statutory authority (Searle 2016), while the appointment of the Greater Sydney Commission is an attempt to enhance this authority in Sydney.[4] These cases aside, the disjuncture between strategic and regulatory functions of the planning system represents a significant barrier to the delivery of the urban strategic vision outlined in MSPs. The most recent round of metropolitan plans have more clearly aligned the delivery of strategic urban visions with broader reforms of the regulatory function of the planning system (Maginn et al. 2016). Thus, the delivery of a regenerated city as presented in MSPs often rests with the reconfiguration of regulatory mechanisms framing development assessment and approval, especially for large-scale regeneration projects (Chapters 6, 7, 8 and 9). Nevertheless, the current generation of MSPs seeks to facilitate implementation in a number of ways.

First, MSPs seek to either provide or set the parameters for finer-gained sub-regional strategic planning. These sub-regional strategies generally provide more detail and direction for the locations of regeneration, the characteristics of desired development and required infrastructure. For Sydney, sub-regional plans[5] are for the "distribution of housing and employment" (NSWDPE 2014: 18). These sub-regional plans represent the middle of a planning hierarchy which seeks to translate the vision of MSPs into development on-the-ground, as they set the context for local planning frameworks established by councils. Nevertheless, the success of this process must be questioned, with the misalignment between local and strategic plans and, in some cases, the active resistance from local governments to the spatial vision and targets set for them acting as barriers to regeneration.

Second, in responding to observations that implementation requires an explicit metropolitan governance structure (Gleeson and Spiller 2012), many states have introduced a metropolitan-wide planning agency. The new delivery agencies are seen to override and coordinate planning and development decisions made at the local level, seeking to facilitate the delivery of the MSP vision which has long been acknowledged as a major challenge (Searle 2013; Bunker and Searle 2009). For example, accompanying *A Plan for Growing Sydney*, the NSW state government announced the establishment of the Greater Sydney Commission. The Commission is "a dedicated new body with responsibility to drive delivery of the Plan" (NSWDPE 2014: 5) and is seen to characterise a "new approach to delivery" (NSWDPE 2014: 5). The Greater Sydney Commission's first major task is the development of sub-regional/district strategies. Likewise, in Victoria, *Plan Melbourne* calls for the establishment of a new Metropolitan Planning Authority with:

> powers to plan state-significant sites and precincts; help to coordinate whole-of-government integrated land use; and provide oversight of the plan's delivery. It will be able to streamline planning.
>
> (VicDTPLI 2014: 17)

The push for new metropolitan-wide planning agencies comes as state governments react to the success of similar agencies in other jurisdictions in delivering planning strategies. Internationally the Greater London Authority has come to represent a model for these agencies, with its role in delivering the *London Plan* no doubt core to the models' appeal. Domestically, the Western Australian Planning Commission (WAPC) also acts as a model for other states. The WAPC, first established in the mid-1950s, has sought to coordinate state and private sector actors to deliver the MSPs for Perth and surrounds.

Finally, many state governments have taken a more active role in delivering, primarily, large-scale regeneration projects. A series of new delivery agencies have been established to facilitate development of key strategic sites. In NSW, UrbanGrowth NSW was established in 2013[6] with the express job of delivering more housing in "designated infill areas" and by "developing surplus or under-used Government land" (NSWDPE 2014: 8), assisting with the delivery of larger urban renewal projects in "strategically important locations" (NSWDPE, 2014, p. 19). Similar agencies, with core objectives of facilitating regeneration as outlined in MSPs, have been established in Melbourne (VicUrban), South Australia (Renewal SA), Western Australia (Metropolitan Redevelopment Authority) and Brisbane (Urban Renewal Brisbane[7]). Despite the intention of facilitating development across public and private sectors, one of the critiques of development agencies has been their tendency to focus solely on government assets, rather than intervening at sites which might better align with the strategic objectives outlined in the metropolitan plan (Fensham 2015).

The emergence of state development agencies represents a reconfiguration of the form of neoliberal urban governance which has been dominant in Australia for the past few decades. It has become apparent that direct state intervention, funding or underwriting is required to negotiate urban change and pursue the strategic objectives outlined in MSPs (Pinnegar and Randolph 2012; Randolph and Freestone 2012). While leaving the delivery of urban regeneration to the market might work in high-value segments of the urban property market, such as inner-city or brownfield sites, regeneration targets are unlikely to be met in low-value areas which fail to attract large-scale developers, resulting in small-scale, piecemeal and ad hoc development which does not necessarily align with strategic objectives (Randolph and Freestone 2012). It is in these locations where state development agencies have the potential to facilitate change. These agencies will need to realign their focus away from larger-scale regeneration projects in the inner and middle of our cities to other geographies and scales of urban change if the urban vision outlined in MSPs is to be enacted.

Conclusion

MSPs outline the state's vision of the future city. While Australian MSPs may have evolved into a unique planning paradigm (Searle and Bunker 2010), so too has a largely coherent vision of the future regenerated city emerged. Australia's metropolitan plans have increasingly sought to redefine the spatial structure of our urban areas to a more dense form via a process of ongoing regeneration, renewal and redevelopment. The regenerated Australian city is one based on infill, densification, compact city, centres and nodes, transit oriented development (TOD) and urban consolidation. A polycentric city constituted by a hierarchy centres distributed across the urban footprint is now the goal. The underlying logic of this form of regeneration rests on claims related to efficiency and cost effectiveness, sustainability, liveability, and affordability, health and ideals of community. These claims have been challenged by many. Nevertheless, these claims endure (or expand).

Despite these intentions, the capacity and tools available to implement these visions in a coordinated manner are limited. There has been much debate about the need to reform urban governance and policy frameworks to ensure the future city visions outlined in MSPs can be implemented. The past decade has illustrated the extent to which the strategies have failed to achieve their spatial and development goals, as growth boundaries have been revised, suburban greenfield development expanded, and infill development languished behind the targets (Bunker 2012). Thus, the regeneration of Australian cities primarily occurs outside those designated spaces identified within strategic planning policy. These regeneration processes and geographies are, thus, largely self-organising, market-led and ad hoc (Newton and Glackin 2014). It is an area's socio-economic status and general housing market position which define regeneration potential, rather than zoning conditions or place in strategic urban planning documents (Rowley and Phibbs 2012). In setting a broad urban vision for our cities, contemporary MSPs fail to adequately respond to the rate of change in urban conditions, fluctuating global economic conditions and political and policy instability (Bunker 2012). Bunker and Searle (2009) advocate for a retreat from the blueprint style MSPs which have dominated the planning of Australian cities and advocate strategic frameworks which acknowledge the fluid, dynamic and uncertain nature of urban change.

MSPs need to be recognised as an overarching strategy outlining the intended form of city-wide regeneration. Yet, planning agencies need to acknowledge, and indeed embrace the fact that they will never be implemented as envisaged. Rather they need to leave space for other regenerative opportunities. There is a significant disjuncture between the urban regeneration vision mobilised by MSPs and delivery on the ground. Yet, many of the projects and sites identified are undergoing significant regeneration pressures. Further, many of those spaces outside the polycentric urban model promoted in MSPs are also undergoing rapid transformation. Our cities are regenerating at an urban scale, just not necessarily aligning completely with the vision endorsed by MSPs. These projects and processes are explored in later chapters.

Postscript[8]

Since the completion of this chapter, a number of developments have occurred in Sydney and Melbourne. In Sydney, following the appointment of the Greater Sydney Commission, a draft amendment to update *A Plan for Growing Sydney* was released in November 2016. *Towards our Greater Sydney 2056* offers a vision of "a metropolis of three cities": Eastern City (centred on the exiting Global Economic Corridor), Central City (centred on Greater Parramatta), and Western City (centred on the future Western Sydney Airport) (GSC 2016a). Central to this new vision for Sydney is the desire to create a "30-minute city", "an equitable, polycentric city" and "a city of housing choice and diversity" (GSC 2016a: 6). The underlying push for urban regeneration and renewal remains at the core of the new plan. Indeed, *Towards our Greater Sydney 2056*, in responding to increased population projections, dedicates two full pages[9] "to accelerating housing opportunities" (GSC 2016a: 8). Urban renewal, medium-density infill development, and new communities in land release areas are identified as the three options for increasing supply. In terms of urban renewal, the strategy re-emphasises the renewal corridors identified in *A Plan for Growing Sydney* and adds criteria for new urban renewal corridors. Medium-density infill is identified as essential housing supply across the established city, with a *Draft Medium Density Design Guide* promoted as a means of "show[ing] how this local scale renewal can promote good design outcomes" (GSC 2016a: 9). Fringe development is recognised as a central element of new housing supply, yet it advocates for more "intense development around centres" in these development (Chapter 17). Simultaneously, the GSC released six district strategies for comment. These district strategies are a form of sub-regional planning outlined above, which seek to link the strategic vision of the MSP with local planning strategies and decisions. Importantly, the district strategies closely align with the vision and detail outlined in *A Plan for Growing Sydney*. One of the key underlying principles of the district strategies is "increasing housing choice in and around all centres through urban renewal in established centres" (GSC, 2016b: 15). The bulk of the district strategies is dedicated to exploring the development opportunities of key centres. A comprehensive review of the regional strategy and final District Strategies is to be released in late 2017.

Meanwhile, in Melbourne, the Victorian State Government released its revised *Plan Melbourne 2017–2050* in early 2017. As with the case in Sydney, the revised strategy seeks to respond to increased population projections. The revised strategy maintains the focus of the 20-minute city, as it seeks to "create mixed-use neighbourhoods at varying densities" and "support a network of vibrant neighbourhood activity centres" (Victoria State Government 2017: 12). The identification of metropolitan activities centres, located along key transport networks, remains central to the spatial vision of Melbourne. Importantly, "development opportunities at urban renewal precincts" (Victoria State Government 2017: 8) is identified as a key direction in the strategy.

Notes

1. Plan was not released at the time of writing.
2. *City of Cities* (2005), *Metropolitan Plan for Sydney 2036* (2010), *Draft Metropolitan Plan for Sydney to 2031* (2013), *A Plan for Growing Sydney* (2014).
3. The permissible ratio of gross floor area (developed floor space) to site area (lot size).
4. By appointing district commissioners to oversee planning panels which will assess significant planning proposals and re-zoning applications and ensure that they align with the vision of the district strategy.
5. Renamed district plans.
6. Replacing the state land development agency, Landcom.
7. A Brisbane City Council agency.
8. Added June 2017.
9. Of a 13-page document excluding front material and font and back covers.

References

ACTDESD (ACT Department of Environment and Sustainable Development) (2012) *ACT Planning Strategy: Planning for A Sustainable City*, Canberra: ACT Government.

Baker, T. and Ruming, K.J. (2015) 'Making 'Global Sydney': spatial imaginaries, worlding and strategic plans', *International Journal of Urban and Regional Research*, 39(1): 62–78.

Bunker, R. (2012) 'Reviewing the path dependency in Australian metropolitan planning', *Urban Policy and Research*, 30(4): 443–452.

Bunker, R. (2014) 'How is the compact city faring in Australia?', *Planning Practice & Research*, 29(5): 449–460.

Bunker, R. and Searle, G. (2009) 'Theory and practice in metropolitan strategy: Situating recent Australian planning', *Urban Policy and Research*, 27(2): 101–116.

Burke, M. (2016) 'Problems and prospects for public transport in Australian cities', *Built Environment*, 42(1): 37–54.

Charmes, E. and Keil, R. (2015) 'The politics of post-suburban densification in Canada and France', *International Journal of Urban and Regional Research*, 39(3): 581–602.

Cumberland County Council (1948) *County of Cumberland Planning Scheme*, Sydney: Cumberland County Council.

Davidson, K. and Arman, M. (2014) 'Planning for sustainability: an assessment of recent metropolitan planning strategies and urban policy in Australia', *Australian Planner*, 51(4): 296–306.

Davidson, K., and Gleeson, B. (2013) 'The urban revolution that isn't: The political ecology of the New 'Urbanology', *Journal of Australian Political Economy,* 72: 52–79.

Dodson, J. (2009) 'The 'Infrastructure Tu' in Australian metropolitan spatial planning', *International Planning Studies*, 14(2): 109–123.

Dodson, J. (2016) 'Suburbia in Australian urban policy', *Built Environment*, 42(1): 7–20.

DPMC (Department of Prime Minister and Cabinet) (2016) *Smart Cities Plan*, Canberra: Australian Government.

Forster, C. (2006) 'The challenge of change: Australian cities and urban planning in the new millennium', *Geographical Research*, 44(2): 173–182.

Fensham, P. (2015) 'The Sydney Metropolitan Strategy: Implementation challenges, in J. Buček and A. Ryder (eds), *Governance in Transition*, Dordrecht: Springer.

Gleeson, B. and Spiller, M. (2012) 'Metropolitan governance in the urban age: Trends and questions', *Current Opinion in Environmental Sustainability*, 4(4):393–397.

GSC (Greater Sydney Commission) (2016a) *Towards our Greater Sydney 2056*, Sydney: Greater Sydney Commission.

GSC (Greater Sydney Commission) (2016b) *Draft Central District Plan,* Sydney: Greater Sydney Commission.

Grosvenor, M. and O'Neill, P. (2014) 'The density debate in urban research: An alternative approach to representing urban structure and form', *Geographical Research*, 52(4), 442–458.

Hamnett, S. (2000) 'The late 1990s: Competitive versus sustainable cities', in S. Hamnett, and R. Freestone (eds), *The Australian Metropolis: A Planning History*., Sydney: Allen & Unwin.

Lennon, M. (2000) 'The revival of metropolitan planning', in S. Hamnett and R. Freestone (eds), *The Australian Metropolis: A Planning History,* Sydney: Allen & Unwin.

Maginn, P.J., Goodman, R., Gurran, N. and Ruming, K. (2016) 'What's so strategic about Australian metropolitan plans and planning reform? The case of Melbourne,

Perth and Sydney', in L. Albrechts, A. Balducci, and J. Hillier (eds), *Situated Practices of Strategic Planning*, London: Routledge.

McGuirk, P.M. and O'Neill, P. (2002) 'Planning a prosperous Sydney: The challenges of planning urban development in the new urban context', *Australian Geographer*, 33(3): 301–316.

McKenny, L. (2014) 'Land releases on Sydney's fringe spark fears of urban sprawl', *Sydney Morning Herald*, 6 January, http://www.smh.com.au/nsw/land-releases-on-sydneys-fringe-spark-fears-of-urban-sprawl-20140105-30bwd.html, (accessed: 31 August 2016).

Newman, P. (2009) 'Planning for transient oriented development: Strategic principles', in Curtis, C., Renne, J.L. and Bertolini, L. (eds), *Transit Oriented Development: Making it Happen*, Aldershot: Ashgate.

Newman, P. (2016) '"The 30-minute city": How do we put the political rhetoric into practice?', *The Conversation*, 18 March, http://theconversation.com/the-30-minute-city-how-do-we-put-the-political-rhetoric-into-practice-56136, (accessed: 31 August 2016).

Newton, P. and Glackin, S. (2014) 'Understanding infill: Towards new policy and practice for urban regeneration in the established suburbs of Australia's cities', *Urban Policy and Research*, 32(2): 121–143.

NSWDoP (NSW Department of Planning) (1995) *Cities for the 21st Century: Integrated Urban Management for Sydney, Newcastle and Wollongong*, Sydney: NSW Department of Planning.

NSWDoP (NSW Department of Planning) (2005) *City of Cities: A Plan for Sydney's Future*, Sydney: NSW Department of Planning.

NSWDoP (NSW Department of Planning) (2010) *Metropolitan Plan for Sydney 2036*, Sydney: NSW Government.

NSWDoPI (NSW Department of Planning and Infrastructure) (2013) *Draft Metropolitan Strategy for Sydney to 2031*, Sydney: NSW Government.

NSWDPE (NSW Department of Planning and Environment) (2014) *A Plan for Growing Sydney*, Sydney: NSW Government.

Office of the State Architect (2011) *H.30 Hobart Capital City Plan 2011–2041*, Hobart: Tasmanian Government.

Orchard, L. (1995) 'National urban policy in the 1990s', in Troy, P. (ed.) *Australian Cities: Issues, Strategies and Policies for Urban Australia in the 1990s*, Melbourne: Cambridge University Press.

Pinnegar, S. and Randolph, B. (2012) *Renewing the West: Prospects for Urban Regeneration in Sydney's Western Suburbs*, Sydney: City Futures Research Centre, UNSW.

Pinnegar, S. and Randolph, B. (2013) *Submission on Draft Metropolitan Strategy for Sydney to 2031*, Sydney: City Future Research Centre, UNSW.

QLDIP (Queensland Department of Infrastructure and Planning) (2009) *South East Queensland Regional Plan 2009–2031*, Brisbane: Queensland Government.

Randolph, B. (2006) 'Delivering the compact city in Australia', *Urban Policy and Research*, 24(4): 473–490.

Randolph, B. and Freestone, R. (2012) 'Housing differentiation and renewal in middle-ring suburbs: The experience of Sydney, Australia', *Urban Studies*, 49(12): 2557–2575.

Rowley, S. and Phibbs, P. (2012) *Delivering Diverse and Affordable Housing on Infill Development*, Melbourne: AHURI.

Ruming, K.J., Maginn, P.J. and Gurran, N. (2013) 'Old wine in new bottles? New wine in old bottles? Convergence/divergence in Australian metropolitan planning', *Proceedings of the Planning Institute of Australia's 2013 National Congress*, Canberra: Planning Institute of Australia.

Searle, G. (2004) 'The limits to urban consolidation', *Australian Planner*, 41(1): 42–48.

Searle, G. (2013) '"Relational" planning and recent Sydney metropolitan and city strategies', *Urban Policy and Research*, 31(3): 367–378.

Searle, G. (2016) 'Strategic planning and land use planning conflicts: the role of statutory authority', in L. Albrechts, A. Balducci, and J. Hillier (eds), *Situated Practices of Strategic Planning*, London: Routledge.

Searle, G. and Bunker, R. (2010) 'Metropolitan strategic planning: An Australian paradigm?', *Planning Theory*, 9(3): 163–180.

SADPLG (South Australian Department of Planning and Local Government) (2010) *The 30-Year Plan for Greater Adelaide*, Adelaide: Government of South Australia.

VicDOI (Victorian Department of Infrastructure) (2002) *Melbourne 2030*, Melbourne: Victorian State Government.

VicDTPLI (Victorian Department of Transport, Planning Local Infrastructure) (2014) *Plan Melbourne*, Melbourne: Victorian State Government.

VicPCD (Victorian Department of Planning and Community Development) (2008) *Melbourne @ 5 Million*, Melbourne: Victorian State Government.

Victoria State Government (2015) *Plan Melbourne Refresh*, Melbourne: Victorian State Government.

Victoria State Government (2017) *Plan Melbourne 2017–2050: Summary*, Melbourne: Victorian State Government.

WADOP (Western Australian Department of Planning) (2010) *Directions 2031 and Beyond*, Perth: Western Australian Planning Commission.

Whitzman, C. (2015) 'Can zombies become human again? Plan Melbourne, zombie institutions, and citizen dissent', *Planning Theory & Practice*, 16(3): 441–446.

3 Urban regeneration and planning regulations

Nicole Gurran and Peter Phibbs

Introduction

Urban regeneration and renewal processes are typically catalysed by 'special' regulatory arrangements which excise major sites from local control. This, often controversial, process exemplifies the dilemmas and ambiguities about planning for urban regeneration on sites in public or private ownership: very often the development context involves a significant upheaval for existing communities – residents and/or businesses. Justified by regional or state-level economic, environmental, or social objectives; the legal arrangements by which urban regeneration has been enabled in Australia, are the focus of this chapter. These legal arrangements can relate to: the declaration of regeneration areas and responsible planning and development authorities; the 'strategic' planning of land uses; the assessment of particular developments; and the mechanisms for infrastructure financing and provision.

While the term 'regeneration' is used in a generic sense throughout this chapter, in practice other terms are used in various jurisdictions to refer to the process of renewal of sites and precincts on which urban development has already taken place. In Australia, these so-called 'brownfield' contexts have typically included under-utilised industrial sites; areas of private residential and or commercial activities where there is significant potential for more intensive development; and social housing estates. Across these three contexts, it has usually been the case that sites earmarked for regeneration will have strategic locational significance, due to their proximity to a CBD or regional centre, waterfrontage, or accessibility to existing or planned public transport projects. By contrast, the international practice in relation to urban regeneration efforts often relates to areas of housing market failure. In these contexts, significant impediments to market-driven renewal – such as economic downturn or socio-spatial concentrations of disadvantage and stigma – provide the impetus for organised intervention by communities or public authorities. These conditions can also be seen in the redevelopment of social housing estates in Australia. However, for the most part the potential to unlock significant commercial value of a strategically positioned site, for instance by investing in new transport infrastructure, cultural facilities, or sponsoring main street revitalisation efforts, has been the primary driver of regeneration efforts in Australia, rather than assisting distressed housing markets and communities.

Therefore, in examining the regulatory arrangements which have enabled Australian urban regeneration projects in this chapter, a number of tensions arise:

- The ambiguous role of government land development authorities in relation to the strategic planning and development assessment of regeneration projects, vis-à-vis state and local planning agencies;
- The extent and circumstances by which existing (community endorsed) planning controls are able to be suspended, bypassed, or amended to facilitate urban regeneration projects; and,
- The balance between public and private benefits arising from the regeneration process, for instance, in relation to public and community facilities, and affordable housing outcomes.

The question of public benefit is particularly pertinent because planned regeneration activities are intended to improve land values to catalyse wider development efforts. Consequently, without provisions to prevent displacement of existing, lower-income residents such interventions are likely to displace these groups to other areas of the city. It may be argued that organised regeneration will improve housing affordability overall by generating increased and more diversified housing supply. However, as outlined in this chapter, there is little evidence to demonstrate that urban regeneration efforts in Australian cities have improved housing affordability outcomes for lower-income residents, either within the regeneration area or more widely. This reflects a failure to embed regulatory approaches which preserve or produce affordable housing units on a sufficient scale. More widely, it points to a regulatory approach to regeneration planning in Australia which prioritises market, rather than public interests.

The structure of the chapter is as follows. The first section sketches the research and literature on intersections between urban regeneration projects and regulatory-planning systems in Australia and internationally, highlighting tensions between local and regional/metropolitan interests, and approaches to managing risks of displacement and gentrification. The second section provides an overview of the regulatory framework for urban regeneration in Australia, distinguishing between three different contexts or regulatory settings:

1 major renewal sites owned or driven by government;
2 major renewal projects driven by the private sector; and,
3 regulatory attempts to promote significant levels of residential and or commercial redevelopment and intensification within a defined locality or localities.

The third section summarises and compares the regulatory arrangements and roles of different levels of government, illustrated with reference to major regeneration projects in Queensland (QLD), South Australia (SA), Victoria, Western Australia (WA) and New South Wales (NSW). We provide particular detail on a range of renewal projects and their legal arrangements in the state of New South Wales where there has been consistent policy experimentation across all three forms of regeneration since the early 1990s. Even so, the comparison projects are not fully representative of the range of regulatory arrangements applying to major renewal sites in Australia. Rather, they provide a basis for identifying and comparing the ways in which these sites have come about; the roles of government development and redevelopment agencies; and the range of outcomes that have been achieved over time. Similarly, it is not possible to examine all of the public benefit outcomes that might arise from regeneration projects. For this reason we focus specifically on the delivery of affordable housing in the context of a chronic national shortage of homes affordable to low and moderate income earners, particularly in Australia's major cities (NHSC 2014).

Urban renewal, planning, and housing

There is a robust and diverse literature on the subject of urban renewal, reviewed elsewhere in this book. Early contributions problematise the role of the modern (particularly post-World War II) planning project which often worked in concert with public housing projects to undertake widespread 'urban renewal' schemes (Gans 1965, Carmon 1999, Keating 2000). These were delivered by local authorities in the United Kingdom and United States but by state governments in Australia. In Australia, while the Commonwealth government has no formal role in relation to urban planning, from 1945 cheap loans were provided to the states to finance inner-city slum clearance and public rental housing development, on the proviso that they enact modern town planning laws (under the first Commonwealth State Housing Agreement) (Troy 2012). Importantly, these early regeneration schemes were criticised at the time for destroying urban and community fabric. However, in seeking to remove and replace substandard housing, the objectives of these post-war public housing schemes were primarily socially motivated to improve the living conditions of the working poor.

By the 1980s and 90s, regeneration efforts began to address diverging goals. On the one hand the modernist public housing estates had begun to fail, due largely to increasing demand for affordable accommodation which was not met through increased public housing provision or in the wider private market (Chapter 15). This led to intense needs-based targeting of access to public housing, causing a 'residualisation' effect whereby public housing estates became stigmatised by concentrated social disadvantage (Forrest and Murie 1983, Arthurson 1998, Pearce and Vine 2014). The distinctive built form of modernist tower blocks and poor management and maintenance arrangements were seen to exacerbate these social problems, leading to a new and ongoing wave of urban renewal and regeneration within social housing, particularly in the USA and the UK, but also in Australia (Keating 2000, Randolph and Judd 2000, Cameron 2003). A second stream of regeneration was also occurring in the wake of economic restructure and the de-industrialisation of many inner city areas, which became important 'brownfield' sites for redevelopment as part of a wider environmental agenda intended to contain the spread of urban growth (Hamin 1999, Dixon 2007, Adams et al. 2010). A third wave of projects were about city branding and providing redevelopment of large waterfront sites as a mode of city branding in a globalised world (Dovey 2005, Galland and Hansen 2012) (Chapters 6, 7 and 8).

A common thread across planned regeneration projects in both the USA and the UK has been strong commitment to affordable housing inclusion as part of the residential development and redevelopment process (Crook 1996, Calavita et al. 2010). In the USA this has often been delivered through inclusionary zoning requirements, while in the UK, mandatory targets provide a basis for site-by-site negotiation with developers about the quantity and composition of affordable homes to be provided (Whitehead 2007). The notion of capturing value created by the public planning and infrastructure process is thus implicit within these international schemes, which increasingly apply to regeneration sites.

The urban containment agenda emerged in Australian metropolitan planning in the early 1990s. Labelled 'urban consolidation' by state planning authorities, the policy of intensifying residential development within existing suburbs was conceived as a strategy for slowing outer metropolitan sprawl, and capitalising on under-utilised infrastructure (McLoughlin 1991, Searle 2007) (Chapter 2). Urban containment policies have been criticised in the UK for reducing the supply of land for housing development, and forcing developers to use more complicated sites which can be expensive to remediate, thus reducing housing supply overall and exacerbating house price pressures (Hall 1973, Cheshire and Leven 1986, Cheshire 2008). Similar arguments about the lack of sufficient brownfield sites, and inadequate greenfield land supply have been made in Australia (Property Council of Australia 2007, Gurran and Phibbs 2016). However, in contrast to the UK, where containment efforts have focused on under-utilised industrial sites and public housing estates in need of renewal, in Australia urban consolidation efforts have extended to established, low-density residential neighbourhoods with the objective of redeveloping detached cottages as multi-storey apartment blocks (Ruming et al. 2012). This has made the urban consolidation agenda particularly contentious with local communities and their representatives. Often established residents are deeply opposed to metropolitan planning goals which aim to increase housing supply by enabling the redevelopment of low-density homes as apartments. A key tension is the particular regulatory approach to planning for urban regeneration in Australian capital cities, which allows state governments to overcome community opposition to densification and renewal, by enacting special laws and creating new authorities empowered to override local planning controls and which might otherwise prevent high-rise flats and apartments.

Three modes of urban regeneration under Australian planning

Urban regeneration practices in Australia have evolved into three main approaches, each of which is subject to a different set of legislative, planning, and funding arrangements. These approaches include: (1) regeneration on major government sites or on sites whereby a public authority (including a government land authority) has a driving role in the process; (2) redevelopment driven by the private sector (i.e. a major redevelopment project with a high capital investment value); and, (3) regulatory attempts to promote redevelopment and intensification within established and built up urban areas.

The first approach was kick-started by the Commonwealth Government's series of demonstration initiatives under the Building Better Cities (BBC) program (see Neilson 2008). Launched in the early 1990s, these regeneration schemes, involving all three levels of government, helped institutionalise a new model of site-specific planned urban regeneration in Australian cities, driven primarily by state governments but typically involving private partnerships. A variation of these government-driven redevelopment schemes has also been rolled out on public housing estates, driven by state land and housing corporations (Arthurson 1998). These government led regeneration approaches are typically subject to a different set of regulatory arrangements than those affecting wholly private developments, at least for the period of active development.

Prior to the Building Better Cities program, private sector housing development primarily focused on greenfield sites. However, from the mid 1990s, major property developers became active in acquiring and redeveloping underutilised industrial land, particularly for housing development. The scale and capital investment value of these developments, and the need for wholesale revision of the existing controls (often through a master planning process), contributed to the evolution of special regulatory arrangements for these major sites.

For state governments, focused development on major sites represented a significant opportunity to deliver new housing within the existing urban footprint. However, wider strategies were also required to meet housing supply targets. Thus across most jurisdictions, state planning policy articulated through metropolitan strategies began to emphasise a more generalised approach to urban consolidation for implementation through changes to local plans (Bunker and Searle 2009). In practice, this often became a form of incremental regeneration through infill housing development driven primarily by the market, enabled to take up new development opportunities released by changes to prevailing residential and commercial zones. In the same way that major urban regeneration sites have been enabled by special state planning instruments which override local planning controls; increasingly incremental, infill redevelopment processes have been facilitated by state powers, exercised via indirect or direct policy and legal instruments.

These three distinctive approaches to urban regeneration in Australia are summarised in Table 3.1.

Government-driven regeneration projects

Urban regeneration projects driven by governments characteristically involve government land, either land already held which is surplus to requirements; public housing estates; or land which is acquired. Often acquired land will include former industrial sites which need significant remediation before being developed. As shown in Table 3.1, examples of these government driven regeneration projects include the former industrial area of 'Green Square/ Victoria Park' in Sydney's inner East, 'Docklands' in central Melbourne, 'Subiaco' in Western Australia, and 'Bowden' in South Australia.

Regeneration of these strategically located sites (all of which are located within the inner urban ring), has involved a combination of government-owned and privately held land. For the most part, processes have been managed by a special authority operating as a development agency, sometimes with planning powers (as in the case of Subiaco). Typically a special planning scheme will apply to the area, prepared by the development authority or local government with close state involvement. Assessment of development proposals within the area is undertaken either by local government (particularly for developments beneath a certain capital investment or environmental impact threshold); by a state authority under ministerial delegation (for more significant projects); or, in the case of Subiaco, by the development authority itself. Ultimately planning responsibility for the area will transfer to the relevant local government authority once the regeneration scheme is well underway or completed. Overall however, the long timeframe involved in the regeneration of these major precincts means that a combination of different regulatory and planning arrangements may evolve in line with larger changes to planning systems and government.

For instance, the redevelopment of the Green Square area in inner Eastern Sydney was precipitated in the early 1990s by the announcement of a new rail line connecting Sydney's airport and the CBD, to form the central point of a new 'Green Square Town Centre'. In the early years of the scheme, a special-purpose development corporation (the South Sydney Development Corporation) was established by the state government to coordinate the master planning and initial development of the 278ha area. The Development Corporation was disbanded in 2005, and its functions adopted by the government's land developer, then named Landcom. Landcom operated in many ways like a private-sector developer but invested

Table 3.1 Examples of urban regeneration approaches in Australia

Mode	*State*	*Example*	*Planning context*	*Planning authority*	*Funding*
Government-driven regeneration projects	NSW	Green Square/Victoria Park (former industrial area) Bonnyrigg (public housing redevelopment), South Western Sydney Barangaroo	Special schemes prepared	State/local	Public / private
	SA	Bowden (former industrial site)			
	Victoria	Docklands (inner Melbourne)			
	WA	Subiaco			
Redevelopment driven by the private sector	NSW	Carlton United Breweries site (Central Sydney)	Proposals assessed under special planning framework	State Local/state (Central Sydney Planning)	Private
Regulatory attempts to promote regeneration within a defined area	QLD	Brisbane City Plan (abolished height limits in parts of CBD)	Schemes amended to accommodate infill development	Local	Private
	NSW	State Environmental Planning Policy 53 NSW (Metropolitan Residential Development) (1997) (required local councils to prepare strategies to accommodate increased housing supply) Councils now required to ensure local plans can accommodate state defined targets for housing growth, amending local zones if required			

Source: Authors

heavily in public space and landscaping, demonstrating leading practice in water sensitive urban design and aiming to lead the market in high-quality high-density housing development. By 2013, Landcom itself had become a development authority and was renamed Urban Growth. Meanwhile, the local government authority, the City of Sydney, has maintained its role in preparing the statutory land-use plans for the area (land-use zones, development controls on floor space ratios, heights etc.), and assessing individual developments beneath a set capital investment threshold.

The renewal of public housing estates typically involves a slightly different process, stemming from the particular objectives associated with the redevelopment and the particular opportunities and constraints affecting each site and community. Public housing renewal schemes have been underway in Australia since the 1990s, but many of the earlier efforts focused on public realm and housing design upgrades which involved estate 'improvement' rather than wholesale redevelopment (Randolph and Judd 2000). In the context of declining funding for new social housing provision, as well as an ageing asset base in need of significant maintenance and upgrading, recent initiatives have attempted to leverage finance by selling assets or renewing estates via more intense, private-sector development (Darcy 2013). In many cases, the objective is for the renewal of public housing stock to be self-financed through the value created by the private scheme. This means that the strategic location of the site, and the establishment of planning rules to facilitate development of the desired scale, are critical considerations.

The renewal of the 'Bonnyrigg' social housing estate in South-western Sydney, provides an example of the way in which these schemes are governed (Chapter 15). By the turn of the new millennium, Bonnyrigg was characterised by deepening social problems and a poor-quality, low-density housing stock constructed in the 1970s in the 'Radburn' style (Rogers 2013). Several of Sydney's public housing estates suffer similar social issues compounded by a substandard dwelling stock, but Bonnyrigg's strategic location in South-western Sydney with access to both Parramatta and Liverpool employment centres, made a redevelopment scheme more financially attractive.

The scheme itself involved a public–private partnership with private developer Becton, for the staged redevelopment of Bonnyrigg (renamed 'Newleaf') over a 30-year period. The existing dwellings, most of which were public housing, were to be demolished to make way for a combination of 2,332 detached homes, townhouses, and apartments, accommodating around 6,850 people (up from 2,900). The original public housing tenants were relocated on demolition of their homes, although most were given the option to return to the area once the development was complete. Concerns were nevertheless raised about impacts on community cohesion and social networks (Pinnegar 2013).

Major redevelopment projects driven by the private sector

The redevelopment of major sites in private ownership is often facilitated by special planning legislation, designed to overcome local-level constraints (for instance, restrictive zones or height controls). In many cases the legislation activates special provisions in planning law which allow projects or development meeting special criteria to be assessed by state rather than local government authorities.

In New South Wales, however, there has been concern that even these special planning and assessment processes are cumbersome, hindering important major developments. To further shorten the process and provide greater certainty to developers, controversial amendments to the NSW Environmental Planning & Assessment Act 1979 (EPAA), were introduced in 2005. These changes (under the former 'Part 3A' of the Act), enabled developers to proceed straight to the minister for 'in principle' approval on the basis of a 'concept plan', irrespective of the provisions within local plans and the attitudes of local government. The provisions were amended again in 2011, bringing private development back within the umbrella of prevailing planning rules, but special provisions for state significant projects, many of which involve redevelopment or regeneration activities, remain.

The so called 'Part 3A' amendments to the EPAA, were instrumental in facilitating the redevelopment of the 5.8ha 'Carlton United Brewery' (CUB) site, located on the edge of Sydney's CBD and opposite the University of Technology Sydney (UTS) (Chapter 10). Initial approval for a 'concept plan' was granted by the planning minister in 2007, two years after the brewery had ceased operation. The concept plan application was assessed by the NSW Department of Planning and Environment, which considered draft controls proposed for the site by the City of Sydney Council (and the state-appointed Central Sydney Planning Committee), but supported the applicant's preferred scheme. The applicant's scheme involved greater site density (from a floor space ratio of 4.0:1 to 4.05:1) and heights of up to 110 metres for two towers (ten per cent above that envisaged in the locally proposed controls) (Department of Planning 2007). The project approval included provision for five payments amounting to $32 million for the provision of affordable rental housing which was to be delivered offsite in the nearby Redfern Waterloo public housing and former industrial redevelopment area (City of Sydney 2014).

Since its original approval, the concept plan has undergone several modifications and a series of changes to the prevailing planning legislation (REDWatch 2013). In 2008 the project stalled during the period of the global financial crisis, and the developer subsequently sought an extension of its development approval. Under arrangements at the time of writing, while the minister remains the absolute authority for projects on the site, determination powers have been delegated to the NSW Department of Planning and Environment and the state-level independent Planning

Assessment Commission (in case of objection from the local City of Sydney Council). Now renamed 'Central Park' the former CUB area is home to over 2,000 residents, accommodated in a mix of apartments, including special-purpose student accommodation. The project has been lauded for its environmental credentials, claiming the largest on-site trigeneration system for power, heating and cooling used in an Australian urban development, as well as rainwater harvesting and re-use for non-drinking purposes (City of Sydney 2014). The high-quality environmental and public realm outcomes reflect the influence of strong local government advocacy from the City of Sydney, with requirements implemented through strict conditions of approval imposed by the state in its determination role.

The unique Central Park example notwithstanding, when state governments or specially convened development corporations drive the planning of regeneration projects, there is a risk of losing or alienating the deep experience and expertise able to be provided by local government authorities.

Regulatory attempts to promote incremental renewal efforts

The third category of regeneration is that encouraged by planning legislation, pre-empting concrete proposals by private- or public-sector developers. Incremental processes of urban regeneration through infill housing and intensification of residential areas has been encouraged through successive Australian metropolitan strategies dating from the 1990s (Bunker and Searle 2009). The implementation mechanism for these general urban consolidation/renewal policies has primarily been through amendments to residential zones in local planning schemes (Gurran 2011), sometimes assisted by state policy. For instance in 1991 the NSW government introduced State Environmental Planning Policy 32 – Urban Consolidation, which provided a framework for requiring authorities when preparing local plans to consider 'whether urban land is no longer needed or used for the purposes for which it is currently zoned or used, whether it is suitable for redevelopment for multi-unit housing and related development' (Clause 6); and if so, to re-zone the land accordingly.[1]

However, when local governments have exhibited resistance to higher-density housing in their areas, the states have sometimes intervened. In 1997, for instance, the NSW government introduced the controversial planning policy State Environmental Planning Policy (SEPP) 53 – Metropolitan Residential Development, which required local councils in the Sydney metropolitan region to have an approved residential development strategy in place, outlining opportunities for accommodating higher-density housing within their areas. While most local councils ultimately complied with this requirement, some, including the municipality of Ku-ring-gai held out. Consequently, the state government intervened to directly re-zone sites irrespective of local controls. A similar, if less heavy-handed approach to promoting local densification can be seen in the Victorian planning reforms of 2012 (Department of Planning and Community Development 2012), which were designed to encourage Melbourne's suburban councils to identify areas for more intense housing development via a threefold residential zoning scheme colloquially described as 'go-go, go-slow, and no-go' (Lee 2012, also see Rowley 2017 for a full description of these reforms).

Brisbane is in the relatively unique situation of having a single local government authority across the entire metropolitan area. In 2015, Brisbane City Council sought to encourage regeneration by abolishing height limits in parts of the CBD. Absolute height limits are set by state aviation regulations (274 metres), and in practice most projects will still undergo site-by-site assessment. However, the approach is intended to give developers confidence that high-density projects will be approved. Unlike the other major cities (Sydney, Melbourne and Perth), Brisbane City Council remains the planning authority for significant projects and has a dedicated urban renewal team to manage processes of redevelopment (Chapter 8).

Legislation, regulatory arrangements and roles of government in Australian urban renewal

A variety of legislative instruments and authorities to facilitate these urban renewal approaches have arisen across the Australian states and territories. In general terms, although closely related to wider urban planning (particularly metropolitan planning) policy and law, special legislative instruments have been used to implement urban-renewal objectives beyond standard (locally driven) processes of land use planning and development. These laws have typically applied to defined urban renewal sites, precincts, or development contexts meeting set criteria, and are often administered by special-purpose government land-development authorities or development corporations with land acquisition and assembly powers. Table 3.2 summarises the different laws and authorities involved in urban renewal processes across Australia.

As shown, state land development authorities (the ACT Land Development Agency; Urban Growth NSW; the NT Land Development Corporation; the Perth Metropolitan Development Authority, Economic Development Queensland (EDQ) and Places Victoria) feature prominently in identifying, facilitating, and sometimes directly undertaking urban renewal (often as a distinct part of wider land-development roles including greenfield residential projects). Public land development authorities were first established by most Australian states under the Commonwealth's Land Commission Program of the early 1970s (Gleeson and Coiacetto 2007). Generally enabled to undertake land acquisition and assembly, an early objective for the authorities was to maintain steady residential land supply, dampening speculative pressures. However, corporatisation of most authorities during the 1980s saw a shift away from volume land supply to cool the market (which in some jurisdictions, such as Victoria, was regarded to be anti-competitive) (McGuirk and Dowling

Table 3.2 Government land development and urban renewal legislation and authorities, Australian states and territories 2016

State	*Legislation/instrument/authority*	*Description*
ACT	Planning and Development Act (PDA) 2007	S. 31 of PDA provides for the establishment of the Land Development Authority, which develops government land. Undertakes renewal and greenfield projects.
	ACT Land Development Agency	
NSW	Growth Centres (Development Corporations) ACT 1974	Provides for identification of growth centres (rural and urban contexts), and for development corporations to manage development in those centres
	Urban Growth NSW Development Corporation Hunter and Central Coast Development Corporations	Develops projects (minister or local council is planning authority). Most major renewal sites are state significant and development of CIV over $10m goes to Department of Planning and Infrastructure
	State Environmental Planning Policy (Urban Renewal) 2010	Allows sites to be identified and developed as urban renewal precincts. Controls development in potential urban renewal precincts.
	Barangaroo Delivery Authority Act 2009	Development in Barangaroo is state significant if CIV over $10m (and for most subdivision applications). Otherwise planning authority is council.
	Barangaroo Delivery Authority	
NT	Land Development Corporation ACT 2014	Originally established in 2003, delivers land for residential, commercial and industrial purposes, and operates where viability might not be sufficient for private sector operating alone. No specific regeneration focus but undertakes projects on brownfield sites.
Queensland	Economic Development Act 2012	Continues functions previously established by the Urban Land Development Authority Act (ULDA) to facilitate redevelopment of identified sites for growth.
	Economic Development Queensland	Specialist land-use planning and property-development unit within Department of Infrastructure, Local Government and Planning
	Urban Renewal Brisbane	Section within Brisbane City Council
South Australia	Urban Renewal Act 1995	Originally enabled the establishment of urban renewal authorities; now combined under Renewal SA. Reports to Minister for Housing and Urban Development
	Urban Renewal Authority (trading as SA 'Renewal')	Emphasis on renewal and infill; charter includes 'facilitating attractive investment options for the private sector … supported by creating planning certainty and identifying any significant infrastructure and site remediation risks to allow integrated and master planned developments to proceed' (Renewal SA, 2016: 2)

Victoria	Urban Renewal Authority Victoria Act 2003	Establishes Urban Renewal Authority Victoria to carry out or manage urban renewal projects; contribute to implementation of government urban planning and development policies; and to complete the development of the docklands area (§ 1). Can 'purchase, consolidate, take on, transfer or otherwise acquire land in metropolitan and regional areas for development for urban purposes' and carries out urban renewal projects" (§ 7).
Victoria	Urban Renewal Authority Victoria Act 2003	Establishes Urban Renewal Authority Victoria to carry out or manage urban renewal projects; contribute to implementation of government urban planning and development policies; and to complete the development of the docklands area (§ 1). Can 'purchase, consolidate, take on, transfer or otherwise acquire land in metropolitan and regional areas for development for urban purposes' and carries out urban renewal projects" (§ 7). Objectives include to 'promote housing affordability and housing diversity'.
	Places Victoria	Property development agency for the Victorian government. Develops surplus government land.
	Victorian Planning Authority	New planning authority, expanded from Metropolitan Planning Authority. Reports to Minister for Planning. Strategic planning role – subdivision and 'unlocking' of strategic sites (e.g. brownfield sites identified as 'National Employment Clusters' under Plan Melbourne. Functions must be carried out on 'commercial basis' but can undertake actions to financial detriment with agreement of the Treasurer.
Western Australia	Western Australian Land Authority ACT 1992 Metropolitan Redevelopment Authority Act 2011	Objectives of the act include: the identification and development, or urban renewal, of centres of population and the provision or improvement of land for those centres (§ 3). The MRA Act makes the authority, planning body for identified redevelopment areas. In turn these areas are subject to their own 'redevelopment schemes', as well as planning policies, design guidelines, contribution plans, and structure plans. Once redevelopment is substantially completed, planning powers are 'handed back' to the local authority. Planning powers for Central Perth, Armadale, Midland, Scarborough and Subiaco; can write policy/guidelines and assess proposals. Promotes a 15 per cent affordable housing objecive.

Source: Authors

2009). Instead, the focus shifted towards demonstrating new products and leading practice in urban development, including through urban renewal projects, in established areas rather than on greenfield development sites. By 2015, the NSW land developer 'Urban Growth' (formerly 'Landcom'), had announced that it was withdrawing from new greenfield projects altogether in favour of urban renewal schemes. Similarly, the South Australian government's development authority 'Renewal SA' also has a sole focus on urban renewal and infill settings. One argument for government intervention in urban regeneration contexts as opposed to greenfield settings is the particular hurdles facing private-sector developers such as the need for complex and costly infrastructure upgrades and site remediation costs.

It is also the case that site- or project-specific development authorities are established by government to oversee or coordinate development of major renewal sites on public land or as a public–private partnership. These authorities are intended to play an important role in coordinating infrastructure provision and planning approvals, and operate somewhere between local and state governments in recognition of the wider significance of the overall precinct or project.

The former South Sydney Development Corporation (formed to facilitate the regeneration of Victoria Park/Green Square as outlined above) and the current Barangaroo Development Authority (BDA) in NSW are examples of this model (Chapter 6). The BDA is guiding the waterfront renewal of the western edge of Sydney Harbour, a 22ha former shipping site which is on government-owned land. The authority coordinates and guides the development of the site. This includes providing public infrastructure, and managing private-sector bids to deliver the residential and commercial components of the development. In return for the right to develop and then sell or own 99-year leasehold agreements for the built form (with completed public domain spaces returned to the authority), successful tenders carry out these developments in line with the vision and requirements set out in the tender documents. Responsibility for assessing specific development proposals within the wider scheme is shared between the local City of Sydney Council (for developments with a capital investment value (CIV) of less than $10m, and by the NSW Department of Planning and Environment, for higher value projects defined to be of state significance).

One obvious conflict with this planning and assessment process is that the state government is both a landowner and the consent authority. The government has an agreement with Lendlease, the developer for the site, to profit share from any additional floor space developed in the project beyond the original contract agreement. With this framework in place, it is not surprising that the development has added significant floor space as the project has been developed.

Private development bids are assessed by the BDA against set criteria, as illustrated below with reference to the Barangaroo Central site tender document (listed in order of importance):

1 Finance: Maximum financial offer
2 Risk: Highest certainty of outcome for the authority and greatest acceptance of risk by bidder
3 Development strategy and delivery plan
4 Excellence in urban planning and design outcomes
5 Public benefit: Maximum public benefit, both short and long term.

(BDA 2015: 6)

The BDA tender document provides no specific guidance on the definition of public benefit, but indicates that 'cultural, civic and community uses' should be accommodated, and that these might include 'arts enterprises, key worker housing, affordable office space, low rent creative studios, life-long education and community meeting spaces and recreational facilities' (BDA 2015: 25). In practice, and in relation to Barangaroo South, the first part of the site to be developed (in partnership with Lendlease), public benefit appears to have been interpreted as the creation of public open space and landscaping.

Other government development authorities with responsibility for urban renewal have had more explicit charters to developer specific public benefits such as affordable housing, as shown in Table 3.2. For instance, Perth's Metropolitan Redevelopment Authority (MDA), which takes on planning powers (preparing redevelopment schemes, planning policies, design guidelines, development contribution plans and structure plans) for identified redevelopment areas, aims to ensure that 15 per cent of all housing delivered is affordable for those on low or moderate incomes. Similarly, South Australia's 'Renewal SA' operates within a planning framework which requires that 15 per cent of homes delivered in new and redeveloping residential areas, be affordable (Davison et al. 2012).

Comparison of major regeneration projects

The range of laws, models, and approaches involved in delivering, facilitating and or regulating different types of urban renewal projects and sites in Australia, combined with frequent changes to these arrangements in line with changing state government regimes, makes it difficult to draw conclusions about which approach might represent leading practice. However, given that one of the most important rationales for government to support and facilitate urban renewal projects, is to facilitate the provision of more diverse and affordable housing supply in well-located areas, examining a series of major renewal projects in relation to housing outcomes seems an appropriate starting point for evaluation. In particular, urban renewal initiatives typically involve heavy government investment through the provision of infrastructure, public land, or the relinquishment of prevailing local planning controls and standards in order to facilitate more intense and high-value forms of private residential development. As noted, one of the drivers of market-led renewal processes is increased property values, so provisions which seek to ensure that lower-income renters are not disadvantaged or displaced by rising prices or rents, seems one important basis for assessing the public benefit outcomes of Australia's urban renewal schemes.

Numerous policy and planning mechanisms are able to be introduced to secure public benefit in general and affordable housing outcomes in particular, through urban renewal processes as demonstrated internationally (Davison et al. 2012). When the renewal sites consist of public land, in theory it is straightforward for governments to require a proportion of affordable housing (or to use the whole site for this purpose) subject to demonstrated community need, desirable tenure mix, and the costs of financing the overall development scheme. Local governments (such as the City of Port Philip in Victoria) have sometimes used their land reserves for affordable housing in this way, but state and Commonwealth governments have been reluctant to do so, perhaps because public land authorities have needed to deliver a commercial return to government. However, in the case of South Australia, the state has been prepared to accept lower returns on government sites in renewal locations, in order to meet its affordable housing target of 15 per cent (Gurran et al. 2008, van den Nouwelant et al. 2014).

Similarly, in Queensland (under the former Urban Land Development Authority (ULDA) and subsequently, the EDQ), affordable housing goals have remained part of the stated rationale of government renewal activities. For instance, the latest planning guidance for the state's Priority Development Areas, which include urban renewal sites of Woolloongabba, Bowen Hills, and Northshore Hamilton in Brisbane, enforces modest targets (15 per cent, 5 per cent and 5 per cent respectively) for homes delivered in these schemes to be affordable to median income earners (incomes up to $110,855 in September 2015) (Economic Development Queensland 2016). In 2016, this means that between 5–15 per cent of homes produced across these projects need to be available for rent at a maximum of $640 per week, and sold at up to $664,804, which is thought to be the maximum purchase price an investor could pay in order to rent the property at the targeted amount (assuming a 5 per cent yield).

There is very little systematic tracking or public reporting of affordable housing produced in Australia, and even less reporting of affordable outcomes on renewal sites. The most comprehensive study to date (Davison et al. 2012) was in many instances only able to report on targeted percentages for affordable housing delivery, rather than actual outcomes, and this remains the case at the time of writing. Acknowledging these data limitations, Table 3.3 summarises a selection of capstone urban renewal projects across five states, highlighting reported affordable housing outcomes. These renewal projects all fall within the first of our project types – government-driven regeneration projects – and are not primarily on public housing sites where the number of affordable dwellings produced is typically at least sufficient to replace the units lost to redevelopment. It should also be noted that across the jurisdictions, definitions of affordable housing differ. In South Australia, affordable housing includes homes for sale or rent that are affordable to low- and moderate-income earners, whereas in NSW, only rental accommodation is able to be delivered through the planning and development process, although affordability criteria also focus on low- and moderate-income earners. In Queensland, affordability targets are met if properties are able to be purchased at a price point that would allow landlords to set rents affordable for median-income earners.

Even with these caveats in mind, the affordable housing outcomes delivered to date on Australia's government-led urban renewal projects, as summarised in Table 3.3, seem very modest indeed. Despite criticism for inadequate affordable housing delivery at the time (Greive et al. 1999), the Subiaco Redevelopment Scheme achieved a proportion of almost 2 per cent affordable units in perpetuity, almost double the yield anticipated for Green Square (0.9 per cent). It is also higher than Docklands in Victoria (1.4 per cent), although the Docklands project has been extended by an anticipated 522 dwellings supported by the federal government's short-lived 'National Rental Affordability Scheme (NRAS)' which provided $10,000 in tax credits per year for ten years, to investors delivering new homes at up to 80 per cent of market rent. Indeed, in most cases additional subsidy or assistance has been available to help deliver the affordable component (Davison et al. 2014).

On private-sector sites, the affordable housing outcomes have been even more limited, despite the significant public benefits received by developers in the form of land-value uplift from changed planning controls and infrastructure investment. In NSW however, the Affordable Rental Housing State Environmental Planning Policy 2009 (ARHSEPP) includes a density bonus for infill residential projects incorporating affordable rental housing (which must be rented at a 20 per cent market discount for a minimum of ten years). This provides one potential model for securing wider public benefits from incremental urban renewal processes, although the approach is only useful where there is a demand for housing developments which exceed existing height and floor space controls. To date, data on the number and locations of residential development projects utilising this policy has not been reported by the NSW government, so it is difficult to assess its impact as a mechanism for infill renewal or for affordable housing creation.

Summary and conclusion

Urban renewal initiatives have been an important motif in Australian planning, and play a key role in evolving housing outcomes. The early slum clearance initiatives were intended, however misguidedly, to improve the housing conditions of the working poor, and saw the introduction of modern town-planning laws and the first modern public housing schemes. However, urban and housing policy in Australia soon detached, as city building efforts

Table 3.3 Selection of urban renewal projects and affordable housing outcomes, Australia 1990s–2016

State	*Years*	*Name*	*Authority*	*Area*	*Description*	*Dwellings*
New South Wales	1997–2031	Green Square town centre	Initially South Sydney Development Corporation, then Urban Growth City of Sydney Council	278 ha	Former industrial area between CBD and airport. Installation of train line catalysed development decision.	Initially 20,000 dwellings; now 30,500 (by 2031) Affordable housing: 330 dwellings (by 2035); inclusionary zoning requirement
Queensland	2010–	Woollongabba	Initially Urban Land Development Authority, then Economic Development Queensland	10 ha	Government-owned land with low-intensity government buildings, transport interchange; 2 km from CBD	2,000 dwellings anticipated; 15% to be affordable to median-income earners (300)
South Australia	2008–2026	Bowden	Renewal SA (Design Review Panel) reviews applications Charles Sturt Council	16 ha	Higher-density urban infill, 2.5km from CBD	2,400 dwellings 15% affordable, to low/moderate-income earners (360)

Victoria	1991–	Docklands	Docklands Authority (1991), later became part of Places Victoria	44 ha	Former industrial lands. Places Victoria sells development rights via competitive bidding; development agreement made for how and when sites to be developed, detailed proposals submitted to Department of Planning and Community Development before commencement. Ministry for Planning responsible for developments over 25,000 square metres GFA; Melbourne City for other projects	10,000 dwellings (approx.) 142 affordable (in perpetuity) 522 to be delivered as National Rental Affordability Scheme (NRAS) units – 20% below market rental
Western Australia	1994–	Subi-Central	Subiaco Redevelopment Authority 1994 Subiaco Redevelopment Scheme 1996	80 ha	Suburb blighted by railway line; submerged and transit oriented town centre developed to stimulate new housing and commercial development	1,975 dwellings; 33 affordable

Source: The authors, drawing on Gurran et al. (2008) and Davison et al. (2012)

focused on low-density suburbia in the 1960s–1980s. The new wave of urban-renewal initiatives in the early 1990s focused on industrial rather than residential areas, but, stimulated by the Commonwealth's Building Better Cities project, remained cognisant of the need to prevent displacement of lower-income residents when land and property values inevitably rose in response to new government-led development opportunity. This saw the introduction of Australia's first inclusionary zoning schemes and the establishment of special-purpose affordable housing company, City West in NSW. In the other states similar efforts to stimulate urban renewal without displacing lower-income renters have been attempted over time but overall affordable housing outcomes remain limited, with renewal activities seemingly focused more around unlocking development opportunities and demonstrating new templates for residential and high-density living, than on improving the housing conditions of existing residents, or leveraging value creation to support affordable housing supply.

The three modes of urban renewal reviewed in this chapter, along with the distinct and complex regulatory arrangements and authorities which enable major renewal schemes, give rise to a number of tensions. A primary tension is between state and regional interests – for development land, housing, and economic growth, versus local community concerns about neighbourhood character, amenity, and the capacity of local infrastructure and services. Often this tension plays out as an ideological contest between the aesthetic experiences of the city and suburb – with high-rise housing development at one extreme pitted against low-rise garden suburbs at the other. While a gradation between these two extremes may be preferable, in many cases it has proved politically and logistically simpler to concentrate high-density renewal schemes within defined areas near existing centres and transit lines, leaving traditional suburbs intact. However, as the supply of suitable sites diminishes, urban renewal will come to mean the redevelopment of residential suburbs as well as of under-utilised industrial lands. Without a comprehensive inclusionary planning framework to ensure that incremental renewal efforts secure affordable housing as part of the development process, this new supply will continue to serve higher-income earners, doing little to address Australia's chronic housing needs. Rather than addressing an explicit social objective (such as improving the housing conditions of the urban poor, or more efficiently utilising former industrial sites to prevent sprawl), the notion of public benefits in this new wave of urban renewal in Australia seems more distant than ever.

Note

1 State Environmental Planning Policy 32 – Urban Consolidation, was repealed in 2016 as redundant.

References

Adams, D., De Sousa, C. and Tiesdell, S (2010) 'Brownfield development: A comparison of North American and British approaches', *Urban Studies,* 47(1): 75–104.

Arthurson, K. (1998) 'Redevelopment of public housing estates: The Australian experience', *Urban Policy and Research*, 16(1): 35–46.

BDA (Barangaroo Development Authority) (2015) *Request for Development Bids Central Barangaroo*. Sydney: BDA.

Bunker, R. and Searle, G. (2009) 'Theory and practice in metropolitan strategy: Situating recent Australian planning', *Urban Policy and Research*, 27(2): 101–116.

Calavita, N., Mallach, A. and Lincoln Institute of Land Policy (2010) *Inclusionary Housing in International Perspective: Affordable Housing, Social Inclusion, and Land Value Recapture*. Cambridge, MA: Lincoln Institute of Land Policy.

Cameron, S. (2003) 'Gentrification, housing redifferentiation and urban regeneration: "Going for growth" in Newcastle upon Tyne', *Urban Studies,* 40(12): 2367–2382.

Carmon, N. (1999) 'Three generations of urban renewal policies: analysis and policy implications', *Geoforum*, 30(2): 145–158.

Cheshire, P. (2008) 'Reflections on the nature and policy implications of planning restrictions on housing supply: Discussion of "Planning policy, planning practice, and housing supply" by Kate Barker', *Oxford Review of Economic Policy*, 24(1): 50–58.

Cheshire, P. and Leven, C.L. (1986) *On the Costs and Economic Consequences of the British Land Use Planning System*, Reading: Department of Economics, University of Reading.

City of Sydney (2014) *Central Park, Sydney*, Sydney: City of Sydney.

Crook, A. (1996) 'Affordable housing and planning gain, linkage fees and the rational nexus: using the land use planning system in England and the USA to deliver housing subsidies'. *International Planning Studies*, 1: 49–71.

Darcy, M. (2013) 'From high-rise projects to suburban estates: Public tenants and the globalised discourse of deconcentration'. *Cities*, 35: 365–372.

Davison, G., Gurran, N., Pinnager, S. and Randolph, B. (2012) *Affordable housing, Urban Renewal and Planning: Emerging Practice in New South Wales, South Australia and Queensland*, Melbourne: AHURI.

Department of Planning (2007) *Major Project Assessment; Carlton United Brewery Site* (MP 06_0171), Sydney: NSW Government.

Department of Planning and Community Development (2012) *Reformed Zones for Victoria*, Melbourne: State Government Victoria.

Dixon, T. (2007) 'The property development industry and sustainable urban brownfield regeneration in England: an analysis of case studies in Thames Gateway and Greater Manchester', *Urban Studies*, 44(12): 2379–2400.

Dovey, K. (2005) *Fluid City: Transforming Melbourne's Urban Waterfront*, London: Routledge.

Economic Development Queensland (2016) *PDA Guideline no. 16 January 2016*, Brisbane: Queensland Government.

Forrest, R. and Murie, A. (1983) 'Residualization and council housing: Aspects of the changing social relations of housing tenure', *Journal of Social Policy*, 12(4): 453–468.

Galland, D. and Hansen, C.J. (2012) 'The roles of planning in waterfront redevelopment: From plan-led and market-driven styles to hybrid planning?' *Planning Practice and Research*, 27(2): 203–225.

Gans, H.J. (1965) 'The failure of urban renewa'l, *Commentary*, 39(4): 29.

Gleeson, B. and Coiacetto, E. (2007) 'Positive planning in Australia: A review of historical and emergent rationales 1', *Urban Policy and Research*, 25(1): 5–19.

Greive, S., Jeffcote, R. and Welch, N. (1999) 'Governance, regulation and the changing role of central Perth: Is there room for affordable housing?' *Urban Policy and Research*, 17(3): 225–233.

Gurran, N. (2011) *How can Australian Local Governments Use Planning Levers for Housing Diversity, Choice And Affordability? Australian Urban Land Use Planning; Principles, Systems And Practice*. Sydney: Sydney University Press.

Gurran, N. and Phibbs, P. (2016) 'Boulevard of broken dreams: Planning, housing supply and affordability in urban Australia', *Built Environment,* 42(1): 55–71.

Gurran, N., Milligan, V., Baker, D., Bugg, L.B. and Christensen, S. (2008) *New Directions in Planning for Affordable Housing: Australian and International Evidence and Implications*. AHURI Final Report Series. Melbourne: AHURI.

Hall, P.G. (1973) *The Containment of Urban England: The Planning System: Objectives, Operations, Impacts*, London: Allen and Unwin.

Hamin, E.M. (1999) 'Turning brownfields into greenbacks', *Journal of the American Planning Association*, 65(2): 236–237.

Keating, L. (2000) 'Redeveloping public housing: Relearning urban renewal's immutable lessons', *Journal of the American Planning Association*, 66(4): 384–397.

Lee, M. (2012) 'Focus: Zoning a go go?', Allens, available online at http://www.allens.com.au/pubs/env/foenv19jul12.htm (accessed 30 October 2017).

McGuirk, P. and Dowling, R. (2009) 'Neoliberal privatisation? Remapping the public and the private in Sydney's masterplanned residential estates', *Political Geography*, 28(3): 174–185.

McLoughlin, B. (1991) 'Urban consolidation and urban sprawl: A question of density', *Urban Policy and Research*, 9(3): 148–156

Neilson, L. (2008) 'The "Building Better Cities" program 1991–96: A nation-building initiative of the Commonwealth government', in J. Butcher (ed.) *Australia Under Construction* (pp. 83–118). Canberra: ANU Press.

NHSC (2014) *National Housing Supply Council 2013 State of Housing Supply Report; Changes in the way we live*. Canberra: Treasury.

Pearce, J. and Vine, J. (2014) 'Quantifying residualisation: the changing nature of social housing in the UK', *Journal of Housing and the Built Environment*, 29(4): 657–675.

Pinnegar, S. (2013) 'Negotiating the complexities of redevelopment through the everyday experiences of residents: the incremental renewal of Bonnyrigg', State of Australian Cities Conference, Sydney: State of Australian Cities Research Network.

Property Council of Australia (2007) *Australia's Land Supply Crisis; Supply Demand Imbalance and its Impact on Declining Housing Affordability*. Sydney: Residential Development Council, Property Council of Australia.

Randolph, B. and Judd, B. (2000) 'Community renewal and large public housing estates', *Urban Policy and Research*, 18(1): 91–104.

REDWatch (2013) *Frasers Broadway Amended Concept Plan*. Sydney: REDwatch.

Renewal SA (2016) *Urban Renewal Prospectus South Australia 2016*. Adelaide: Government of South Australia.

Rogers, D. (2013). 'Urban and social planning through public-private partnerships: The case of the Bonnyrigg Living Communities Project, Sydney Australia'. In

J. Colman and C. Gossop (eds), *ISOCARP Review 09: Frontiers of Planning: Visionary Futures for Human Settlements* (pp. 142–155). Brisbane: International Society of City and Regional Planners.

Rowley, S. (2017) *The Victorian Planning System: Practice, Problems and Prospects*. Melbourne: Federation Press.

Ruming, K., Houston, D. and Amati, M. (2012) 'Multiple suburban publics: Rethinking community opposition to consolidation in Sydney', *Geographical Research*, 50(4): 421–435.

Searle, G. (2007) *Sydney's Urban Consolidation Experience: Power, Politics and Community*, Nathan, QLD: Urban Research Program, Griffith University.

Troy, P. (2012) *Accommodating Australians; Commonwealth Government Involvement in Housing*. Sydney: The Federation Press.

van den Nouwelant, R., Davison, G., Gurran, N., Pinnegar, S. and Randolph, B. (2014) 'Delivering affordable housing through the planning system in urban renewal contexts: converging government roles in Queensland, South Australia and New South Wales', *Australian Planner*, May: 1–13.

Whitehead, C.M.E. (2007) 'Planning policies and affordable housing: England as a successful case study?' *Housing Studies*, 22(1): 25–44.

4 Funding large-scale brownfield regeneration projects

Glen Searle

Introduction

In this book, regeneration in Australian cities embraces the idea of active governance to bring about desired redevelopment of the existing urban area. A central consideration in this is the question of how the regeneration can be funded. This chapter focuses on the funding of major brownfield projects, a category of regeneration that are often the most obvious and largest examples of regeneration, and in which funding issues have been especially central to the way in which planning and development have been produced. The chapter surveys the different ways by which major Australian brownfield urban-regeneration schemes have been funded, and the governance context of these various methods.

Because regeneration of brownfield sites is sought by governments, since these sites are frequently in areas which are attractive for more intensive development on the one hand, but which require market intervention to address aspects such as transport capacity, poor image, disaggregated land-parcel ownership, and soil contamination on the other hand, governments cannot rely on the market for redevelopment at rates that are desirable from a strategic planning perspective. Thus governments need to provide incentives for developers to fund brownfield regeneration, or else provide funding themselves and then perhaps extract returns from post-development property sales. The ways in which governments intervene to fund brownfield regeneration under these imperatives is the focus of this chapter. These ways are discussed in two basic parts. The first part surveys the basic methods used to provide incentives for developer funding of brownfield regeneration in general: re-zoning and development corporations. The second part surveys methods used to fund key components of brownfield regeneration that often need specially crafted interventions: transport infrastructure, affordable housing and soil remediation.

The fundamental governance frame that has underlain Australian regeneration financing over the last quarter of a century has been that of neoliberalism. By the early 1990s, Australian governments were following the lead of the UK in particular and embracing the use of private-sector investment to replace government funding of infrastructure and utilities, as an enlarged welfare-based public sector proved unsustainable (Gleeson and Low, 2000). Thus initiatives to regenerate older city areas have attempted to create governance arrangements that encourage significant private investment and reduce government fiscal outlays as much as possible. Nevertheless, government involvement has been at the centre of this process, and brownfield regeneration has certainly not involved a 'hollowing out' of the state (Peck 2001). Indeed, the use of Commonwealth Better Cities funding to kick-start several significant early regeneration schemes from 1991 to 1996 in some ways marked the last throes of the welfare state, although unlike that era, the funding was always intended to be a demonstration of how state capital could be used to draw in private investment and create a sustainable urban form. Better Cities funding also exemplified the post-welfare era of government intervention that was differentiated according to local context (Brenner, 2004). This dynamic has continued in variegated governance and associated financing mechanisms for subsequent brownfield regeneration projects.

The chapter starts with an analysis of the most basic method of financing brownfield regeneration projects, re-zoning of land, especially old industrial areas, to higher-value uses. Re-zoning of government land as a special case is also considered. Then the most important governance initiative to bring about brownfield regeneration – the establishment of development corporations – is discussed. The financing of necessary infrastructure is then analysed, followed by an analysis of the special case of rail infrastructure. Financing aspects of the provision of affordable housing in regeneration schemes are discussed in the ensuing section, and a short treatment of regeneration-site remediation follows. The chapter concludes with a brief perspective that considers some factors that have shaped the choice of financing methods used to achieve brownfield regeneration in Australian cities.

Re-zoning

Re-zoning lower-value land uses to uses that have a higher land value has been the most common method of making urban regeneration financially possible in Australian cities. In recent decades this has usually involved re-zoning old inner-city industrial areas and obsolete port sites from industrial and transport zonings to commercial and higher-density residential uses. The generally central location of such areas has meant that they have been in high demand for offices, entertainment and tourist attractions, and apartments. The current proposal for major redevelopment of inner-city Fishermans Bend in Melbourne is just the latest in a long line of such developments over the last three decades (Chapter 7). A parallel approach has involved the up-zoning of lower-density residential zones to more valuable higher-density residential zones, although this has only recently started to become significant as a regeneration tool (as opposed to its older and wider role of promoting urban consolidation for other reasons). A central issue here involves the capacity of developers to negotiate up-zoning outcomes that are favourable to them, as a form of rent-seeking, perhaps beyond desirable planning limits. This section includes examples of the way in which developers have extracted such outcomes.

Early examples of regeneration involving re-zoning were inner-city high-density public-housing projects to regenerate 'slum' precincts in inner Sydney and Melbourne. Inner-city precincts zoned as living areas in Sydney's County of Cumberland plan of 1948, such as in Redfern and Surry Hills, were then subjected to municipal plans that included large swathes of high-density public housing (Murray 2013). Several such schemes went ahead, involving the compulsory acquisition of old houses and shops by state housing commissions. In those cases, funding was shared between the states and the Commonwealth government under former Commonwealth-State Housing Agreements. More recent schemes have regenerated public housing precincts by rebuilding the sites and adding new private-sector dwellings to the reconstructed public housing in public–private partnerships (Chapter 14). These schemes have not required re-zoning, as state government developments are not required to conform to local development controls. In the absence of Commonwealth funding and with minimal state government funding available, such current regeneration of public-housing precincts has been paid for by the return to the government from the site developers. The latter make their profit from the sale of the new private dwellings. While public housing estate regeneration is different in important respects to other forms of brownfield regeneration, it shares the assumption that existing land uses are inappropriate or inefficient, and need to be changed and an opportunity for redevelopment be taken. In each case, up-zoning represents a policy tool to increase the economic viability of redevelopment.

The approved density of re-zoned sites frequently involves a contested and contentious process that does not necessarily produce socially optimal outcomes. Rather, it can generate returns beyond those needed by the market to bring about regeneration, and produce super-profits for developers and/or landowners. In the first place, developers may pay above-market values for sites prior to re-zoning, in the expectation that re-zoning will allow them to achieve good profits from projects on those sites. In this case, landowners receive unearned windfall gains that increase the costs of regeneration. Second, after purchasing a site, developers may try and negotiate controls allowing higher densities than proposed by the state or local government. Such increased densities may be needed to return developers an acceptable profit because they have paid a premium to buy the site in the first place. This has two aspects. Under most Australian planning systems, the proposed state or local controls require some degree of community consultation. Overturning controls that the community has been involved in setting results in reduced procedural fairness. The second aspect is that various social, economic or environmental considerations that framed the initial development controls are being bypassed, resulting in sub-optimal outcomes for society. In particular, the resulting density may be too high in terms of those considerations, but acceptable to state governments because metropolitan urban consolidation targets get closer to being realised (Chapter 2).

A case study of the way in which developers have achieved denser, more profitable outcomes than required to kick-start regeneration is the story of the ACI glassworks site in Waterloo, inner Sydney (Searle 2003b). In 1997 the local council adopted a South Sydney-wide Local Environmental Plan (LEP) and Development Control Plan (DCP) in which key standards such as floor space ratios (FSRs) were placed in the DCP which, under the NSW Environmental Planning and Assessment Act 1979 (EPA Act), set out controls that were only quasi-enforceable. The council then took this same approach in an LEP and DCP for the major Green Square redevelopment area. The Green Square DCP also contained floor space bonuses on development sites in return for community benefits beyond those mandated by council under Section 94 of the EPA Act. The first major development under the Green Square LEP/DCP was Meriton's proposal for 2,300 flats on the 11.25ha ACI site. As allowed in the DCP, Meriton received a floor space bonus of 1:1, or a density 67 per cent above the 1.5:1 standard in the DCP, in return for extra open space, 80 affordable housing units, public roads and other benefits (Harrison 2000). Meriton eventually decided to contribute $16.6 million instead of building the affordable housing (Chandler 2000), but at then land values of around $1,500 per square metre (Harrison 2000) the remaining benefits provided for the bonus were worth around $136 million to the developer. These benefits included public roads already shown in the DCP, and which are a standard inclusion in Section 94 plans[1] for local infrastructure of outer suburban councils. The floor space bonus in this case was based on a guess by the council about the value of the benefits, which in the main merely seemed to reflect good design principles in the DCP. But here the council was also negotiating with Australia's largest apartment developer, which had extensive experience in such negotiations. The council later employed consultants to develop a policy formula for estimating the worth of floor space bonuses in each area to provide a firmer basis for negotiating benefits in return (Harrison 2003).

The potential of floor space limits in DCPs to be varied in negotiations to the financial advantage of developers is illustrated in another South Sydney case, the St Margaret's Hospital site redevelopment (Searle 2003b). A site-specific DCP was developed by the hospital's consultants in consultation with the community, which set the maximum height at 20 metres (7 storeys) and a maximum FSR of 2.1:1 (Lawson 2003). On this basis, the site was sold to developers. Council planners accepted the developer's position that the maximum height should be increased to reflect the previous situation where one of the demolished buildings was around 14 storeys. In addition, the council agreed to an FSR bonus of 0.25:1 in return for community benefits. These would mainly comprise a basement area of 900 square metres for a new community library, public art valued at $200,000, and a public plaza (Harrison 2003; Lawson 2003). The bonus for community facilities was required because the council had not included these in its Section 94 plan or in other budgeting (Moore 2003). In the end though, the council approved a development in which the tallest block would be 51 metres high (19 storeys), and with an overall FSR of 3.25:1 (Lawson 2003). The discussions leading up to the final approval involved developer-facilitated networking with senior interstate and national identities within the ALP, which controlled the council, according to one experienced consultant (Anon 2003).

A special case of up-zoning old industrial sites and obsolete transport facilities occurs where the land involved is owned by the state government. This is a common situation, arising from Australia's historic provision of regional infrastructure by state businesses (Searle and Bunker, 2010). Thus a range of government-owned inner-city sites have become available for redevelopment as their original infrastructure function has become obsolete, such as port areas (Sydney, Melbourne, Brisbane, Adelaide and Newcastle) and associated railyards, railway workshops (Sydney and Perth), power stations (Sydney and Brisbane), and gas works (Perth).

At Honeysuckle in Newcastle, the cost of relocating port functions to free up land for regeneration was borne by the Commonwealth government's early 1990s Better Cities program (McGuirk et al. 1996), an $816 million program of investment in 26 projects around Australia intended to be a catalyst for efficient and equitable city development. More recently, the state government has closed the last section of central Newcastle's rail line to make the rail corridor land available for development. Instead, it has proposed to build a light rail line from the new rail terminus down the main retail street to promote high-rise redevelopment of major sites along the street via a joint developer–UrbanGrowth (the government's urban regeneration agency) project (Ruming et al. 2016). In Perth's Subi Centro, the rail line was sunk underground to free up surface land for residential and retail uses and community open space.

State government sites in inner suburbs used since the nineteenth century as showgrounds for the annual shows of agricultural societies have been redeveloped as governments realised their regeneration potential (Sydney and Brisbane). Old sites from the era of state ownership in some industries have also become available, such as the former state abattoir and brickworks sites at Homebush, Sydney. In nearly all cases where governments have sought to redevelop major inner-city sites owned by them, they have set up special development corporations, or used existing ones, to oversee the process.

Development corporations

To facilitate the capture of land-value increases from the regeneration of government-owned sites and to coordinate and encourage private investment on those sites and adjacent non-government land, state development corporations have been set up. These have been given development powers through special acts of parliament that enable them to bypass local planning controls and coordinate the provision of infrastructure, particularly that provided by state government agencies (Craik 1992; Daly and Malone 1996; Searle and Byrne 2002). This is often carried out within the framework of a master plan that is formulated by the corporation. The corporation members are appointed by the state governments, and usually consist of a mix of private-sector and public-sector representatives. Where state government land is involved, the development corporations are the vehicle for selling or leasing that land to developers, and using the proceeds to provide infrastructure for the precinct and to return the surplus to the government. Where specific developments are desired by the government, formal public–private partnership agreements are made where the government wants to fund the development with private-sector funds (with public land included as government collateral). A current example is Darling Harbour South, branded as Darling Harbour Live, discussed below.

Development corporations were given powers to make statutory plans in major regeneration schemes such as Darling Harbour and City West (including Pyrmont-Ultimo) in Sydney, Docklands (Melbourne), South Bank, Bowen Hills and Woolloongabba (Brisbane), and East Perth, Subiaco, Midland and Armadale (Perth), removing them from the normal local-government plan-making process. In the latter three cities, the development agencies also had development approval powers (now reverted to the relevant city or municipal council in the cases of Melbourne, South Bank Brisbane and the Perth schemes). The Perth redevelopment authorities also had compulsory land acquisition powers. These arrangements enabled the states to maximise development and the return to the government from property sales (Chapters 6, 7, 8 and 9).

Development corporations for redevelopment and regeneration in Australian cities date back to 1968. In that year the Sydney Cove Redevelopment Authority (SCRA) was established to capture value from the redevelopment of government land on the western side of Circular Quay at the Rocks in Sydney's CBD. The SCRA would prepare a redevelopment plan, manage and lease land for development, provide infrastructure, and receive revenue from development to fund the plan (Webber 1988: 132). Following a green ban initiated by the Builders Labourers Federation, a revised plan was prepared that greatly reduced the scale of development and preserved the historic core of the Rocks precinct.

The next regeneration corporation, Sydney's Darling Harbour Authority, took its cue from the development of Baltimore's Inner Harbour by the Rouse Corporation, which was invited to submit a master plan for Darling Harbour's redevelopment (Lang 2005). Government port land in Darling Harbour had become available for redevelopment following the construction of a new port at Botany Bay. Darling Harbour redevelopment would be firmly oriented at capturing more of the already expanding global tourism numbers coming to Sydney. The project would have major exhibition and convention buildings, a harbourside market and a hotel along the foreshore, as well as the Chinese garden and maritime museum. A casino would be developed, with the state licence fee intended to recoup much of the project's development costs. In September 1984 the Darling Harbour Authority (DHA) was set up by an act of state parliament. The authority's board contained appointees of the government, including several from the business sector. The authority itself had sweeping powers, able to make its own planning controls outside the existing state planning act (the Environmental Planning and Assessment Act) and outside the Heritage Act. It proceeded to produce its own development strategy. The design and development responsibilities of the authority were given to private companies which reported to the authority's executive (Young 1988: 195).

The Darling Harbour scheme saw the completion of most of the planned facilities by the target Australian Bicentennial date of 1988. A similar opportunity arose in Melbourne in the 1980s as port activity continued to relocate west, away from the old government-owned Yarra River docks adjacent to the CBD. In 1991 the Docklands Authority was established to oversee the regeneration of the old docklands. By 1995 a plan for the area had been produced. Unlike Darling Harbour, new development in Docklands was focused on office and residential development, as well as a major privately financed football stadium. Separate development agreements were signed with developers for each precinct or major site within the scheme. These required development in accordance with the plan, and payment for the land and for infrastructure costs. To maximise the speed of development, the government sold docklands land at very low prices (Chapter 7). In one of the first land deals made by the state government in 1999 in an effort to kick-start the Docklands scheme, the 136,970 square metre New Quay precinct, roughly five city blocks in size, was sold to developers for $3 million (Johanson 2012). That priced the land, which had harbour access, at $22 per square metre. When the infrastructure payment is included, the government's return from the site was about $79 per square metre at a time when a typical inner-city worker's cottage was costing about $1,000 per square metre (Johanson 2012).

A regeneration development that was more similar to Darling Harbour was Brisbane's South Bank scheme. The 42 hectares of land for this scheme, along the Brisbane River opposite the CBD, had originally been used to stage the world Expo in 1988. This mainly industrial area was largely derelict (Chapter 8). The state government passed the Expo 88

Act and established the Brisbane Exposition South Bank Redevelopment Authority (BESBRA) to compulsorily acquire and develop the site for Expo 88, organise the event and dispose of the site afterwards. The authority was required to achieve a satisfactory return on the government's investment in Expo (Fagence 1995: 80). At the time the Expo 88 Act was passed (1984), the long-standing state government of premier Bjelke-Peterson was continuing to pitch itself as very pro-development, and the act could be regarded as reflecting the government ethos. However, public outcry in 1989 over BESBRA's preferred post-Expo scheme (involving a world trade centre, casino and international hotel) (Craik 1992; Fagence 1995) and the replacement of Bjelke-Peterson by an interim premier after the findings of a corruption inquiry shifted the balance of power. Another development authority, the South Bank Corporation (SBC), was established to develop the site. Its remit was the less authoritative one of ensuring that development met the highest possible standards and was in the public interest (Craik 1992). Wide consultation and planning processes then took place in the preparation of a new plan, which incorporated extensive public open space in addition to a state convention and exhibition centre, cultural centre, university buildings, restaurants and other retail outlets, and apartments and offices. The SBC raised revenue from site leases for restaurants and other commercial uses, and land sales for apartments. Up until 2012, when the new state government decided to transfer SBC functions to Brisbane City Council with long-term development of South Bank substantially complete, the corporation was receiving a state grant of around $10 million each year, or about ten per cent of its total income (South Bank Corporation 2010). This could be attributed to recouping costs of maintaining the parklands, a major regional recreation resource, that could not be met from SBC land lease/sale revenue or from its car parking and convention and exhibition centre revenue.

Variants of the basic development corporation model have been used and modified to accommodate specific institutional and political contexts. In Sydney's Green Square scheme, initial planning was carried out by South Sydney Council for the South Sydney Development Corporation, a partnership between the council and the state government under a joint memorandum of understanding. This was partly in response to concerns that there had been inadequate community involvement in Pyrmont's redevelopment under the City West Development Corporation (see below). The local council retained basic development control powers, but large development applications were reviewed by the corporation (Searle 2005). In other cases the planning of regeneration precincts has been done by a stand-alone process, with the management of ensuing development then given to a more circumscribed development authority. In Barangaroo (Sydney), being developed by the Barangaroo Delivery Authority, the government ran an international master plan competition to ensure a global-standard outcome on Sydney's most valuable brownfield site (Chapter 6). The master plan and development controls were set down prior to the establishment of a development agency, which is responsible for implementing the plan. In this case the new harbourside park and community facilities are being paid for by the developer, Lendlease, under agreement with the government. When the original master plan was changed to incorporate a more expensive harbourside park, the developer was given extra floor space and height allowances for new buildings. The state government received revenue from the scheme under a fixed payment plus profit share agreement with Lendlease for the southern office precinct. However, the government will now receive $400 million less than budgeted from its share of developer profits after Lendlease won a long court case against it (Ruehl and Harvey 2014).

The deficit in transparency and accountability of development corporations, and the imperative of the corporations to bring about timely development, have meant that developer attempts to make excess profits by exceeding development controls have increased their chances of success. This is illustrated by the way in which Lendlease was able to obtain higher apartment buildings in the redevelopment of Pyrmont under the City West Development Corporation. Lendlease had bought the old CSR sugar refinery site. It obtained approval for development which replaced the intended 'urban village' scale of the Regional Environmental Plan (REP) with more profitable high-rise towers. The REP allowed owners of large sites – specifically Lendlease and the government – to make master plans for their sites that required only ministerial approval. In the CSR site master plan, a single 21-storey tower permitted by the REP was replaced by four towers, including one of 30 storeys. The master plan set out two more towers at other locations on the refinery site. It also allowed the height of adjacent buildings to be increased from the regional plan's five to eight storeys to ten storeys. This ensured extra profits for Lendlease by allowing more apartments with harbour views to be built (Searle and Byrne 2002).

Even less transparent have been developer gains in South Darling Harbour, where there are joint development agencies (the state's infrastructure coordination agency, Infrastructure NSW, and Sydney Harbour Foreshore Authority). A public–private partnership was signed between the government and the development consortium, Darling Harbour Live, to rebuild the convention and exhibition centres and the entertainment centre, at a total cost of $1 billion, financed by the consortium (Lendlease 2013). Planning and development is subject to state guidelines and specific negotiation between the agencies and development consortium. This gives the consortium more freedom to negotiate the terms of development, particularly the scale of mixed-use developments supplementing the major facilities, and thus to recoup costs of the contracted rebuilding. The requirements for redevelopment set out by Infrastructure NSW for bidding consortia were specific in terms of size – principally an exhibition centre of 40,000 square metres, a convention centre able to accommodate 10,000 people, a meeting-room space of 8,000 square metres between the two centres, and a premium entertainment centre holding up to 8,000 people. Urban design guidelines for redevelopment were also set out. These covered 'reorientation', connectivity, activity, and open space and public realm (Sydney Harbour Foreshore Authority 2012 and Infrastructure NSW 2012). These were mostly very general, such as 'Building a place that invites tourists, visitors and locals to explore', though some were a little more specific such as requiring new east–west and north–south connections, and no reduction of open space. These requirements allowed bid consortia considerable freedom in master-planning and design, especially in the Haymarket precinct where building size and street layout were very much developer-determined. This resulted in a virtually privatised planning process in which the successful bid was selected mainly on financial criteria and to some extent the architectural quality of the new convention and exhibition facilities. Eminent planning lawyer John Mant summed up the process as: 'a government authority did a deal with a developer, giving them free rein over swathes of land in exchange for new public facilities' (Robertson 2016).

Where regeneration schemes have involved large buildings or extensive open space on land that was not all publicly owned, state governments have compulsorily purchased the private land within the scheme area to increase the attractiveness and financial viability of redevelopment. The Darling Harbour, Brisbane South Bank and Sydney Olympic Park regeneration projects all required some state purchase of private land holdings. Similar powers have also been used by the Perth redevelopment agencies. At Fishermans Bend, the purchase by the state government of the GM Holden site is expected to act as a catalyst for 'high tech, high amenity employment' (Lenaghan 2017).

Infrastructure provision

While redeveloping older parts of the city instead of pursuing greenfield development can save on infrastructure costs, significant regeneration of brownfield sites usually requires enhanced infrastructure to be provided. This is because infrastructure in these areas can be at the end of its functional life (such as with older clay sewerage pipe networks), and because regeneration usually involves an increased level of overall development. Some infrastructure costs can be avoided where spare capacity exists, although steadily increasing populations in inner-city areas where the main regeneration efforts are focused mean that much of the excess infrastructure capacity generated by post-war population decreases in these areas has now disappeared. Nevertheless there are still some opportunities to save money by making more intensive use of existing infrastructure. One example is open space. While this is usually under-funded and under-provided in regeneration areas (Searle 2011), there is potential to intensify the use of centrally located regional open space in some places. At Sydney's Moore Park, the city council is proposing to reduce the length of the public golf course so that more of the open-space needs of the expanding population in the adjacent regeneration area of Green Square can be better met.

Finance for the required infrastructure comes from a variety of sources. Local infrastructure for new development, such as water and sewerage connections, drainage, and local roads, is legally required to be provided by developers under provisions such as Section 94 in the New South Wales Environmental Planning and Assessment Act 1979. These costs are generally passed on to the first purchasers through higher sales prices, although in theory they may result in lower prices being paid by developers for sites (Gurran et al. 2009). But the major infrastructure costs for general urban development involve items provided at a regional level, such as main roads and rail lines, water and sewer mains and pumping stations, power lines and human services infrastructure such as educational and health facilities. Such items have traditionally been funded by state governments (with the exception of Brisbane City Council for most regional infrastructure there). Nearly all recent regeneration schemes in Australian cities have been too small to necessitate new regional infrastructure, with the main exception of new rail lines in Sydney and tram lines in Melbourne (see below). In Pyrmont-Ultimo (Sydney), new water-main capacity has had to be provided by Sydney Water (a government business), but would have been recouped from the government's development agency under Sydney Water policy. In Brisbane's inner north regeneration area across New Farm and Teneriffe, Better Cities funding of $22 million was used to upgrade the main sewer line, which was seen as the main infrastructure constraint to redevelopment (Urban Renewal Task Force 1996; Duncan 2004). At Brisbane's South Bank and Sydney Olympic Park, new regional open space was created as a central component of the regeneration.

Rail and tram infrastructure

Regeneration invariably involves intensification of development in the areas concerned. In turn, this increases demands on local transport infrastructure. In a number of regeneration areas, schemes have taken advantage of existing rail/light rail infrastructure – Rhodes, Pyrmont and Eveleigh (Sydney); Bowen Hills (Brisbane); Honeysuckle (Newcastle); and Subiaco, Midland and Armadale (Perth). In the case of Pyrmont, this required conversion of a disused government-owned goods rail track to a light rail passenger line. Commonwealth Government Better Cities funding of $24 million, part of its financing of Pyrmont-Ultimo and other regeneration projects around Australia, was granted toward the total cost of $70 million. The remainder was funded by City West Development Corporation loans, intended to be recouped from sales of government land to developers. In Honeysuckle,

Better Cities funding was used to construct a transport interchange at Newcastle Station (Roberts 2009, cited in Bryant 2016). Also funded by the Better Cities program was the redevelopment of Subiaco station, with the undergrounding of the rail station enabling land to be freed up for the Subi Central regeneration project. The Woolloongabba regeneration area in inner Brisbane is served by an existing station on the city's busway network.

New rail and tram lines were built to provide necessary transport capacity for regeneration on old industrial and port lands in Sydney and Melbourne. In Sydney, a new rail line to serve Olympic Park was constructed from the existing main western line. This was funded by the state government through a capital grant to its rail track authority. In Melbourne, the Collins Street tram was extended 450 metres to serve Docklands development. This was part of a $20 million Collins Street road, pedestrian and tram network extension project that was funded by Lendlease, the developer of Docklands (Public Transport Victoria 2014). In Sydney's Barangaroo, where 13,000 workers are expected by 2021, there was no planning provision for rail access at the time the size of the scheme was agreed with the developer, Lendlease; extra bus services were not feasible as central city bus routes were already extremely congested. However, a decision was made in 2015 to build a station there on the planned cross-harbour metro rail line that will link the new North West rail line and a rebuilt Bankstown line in the south-west. The decision caused the tender process for the central section of Barangaroo to be re-opened to reconsider the precinct's 'optimal scale' (Ruehl 2015). It seems that some revenue from the extra development thus enabled will help pay for the new station, although 'commercial-in-confidence' restrictions mean Barangaroo financial arrangements are difficult to discern.

The construction of a new rail line was critical to the regeneration of Sydney's Green Square area. This was the New Southern Railway (NSR), connecting Central Station to Sydney Airport. A proposal was being developed by Transfield and another construction company, CRI, in 1993 when Sydney was awarded the year 2000 Olympics. This meant that the ability of the proposed line to handle some of the anticipated Olympic Games airport traffic added further to other government arguments in favour of the line. Connecting the NSR to existing lines to the west at North Arncliffe would obviate the forecast need to expand track capacity north of that point. The line would also allow significant urban consolidation to be generated. A feasibility study forecast that if the line was constructed, 13,000 to 14,000 new dwellings would be built on old industrial land around the proposed Green Square Station (Airport Link Association, 1994). The Transfield–CRI consortium was selected to build and operate the railway and its stations, and also own the latter. The consortium's $131 million share of construction costs was recouped through a fare surcharge on passengers using the new stations on the line, including the two new airport stations. The government and its rail track authority, Rail Corporation NSW, contributed $542 million of the total cost of $673 million (NSW Treasury 2016b). This was funded by government capital grants. Patronage on the line was below forecasts and the operating company went into receivership. The government made compensation payments to the company and the stations were sold to Westpac Bank, but the margin on construction costs of the line had already ensured satisfactory overall profits for Transfield–CRI.

Affordable housing

The question of how regeneration should be funded has often downplayed the issue of whether some of the funding should be allocated to provide affordable housing. As Australian city regeneration schemes invariably involve an uplift in land value, there is clearly potential for a portion of the uplift to subsidise some affordable housing. But developers have been reluctant to include such housing because this would reduce the yields they can achieve within the relevant development controls over scale and density, and because they consider that an affordable housing component would reduce the attractiveness of their product to higher-income buyers or renters (c.f. the Meriton example in Green Square discussed above).

There are two main exceptions to this in Australian city regeneration over the last two decades or so. The first are NSW and Perth schemes with Better Cities funding. Bunker (2015) argues that it was only strong pressure from the Commonwealth government that led to the inclusion of an affordable housing component in Sydney's Pyrmont-Ultimo scheme; it is likely that similar pressure might have been exerted in the Perth schemes. A City West Affordable Housing Program (CWAHP) was started in Pyrmont-Ultimo in 1994, with funding provided by developer contributions mandated under the City West regional plan. A total of 600 affordable housing units were to be built for very low- to moderate-income households over a 20–30 year timeframe (Department of Planning 2010). This housing provision is being administered by a non-profit rental housing company. It has been supplemented by 100 units of public housing constructed by the Department of Housing. Over the life of the City West plan, approximately 8–10 per cent of total housing will be affordable or public rental housing. In 1998 the CWAHP was extended to the rest of the City of Sydney, including the Eveleigh and Green Square regeneration areas. The Honeysuckle scheme has likewise included an affordable housing component of 170 dwellings in community ownership. Developers in later Sydney schemes have been reluctant to include affordable housing. At Barangaroo, developer Lendlease committed to build 89 affordable dwellings, equivalent to three per cent of Barangaroo's main South precinct (Hutchinson 2016). However, it has proven reluctant to build the dwellings at Barangaroo, preferring instead to build off-site up to five kilometres away (Hutchison

2016). This would 'potentially allow ... it to reap a greater profit from the Barangaroo residential floor space' (Hasham and Russell 2014).

In Perth, the affordable housing component has been higher. Both the East Perth and Subiaco schemes received Better Cities funding. In both, the target for affordable housing units is 10–15 per cent of total dwellings. Even this level is not enough to prevent the schemes from becoming gentrified precincts.

The other main exception to deficient provision of affordable housing in regeneration developments has been in Queensland. Concern about the escalating cost of housing in the south-east led the government to release a Queensland Housing Affordability strategy in 2007. A major resulting initiative was the creation of the Urban Land Development Authority (ULDA) in the same year to plan, carry out, promote or coordinate and control land development in nominated urban-development areas (England 2010: 68). The ULDA would have power to make local plans and determine development applications. The goal was to bring housing onto the market quickly at the lowest possible cost. The nominated urban development areas were centred on state-owned land. Three involved regeneration of brownfield sites – Bowen Hills, Northshore Hamilton, and Woolloongabba. A small subsidy from the state government was required to achieve the ULDA's goals in its later years before its functions were transferred to a government department in 2013. In 2010–11, a government grant of $7.5 million was added to development profit (land sales to developers minus land costs) of $10.3 million to produce an overall operating surplus of $3.5 million (ULDA 2013).

Older public-housing estates represent a converse regeneration imperative, whereby reducing isolated concentrations of residents with social problems is seen as a means of improving social sustainability. The public-housing fabric is also frequently in need of upgrading. This has seen public–private partnerships being used to fund regeneration. Private investment is invited to rebuild the public housing and to construct new dwellings for sale on the rest of the estate, thus also addressing urban consolidation planning goals (Chapter 15). The NSW government has begun to regenerate its public housing estates by rebuilding them with a majority of new dwellings, around 70 per cent, being private housing, and new or upgraded social-housing units to re-house the existing population, as well as new community facilities. The schemes are funded via public–private partnerships involving the sale of estate land to private sector partners. In the Bonnyrigg renewal project in south-west Sydney, the private-sector involvement is stated as including 'the responsibility for financing, planning, developing, designing, construction, refurbishment and maintenance of the public housing, as well as providing tenancy management services over a 30-year period' (NSW Treasury, 2016a). The private-sector consortium includes a social housing cooperative. The government achieves rebuilding of the public housing by the consortium through profits the consortium makes after buying the estate land and then building and selling the private housing.

Remediation costs

Obsolete brownfield sites can have significant soil contamination stemming from the operation of former land uses. Such contamination imposes expensive remediation costs on the regeneration of these sites. The costs are often borne by the developer in the expectation of significant profits from the sale of the final development. However, on sites identified by the state as strategic regeneration precincts (as are the three examples discussed below), the state subsidises these costs either through paying for the remediation or negotiating planning conditions that limit the impact of remediation costs on the developer.

Remediation of contaminated land on regeneration sites has been a significant issue in Sydney on two old industrial sites. At Olympic Park, the remediation involved the city's largest contaminated zone, covering 160 hectares at various sites around Homebush Bay (Searle, 2003a). Waste was consolidated into four containment areas and capped with topsoil. Special drains collect leachate, which is treated at the adjacent liquid waste treatment plant. An impermeable membrane prevents runoff into the creek running through the area. The cost was borne by the government as part of its financing of Olympic Games capital works, which was in turn partly funded by cutting back on annual maintenance expenditure on state health and education facilities (Searle 2003a).

More generally, Sydney Olympic Park is distinctive because it is a significant Australian example of how a major event can enable the regeneration of contaminated land with obsolete uses. The year 2000 Olympic Games were the catalyst for state government funding, with additional private-sector funding for stadiums and Olympic Village housing, of transport and public-domain infrastructure that could be the basis for intensified post-Olympics development under a restructured development corporation with full planning powers. The current draft *Master Plan 2030* provides for 10,700 dwellings and 34,000 jobs. Government ownership of the precinct means that the sale of land for this new development will return dividends to the State Treasury that will help offset expenses incurred in operating sporting and other facilities at Olympic Park and adjacent Millennial Park: in 2014 there were losses of $12m in addition to government grants of $35m (Sydney Olympic Park Authority 2014).

Regeneration of the heavy-industry area of Rhodes peninsula on the Parramatta River necessitated remediation of two old chemical works sites. One of the sites was owned by the government's Waterways Authority, and required major remediation at a cost of $100 million to deal with on-site contaminants generated by previous chemical production of Union Carbide (Bureaucrat A, reported in Searle 2003b). To enable this, development was allowed to proceed at a significantly higher density than on the other sites in the regeneration precinct. But the density trade-off was dictated by the developer: 'The government hasn't got the experience to know how to deal with developers, how far it should push' (Bureaucrat A, in Searle 2003b).

For Newcastle's Honeysuckle regeneration, the land on which the old port facilities stood required decontamination. Better Cities program funds were used to decontaminate and remediate the site (Roberts 2009, cited in Bryant 2016).

Overview and conclusion

One striking aspect of brownfield-regeneration funding in Australian cities is the relatively limited range of methods used. The uplift in values from re-zoning has been a key mechanism. This uplift has, in large part, been captured by developers, especially on non-public land, but also in part by private land-owners. Governments have offset infrastructure costs through developer levies and enhanced sale prices for land that they own. There has also been significant 'free' provision of infrastructure such as rail/light rail lines and via Better Cities funding. The setting up of development corporations has allowed more favourable development controls to be set and has facilitated the coordination of development and infrastructure provision.

From an international perspective, this menu of methods omits approaches that might have returned greater revenue to the state governments and/or lowered the need to give planning concessions to developers (see Kirwan 1989). One approach that has been missing is value capture, whereby a proportion of the increase in land value consequent on re-zoning and/or on the provision of a new rail/light rail line is paid to the government by land owners to offset infrastructure costs. This could have been used to help pay for the airport rail line through Green Square, for example. Another approach that could have been used more extensively is inclusionary zoning. This includes provisions in re-zoning that require developers to contribute to public-interest items such as affordable housing. Such provisions have produced inadequate levels of public housing in regeneration schemes in Sydney, and are absent in Docklands, Melbourne and the Better Cities-funded New Farm-Teneriffe area of Brisbane. The result is that regeneration schemes are reinforcing wider inner-city gentrification in these cities.

Thus opportunities have been missed to ensure that developers pay a fairer share of appropriate regeneration costs. Conversely, there is evidence of rent-seeking by developers that has produced excessive returns in some cases, or of excessive prices paid to land-owners that require development beyond existing planning controls. It is possible that governments might be gradually becoming more adept in negotiating with developers: the current profit-sharing arrangement at Barangaroo is very different to the first stages of

Docklands where government land was sold off at fire-sale prices. Nevertheless, recent experience at Barangaroo shows that unless there are tight contracts and multiple developers, extracting decent revenue from government sites and getting adequate public benefits will continue to be difficult.

The broader reasons why developers continue to hold the upper hand are several. One is the rapid growth of Australia's biggest four cities and the concomitant policy imperative to accommodate the expanding populations. The regeneration schemes are seen by governments as important providers of housing or jobs for new residents, and are reluctant to impose conditions that might reduce the rate of development in those schemes. An equally important reason for this situation is the powerful institutional position of the development industry in Australia. Australia's population growth rate has been one of the fastest in the Western world for some time, and this has generated a large development industry that is powerful and politically well-connected. Strong developer lobby groups have emerged to reinforce the sector's influence over governments (Gurran and Phibbs 2015), and the sector is now being seen by the states as a major contributor to their economic development. The evidence for the existence in Australian cities of an urban 'growth machine' (Molotch 1976), in which business and government act together to make it a priority to maximise urban growth, seems strong. In this circumstance, attempts to make developers pay more for regeneration will first require a new public discourse to change the prevailing mindset.

Note

1 Plans outlining compulsory development contributions.

References

Airport Link Association (1994) *Airport Link Feasibility. Summary Report on Detailed Feasibility Studies*. Sydney: NSW Department of Transport.

Anon. (2003) Personal communication, 21 October.

Brenner, N. (2004) *New State Spaces: Urban Governance and the Rescaling of Statehood*. Oxford: Oxford University Press.

Bryant, L. (2016) *Investing in Australian Cities: The Legacy of the Better Cities Program*. Brisbane: Queensland University of Technology.

Bunker, R. (2015) 'The changing political economy of the compact city and higher density urban renewal in Sydney', *Planning in a Market Economy*, ARC Discovery Project Working Paper, No. 1. Sydney: City Futures Research Centre, Faculty of Built Environment, University of NSW.

Chandler, M. (2000) 'Meriton unveils ACI project', *Australian Financial Review*, 4 September, p. 14.

Craik, J. (1992) 'Expo 88: Fashions of sight and politics of site'. In T. Bennett, P. Buckridge, D. Carter, and C. Mercer (eds), *Celebrating the Nation: A Critical Study of Australia's Bicentenary*, St Leonards, NSW: Allen & Unwin.

Daly, M., & Malone, P. (1996) 'Sydney: the economic and political roots of Darling Harbour'. In P. Malone (ed.), *City, Capital and* Water, London: Routledge.

Department of Planning (2010) *Revised City West Affordable Housing Program (June 2010)*. Sydney: NSW Government.

Duncan, T. (2004) 'Revitalising Brisbane's Newstead-Teneriffe waterfront', paper presented at the International Cities, Town Centres and Communities Society Conference, Fremantle, West Australia, 4–7 May.

England, P. (2010) 'The legislative challenge'. In B. Gleeson and W. Steele (eds.), *A Climate for Growth: Planning South East Queensland*. St Lucia, QLD: University of Queensland Press.

Fagence, M. (1995) 'Episodic progress toward a grand design: Waterside redevelopment of Brisbane's South Bank'. In S.J. Craig-Smith and M. Fagence (eds), *Recreation and Tourism as a Catalyst for Urban Waterfront Redevelopment*, Westport, CT: Praeger.

Gleeson, B. and Low, N. (2000) *Australian Urban Planning: New Challenges, New Agendas*. St Leonards, NSW: Allen & Unwin.

Gurran, N. and Phibbs, P. (2015) 'Are governments really interested in fixing the housing problem? Policy capture and busy work in Australia', *Housing Studies*, 30(5): 711–729.

Gurran, N., Ruming, K., and Randolph, B. (2009) *Counting the Costs: Planning Requirements, Infrastructure Contributions and Residential Development in Australia: Final Report*, Melbourne: Australian Housing Research Institute.

Harrison, J. (2000) Presentation to UTS Master of Planning students, South Sydney Council offices, Surry Hills, 23 March.

Harrison, J. (2003) Interview, 29 October.

Hasham, N. and Russell, M. (2014) 'Lendlease baulking at providing affordable homes in Barangaroo', *Sydney Morning Herald*, 15 August. http://www.smh.com.au/nsw/lend-lease-baulking-at-providing-affordable-homes-in-barangaroo-20140814-1049rr.html (accessed 30 January 2017).

Hutchinson, S. (2016) 'Barangaroo sales to help charity housing', *The Australian*, 9 August, http://www.theaustralian.com.au/business/property/barangaroo-sales-to-help-charity-housing/news-story/06d31f35db5b627bd6cb783efe5f76ca (accessed 3 July 2017).

Johanson, S. (2012) 'Docklands: Kennett was almost giving the land away', *The Age*, 3 March.

Kirwan, R. M. (1989) 'Finance for urban public infrastructure', *Urban Studies*, 26(3): 285–300.

Lang, J. (2005) *Urban Design: A Typology of Procedures and Products*. Oxford: Architectural Press.

Lendlease (2013) 'Contracts signed and construction to commence on Sydney's new convention, exhibition and entertainment hub at Darling Harbour', Media Release, 6 December.

Lawson, V. (2003) 'Gimme shelter', *Sydney Morning Herald*, 27 June, p.9.

Lenaghan, N. (2017) 'Equinix gets council nod for Fishermans Bend', *Australian Financial Review*, 15 March, p. 32.

McGuirk, P. M., Winchester, H. P. M. and Dunn, K. M. (1996) 'Entrepreneurial approaches to urban decline: The Honeysuckle redevelopment in inner Newcastle, New South Wales', *Environment and Planning A*, 28(10): 1815–1841.

Molotch, H. (1976) 'The city as a growth machine: Toward a political economy of place', *American Journal of Sociology*, 82(2): 309–332.

Moore, C. (2003) 'Public debate for public information', *Bligh eNews*, http://cloverarchive.com/archive/idx.htm?http://cloverarchive.com/archive/enews/2003/, 158, 1 August, (accessed 30 January 2017).

Murray, L. (2013) 'From slums to Sunny Hills', *Architecture Bulletin*, Autumn: 14–17.

NSW Treasury (2016a) 'Bonnyrigg Living Communities Project (Social Housing PPP)'. Sydney: NSW Treasury. www.treasury.nsw.gov.au/ppp/nsw_projects/bonnyrigg_living_communities_project_social_housing_ppp (accessed 20 August 2016).

NSW Treasury (2016b) 'New Southern Railway Stations Agreement'. Sydney: NSW Treasury. www.treasury.nsw.gov.au/ppp/nsw_projects/new_southern_railway_stations_agreement (accessed 18 August 2016).

Peck, J. (2001) 'Neoliberalizing states: Thin policies/hard outcomes', *Progress in Human Geography*, 25(3): 445–455.

Public Transport Victoria (2014) 'New timetable delivers more than 1,200 extra trams to Docklands: effective from Sunday 26 January'. 10 January. https://www.ptv.vic.gov.au/news-and-events/news/new-timetable-delivers-more-than-1-200-extra-trams-to-docklands-effective-from-sunday-26-january/ (accessed 18 August 2016).

Roberts, J. (2009) *Building Better Cities/Newcastle: A case study in renewal,* Newcastle: Hunter Development Corporation.

Robertson, J. (2016) '2016: The year of a new Darling Harbour', *Sydney Morning Herald*, 15 January.

Ruehl, M. (2015) '"Choo Choo": Barangaroo Central back at square one with new train station', *Australian Financial Review*, 24 June.

Ruehl, M. and Harvey, R. (2014) 'Lendlease wins $1bil appeal against Barangaroo Delivery Authority', *Australian Financial Review*, 21 August.

Ruming, K., Mee, K. J. and McGuirk, P. M. (2016) 'Planned derailment for new urban futures? An actant network analysis of the "great [light] rail debate" in Newcastle, Australia'. In Y. Rydin and L. Tate (eds), *Exploring the Potential of Actant Network Theory*, London: Routledge.

Searle, G. (2003a) 'The urban legacy of the Sydney Olympic Games'. In M. de Moragas, C. Kennett and N. Puig (eds.), *The Legacy of the Olympic Games: 1984–2000,*Lausanne: International Olympic Committee.

Searle, G. (2003b) 'Planning under-empowerment and urban over-development in inner Sydney'. Paper given at State of Australian Cities Conference, Parramatta, 3–5 December.

Searle, G. (2005) 'Power and planning consent in Sydney's urban consolidation program'. In D. Cryle and J. Hillier (eds.), *Consent and Consensus: Politics, Media and Governance in Twentieth Century Australia*, Perth: API Network.

Searle, G. (2011) 'Urban consolidation and the inadequacy of local open space provision in Sydney', *Urban Policy and Research*, 29(2): 201–208.

Searle, G. and Byrne, J. (2002) 'Selective memories, sanitised futures: Constructing visions of future place in Sydney', *Urban Policy and Research*, 20(1): 7–25.

Searle, G. and Bunker, R. (2010) 'Metropolitan strategic planning: An Australian paradigm?', *Planning Theory*, 9(3): 163–180.

South Bank Corporation (2010) *Annual Report 2009–2010*. Brisbane: SBC.

Sydney Harbour Foreshore Authority (2012) *Community information session briefing*, Sydney: Sydney Harbour Foreshore Authority.
Sydney Olympic Park Authority (2014) *Annual Report 2013–14*. Sydney: SOPA.
ULDA (Urban Land Development Authority) (2013) *Final Report 1 July 2011–31 January 2013*. Brisbane: ULDA.
Urban Renewal Task Force (URTF) (1996) *Brisbane Urban Renewal 1996 Report and Five Year Overview*. Brisbane: Brisbane City Council.
Webber, P. (1988) *The Design of Sydney: Three Decades of Change in the City Centre*. Sydney: Law Book Company.
Young, B. (1988) 'Darling Harbour: A new city precinct'. In P. Webber (ed.), *The Design of Sydney: Three Decades of Change in the City Centre*, Sydney: Law Book Company.

5 Experiencing density

The implications of strata titling for urban renewal in Australian cities

Hazel Easthope and Bill Randolph

Introduction

In the twenty-first-century Australian city, urban renewal has become synonymous with higher residential densities. Since at least the 1960s, the renewal of older low-density housing, as well as the more recent redevelopment of extensive areas of redundant industrial and transport-related land in inner- and central-city areas, has been increasingly focused on the development of multi-unit multi-owned residential dwellings. This process has steadily accelerated over the last half century, since the development of strata titling in the early 1960s enabled a rapid escalation of multi-unit housing to be constructed and sold on an individual title. This chapter considers some of the consequences of the emergence of this form of property ownership and its implications for the contemporary urban renewal process with a focus on the communities who have experienced the outcomes of this process and who live in or own these multi-unit properties. Arguably, without strata titling, much of what currently passes for urban renewal activity would not have been possible, at least not on the scale and with the outcomes we are now witnessing. Importantly, as we have argued elsewhere (Randolph and Easthope 2014), the emergence of this form of property ownership was not coincidental or somehow a fortuitous function of purely demand-driven processes. On the contrary, it was driven by the deliberate intervention of the development industry to enable an intensification of residential development, thereby recapitalising land values and with it, speculative land development.

But in creating this new market in multi-unit property ownership suitable for individual mortgage funding, the progenitors of strata titling created important consequences both for those who have come to live in this form of housing and on the form and outcome of the urban renewal process that has accompanied the implementation of compact city planning policies (Easthope and Randolph 2016). Dredge and Coiacetto (2011: 428) aptly note:

> Australia is one of the most heavily suburbanised countries in the world and is moving into a new phase of urban development principally as a

> result of financial and infrastructure-related pressures, and to a lesser extent environmental and social issues. Yet we are moving in this direction based on the ideological and intuitive influence of planning practitioners, and without a clear understanding where it is taking us.

The contemporary process of strata-driven urban renewal will leave an indelible imprint on our cities in the form of the newly emerging community of strata owners, residents, funders and managers. How we resolve the conflicts and tensions inherent in this form of property ownership will determine the success or otherwise of the urban renewal agenda that is reshaping our cities into new higher-density communities, long after the developers and planners have left the scene.

From its modest beginning just over 50 years ago, strata title and its close cousins across the urbanised and urbanising world, some of which predate the Australian model,[1] now represent a major form of property ownership globally. So what are the main consequences that have emerged with the development of this form of property ownership and their implications for wider urban renewal processes? Most of the issues exemplified by the Australian case are broadly replicated in comparable higher-density developments elsewhere. First is the potential for conflict created by a semi-communal form of property relations based on the principles of individual ownership. Within this context tensions arise because of the complexity of relationships within each scheme and the varying interests of stakeholders in the property. The range of stakeholders involved in a typical strata scheme – including resident owners, investor owners, tenants, commercial owners and businesses, building managers and others, including mortgage lenders – raises the potential for significant conflicts over intentions and aspirations. The potential for conflict is growing as the size and complexity of strata developments, especially the requirement for mixed use development in many urban renewal areas, increases.

Second, the proliferation of strata development has led to a range of problems of building design, build quality and defects and ongoing problems with major repairs and maintenance. The prevalence of defects is a growing cause for concern in many countries worldwide (e.g. Christudason 2007; Noble-Allgire 2009; Murphy 2011). In part, this is a consequence of the way projects are proposed and the development managed on the ground. The fragmented supply chain for large-scale strata development with subcontracted lines of delivery generates conditions where poor build quality can occur. This has long-term implications for those living in the sector, as well as for the character of the neighbourhoods being created under urban renewal, especially if failure to police build quality results in poor building performance and expensive remediation.

Third, in effectively facilitating the subdivision of property ownership into multiple parcels of land, strata ownership – which facilitated higher density urban renewal when it was introduced – now poses a formidable barrier to further renewal in areas that otherwise would be prime sites for redevelopment and further densification. This is because of the difficulty in getting multiple owners to agree to the redevelopment of their building. Recognition of this challenge in a number of jurisdictions worldwide has led to changes in law to allow the vote to redevelop a building to be made by a majority of owners rather than unanimously, in the process challenging long-held understandings about the nature of property ownership.

Finally, at the broader metropolitan level, the potential for a new form of socio-spatial segregation based on dwelling type is emerging, with new high-density development in neighbourhoods undergoing urban renewal attracting clearly demographically differentiated sub-markets. How this will play out over time will become a major issue in the future, but one that strategic planning for higher-density renewal in Australian cities has largely ignored, instead focusing attention on housing supply.

The chapter will explore these issues in turn. In doing so, we draw on a series of interlinked and ongoing research projects undertaken by the authors and colleagues over the last decade looking at aspects of the emerging strata title sector in Australia. The research is well documented elsewhere (Easthope et al. 2009; Easthope et al. 2012; Troy et al. 2015) but has involved analyses of census data, surveys of strata residents (owners and tenants), surveys of strata executive committees and strata managers, stakeholder interviews and focus groups in strata communities in the state of New South Wales (NSW).

Tensions arising from communal ownership and multiple stakeholders

For many, the promise of property ownership in the compact city, or more precisely the promise of owning a strata-titled apartment, is a promise of convenience. Apartments are sold as a more affordable option than a house where people might see themselves trading size for convenience and in some cases shared facilities such as pools and gyms. Apartments are also marketed as a carefree option allowing minimal maintenance responsibilities and increased security (Fincher 2004; Kern 2010), while developers promote apartments as part of a broader urban, high-rise lifestyle (Shaw 2006; Lehrer et al. 2010 speaking about Australia and Canada respectively). Yet in practice, Australian research has identified a common 'disconnect between marketing of the [strata] lifestyle and the lived experience' (Reid 2015: 445).

There is also another way that strata-titled apartments are marketed that causes great difficulties for strata schemes later on – the individual ownership of the unit is marketed heavily, while the joint ownership of the common property (the building and grounds) is often underplayed. In effect, people are told that what they are buying is the equivalent of buying a house, when it is not. They are buying into joint ownership of the building and its common areas as well as the mandatory membership of the governing body (the 'owners corporation') charged with the maintenance and management of that shared property.

Many of the disputes and tensions in strata schemes arise because of this fundamental tension between individual and collective ownership and owners' misunderstandings about this. This is especially the case in societies like Australia, where the notion of home ownership is synonymous with control and autonomy over one's property (Easthope 2014). A strata-owner put this well in an interview as part of one of our earlier projects on this issue:

> I think people still have this idea it's my home and I can do whatever I want in it, and I think if just that idea were better explored and people were to understand that your rights to enjoy your home as you want to does have a limit and it's actually probably a lot earlier than you would hope or like to think, then I think people might live better together.
>
> (Owner interview, respondent 1004 in Easthope et al. 2012: 96)

This tension between individual ownership (and assumed autonomy) and collective ownership (and associated responsibility) can play out in a number of ways in a strata scheme. First, and most obviously, it can manifest in owner-residents engaging in activities that negatively affect other residents and refusing to change their behaviour because they can do as they please 'in their property'. Common examples resulting in disruption and disputes include smoking and noise (Easthope et al. 2012). The smoking example is a particularly interesting one because owners' corporations have the ability to ban smoking within a building, and to ban owners from smoking within their own property where that smoke affects people in other units or the common property (Cancer Council NSW 2015). This is further complicated by the fact that in strata properties in Australia it is not possible to pick or vet new residents as it is in cooperatives or company-titled schemes, for example. This is because the legislation specifies that 'no by-law is capable of operating to prohibit or restrict the devolution of a lot or a transfer, lease, mortgage, or other dealing relating to a lot' (Strata Schemes Management Act 2015 (NSW) §139.2). This part of the legislation was specifically developed to protect the individual property rights associated with the ownership of the lot, and to ensure that banks were still willing to issue mortgages against strata-title properties on the same conditions as those they were issuing to detached houses – as this was the primary driver of the introduction of strata title in Australia (Randolph and Easthope 2014).

The second way that this tension plays out is related to the first, in the frustrations of some owners about other people telling them what they can or cannot do. This relates to the battle that sometimes manifests itself in strata schemes between the strata committee (known as the board in other jurisdictions) and other owners (Guilding et al. 2014). In essence, this is the manifestation of the tension between individual property (represented by the individual property owner) and collective property (represented by the members of the strata committee). Certainly apathy on the part of owners is an issue that has often arisen in our research on strata schemes in Sydney with results ranging in seriousness from an inability to reach a quorum at meetings through to poorly managed schemes (Easthope et al. 2012). Yet it is not always as simple as an individual owner choosing not to join the committee when others have chosen to altruistically represent the collective interests of the scheme. Often owners do not get involved in their strata committees or in the decision making of their schemes because of a perceived inability (rather than a lack of willingness) to do so (Blandy et al. 2006). This might be because they do not know what their rights and responsibilities as owners are; they have been bullied or intimidated by other owners; or they feel that their vote would not make a difference (Easthope et al. 2012), sometimes because of a concentration of power by an individual or small group (Johnston and Too 2015).

So what does all of this infighting within buildings have to do with the bigger picture of urban renewal? It is critical, as it affects the long-term social fabric of high-density urban renewal areas and the quality of the built environment. In regards to the first point, it is difficult to have a 'vibrant' local community when neighbours within buildings are in dispute or have an acrimonious relationship. Regarding the second point, if there are difficulties in reaching decisions in strata schemes then it will be difficult for their strata committees to properly carry out their major function, which is to ensure the proper maintenance and upkeep of the shared asset (the common property). Essentially the situation is that untrained volunteers are being expected to manage assets, each of which may well have a combined market value in the multiple millions of dollars, without any training, while at the same time managing the needs and expectations of a diverse group of owners who have not chosen to live together and who may have very different needs and expectations about their building (Guilding et al. 2005; Ngai-Ming and Forrest 2002). The result is that increasingly large areas of renewed urban areas are being left in the hands of poorly qualified amateur strata committees, who are often not recognised by land-use planners and policy-makers (Leshinsky and Mouat 2015). The risk is that the lack of recognition of, or engagement with, these committees – who act in effect as local-scale urban managers – has the potential to result in poorly maintained neighbourhoods with little recourse to public remedy.

The surprise really is not that people in some apartment buildings run into trouble managing their strata schemes, but that so many do not, most of the time. Where people understand what they are buying and moving into, many of these problems can be avoided and if managed properly, joint responsibility of property owners for managing shared assets can have many benefits resulting from economies of scale and the pooling of resources. Similarly, living in close proximity with one's neighbours need not be a nightmare if those neighbours are aware of their rights and responsibilities as co-owners or co-tenants of the building (rather than prioritising their dominion over their individual unit property). Our primary concern then lies not with the strata committee members, nor with the other owners or their tenants, but with a system of housing development and marketing that is selling a lie, or at least a half-truth, to new owners and in the process setting up expectations that cannot be met in practice. Moreover, the fact that the financial institutions lending to the strata development sector require pre-sales of a large proportion of units prior to development finance being committed has meant that developers have fixed their focus on the investor market as a predominant source of demand (discussed further below). The potential tension between resident owners, tenants and their non-resident investor landlords is likely to become a major source of ongoing conflict into the future.

Design and construction quality

When people buy a newly built home they might expect that it has nothing major wrong with it and that if it has, they have some consumer protection when a fault is found. In NSW this is very often not the case when it comes to strata-titled properties. Building defects are common, with recent research suggesting that around three-quarters of all new strata properties in the state have had some type of defect (Easthope et al. 2012). Moreover, there are limited protections for new owners. In NSW, for example, building warranty insurance is not required for buildings over three storeys (NSW Fair Trading 2012) and owners have only two years to find and then require general defects to be remedied by the developer (the period is six years for major defects) (NSW Fair Trading 2016). This assumes that there is a developer or builder to claim against, which is not always the case. One strategy is for developers to form a one-off company to undertake a single development and then dissolve the company on completion. If defects are later found, there is effectively no one to undertake the repair (Cooper and Brown 2014). Building defects can also have a negative impact on ongoing insurance cover. At the time of writing, a recently completed block of strata units in the suburb of Lidcombe in inner Sydney had its roof blown off in a gale. Residents claimed against their insurance policy for repairs, but the insurers refused to meet the claim as they argued the roof had been inadequately built (Han 2016).

The problem of building defects in apartment buildings is not unique to NSW or Australia. Building defects in apartment buildings are common in many jurisdictions internationally (e.g., Christudason 2007; Noble-Allgire 2009; Murphy 2011; de Silva 2011). There are multiple possible reasons for this. These include: a withdrawal (in the case of private certification) or reduction of government oversight of building quality (Cooper and Brown 2014); oversights resulting from a cost-based rather than quality-based project-management approach to construction (Drane 2015); and increasing innovation in the design of buildings made possible with the introduction of computer-aided design and the rapid evolution of new and potentially untested building materials (Agola and Kashiyani 2015). A notable example of the latter issue was the recent fire in the façade materials of a tall block of flats in Melbourne (Dow and Johanson 2015). More generally, effective identification and remediation of building defects is also hampered by market disincentives for buyers to conduct due diligence before purchase (the cost of inspecting an entire apartment building before purchase can be prohibitive), the costs of paying for legal advice to properly review legal documents relating to complex developments for off-the-plan sales, a lack of incentives for developers to trade on reputation in heated housing markets where anything sells, and a lack of effective accountability for developers and builders who deliver defective buildings, as noted above. The problems faced by owners who have bought 'off-the-plan' before construction can be even greater as they have not had a chance to inspect the product they are buying before purchase. The incentive for developers and builders to complete the product to the promised standard diminishes when large numbers of the units have already been presold (Ong 1997).

So while purchasers of new apartments might expect them to be defect-free and under warranty, often they are not. Most purchasers will take possession of a unit in a building with at least some defects and at the same time become part of an owners' corporation that has responsibility for seeking remedy from the developer for those defects. The process of seeking to remedy defects is complicated by the dualistic nature of property ownership in strata schemes: the existence of both individual lots and common property and therefore multiple interests (individual owners and the owners' corporation) means seeking remedies from the developer or builder for defects that might be in lot property, common property, or extend across both.

Where owners' corporations, usually represented by their executive committees, are in the position of having to seek remedies for defects from builders or developers, they play on a very uneven playing field. The scenario typically plays out like this: a group of new property owners who have not met each other before, and likely have no experience of building defects or dealing with developers and builders, find themselves members of an owners' corporation that has the responsibility of chasing a developer or builder to fix the defective work. At the least they will likely need to hire and pay for a lawyer and building surveyor or engineer to assist them in this task and decide whether the strata manager who was appointed by the developer can be trusted to act in their best interests in this matter or whether they need to be replaced. This is a steep and expensive learning curve for the new owners. On the other hand, the developer and builder have seen this all before and know exactly how to respond. If they are reputable, they will respond by negotiating directly with the owners' corporation around returning to remedy the defects or offering a cash settlement. However, they also have the option of fighting the claims and escalating the issue to legal action. Or they might just go into voluntary liquidation and walk away. At this point, it is likely to become a more expensive proposition for the owners to pursue legal action than to just pay for the defects to be remedied themselves.

When we consider the bigger picture of urban renewal, the question becomes why governments have pushed for increasing numbers of higher-density developments as part of their renewal plans without simultaneously introducing reforms to ensure that the quality of those new buildings is improved. Instead, we have seen a steady *reduction* in the protections for owners against defective building works and subsequently reduced incentives on the part of builders and developers to design and construct properties without major defects. Again in NSW, the switch to private certification, scaling back of home warranty periods and coverage and changing definitions of different types of defects, have all led to a reduction in consumer protections for apartment owners (Cooper and Brown 2014). And all these issues are potential sources of conflict even before we consider the longer-term provisions for repair and maintenance through adequate sinking funds, proper property maintenance scheduling, communal planning and budgeting for cyclical and major maintenance and other operational costs over the lifetime of the strata scheme. This is made all the more difficult where owners intending to 'flip' their properties in the short term are reluctant to contribute to sinking funds that might benefit subsequent owners of their building.

Overcoming strata as a barrier to renewal

All buildings have a useful life. But at the end of this life, for either physical or economic reasons, buildings need to be replaced. However, the drafters of the original strata legislation in Australia did not factor in the need to make provision for what happens when a strata-titled building needs to be redeveloped, other than to assume all those with a legal or financial interest in the building would simply simultaneously agree to terminate the scheme at the appropriate time. This proved relatively unproblematic until the early part of the current century when the potential barrier to these schemes created for urban renewal was recognised. During the 1960s and 1970s the introduction of medium-density zonings to facilitate strata development were often focused on larger single dwelling residential blocks close to town centres or rail stations (Cardew 1980). This first wave of higher -density urban renewal resulted in large areas around suburban town centres as well as higher-value beachside suburbs transforming into medium-density housing replacing older single dwellings. Many of these locations are precisely the locations now favoured by planners and urban designers to promote planning schemes based on transport-orientated development principles (Dovey and Woodcock 2014).

In Sydney, where the scale of strata development is much greater than in other Australian cities, this issue came to the fore in the 2005 Sydney Metropolitan Strategy, which recognised the difficulty of redeveloping existing strata buildings in the planned renewal of existing town centres and transit corridors (NSW Department of Planning 2005). However, it took another decade to enact legislation which reduced the need for 100 per cent of lot owners to agree to terminate their scheme before it can be demolished. The Strata Schemes Development Act 2015 (NSW) established a much reduced requirement for just 75 per cent of lot owners in a building to agree to termination, with the remaining 25 per cent required to comply. Similar legislation has been recently enacted in the Northern Territory and other jurisdictions are considering similar reforms.

This move highlights the tension between protecting the rights of the individual to 'quiet enjoyment' of their title in a property, and the rights of the majority of owners to maximise the potential value inherent in the land they collectively own, as well as the wider community's need to periodically renew the urban fabric. With the NSW legislation only enacted while this chapter was being written in late 2016, it is too early to assess the impact of this legislation in generating a significant level of urban renewal. However, recent research by the authors and colleagues (Troy et al. 2015) indicated that widespread renewal of strata will be largely determined by the added value that might be created through the renewal process. Analysing the redevelopment potential of 17,000 pre-1990 strata blocks in Sydney using a simplified development feasibility model applied to each scheme, the study found that 17 per cent of these schemes could be profitably redeveloped with new buildings up to three storeys (i.e. not requiring re-zoning). A further 33 per cent could be redeveloped if additional height was permitted up to 10 storeys (which may require re-zoning in some areas). However, the remainder were either unfeasible without re-zoning to a much greater height or without site amalgamations (which were not modelled). In some localities, potential land values were simply not sufficient in themselves to support any renewal process. The three outcomes are summarised below. Notably, these had a distinctive geography with clear implications for urban renewal policy.

1 *Gentrification*: The replacement of low-rise schemes (up to three storeys) with new low-rise schemes built for a higher-value market. These were prevalent in more expensive locations (including the eastern suburbs and northern beaches) where the resulting units could be sold at a significant premium to existing values, in effect displacing existing residents.
2 *Densification*: The replacement of low-rise schemes with new higher-rise schemes (up to 10 storeys). This was the most feasible outcome in locations where sales values were closer to the metropolitan average but the additional value uplift from denser development would make the scheme profitable, including many of the middle-ring suburbs of Sydney.
3 *Residualisation*: Areas where the replacement of existing strata schemes was not economically feasible. In these locations, the sales values achievable were insufficient to make a redevelopment profitable, even if additional height was permitted. These were concentrated in areas already experiencing multiple disadvantage, largely in the outer suburbs of western Sydney.

The implications for the renewal of older strata blocks, both in Sydney and elsewhere, are obvious. Attempts to undertake renewal of strata schemes in higher value areas are likely to be successful and in many cases without the need for re-zoning, but to the detriment of the communities displaced. In the older town centres and corridors in the middle-value suburbs, re-zoning to significant new height will often be required to provide the additional value uplift to incentivise renewal. But in many town-centre locations in the lower-value western suburbs, renewal seems largely unlikely if the market alone is the driving force, unless it leads to a process of gentrification and, by implication, social displacement. Without an effective market, these areas are destined to become even more disadvantaged as the stock deteriorates.

This analysis highlights how the variation in property markets and property values that can be generated by renewal interact with the local planning system to effectively determine both the scale of renewal of strata developments as well as the likelihood of a neighbourhood being renewed. While market factors constrain all forms of renewal, the problems of renewing multi-owned strata buildings have necessitated the passing of laws which place a significant limitation on the rights of existing individual lot owners to exercise the full rights of possession enjoyed by other property owners (Troy et al. 2017). Whether this proves to be a socially divisive change or one that successfully liberates sites for renewal remains to be seen.

A new socio-morphology of the high density city?

As we noted above, driven by a requirement for rapid pre-sales of planned developments before full development finance is provided, apartment developers are keen to sell to whomever they can. Investors are a favoured target and as a consequence, drive the demand for such housing. The investor market can be relied upon to consume new apartments without being too fussy about the actual product, as long as it returns an income and, more importantly, can promise a significant capital uplift over time. Investors comprise over half of apartment owners in Sydney and elsewhere in Australia (Troy et al. 2015). The recent emergence of a strong offshore investor market in residential real estate in Australia compounds that trend (Australian Government Foreign Investment Review Board, various years). The Australian apartment market has become so attractive that direct involvement by leading offshore developers has also become more common, often to target an overseas market. This has led to much controversy and no shortage of newspaper headlines (see Rogers 2015).

This focus on investment value has therefore had a significant impact on the dwelling form on offer in urban renewal developments (Sharam et al. 2015). In particular, the developments on offer overwhelmingly consist of small one- and two-bedroom properties (Australian Bureau of Statistics 2013). Indeed, the new urban-renewal market could be seen to be developed and marketed explicitly without families with children in mind (Fincher 2004;

Woolcock et al. 2010; Easthope and Tice 2011). As Birrell and McCloskey (2016) have argued, the recent inner-city apartment boom in Melbourne has ignored the demand for family-sized housing, a pressing need in a country whose population is being driven by immigration of people at critical child-rearing-age cohorts.

As a result, the market for new apartments generates its own outcomes. Predominant among these is the demographically selective character of those who end up living in the new vertical communities. A recent analysis of more recently built central Sydney apartments showed that the market here was overwhelmingly skewed towards the young and childless, with a predominance of renters (van den Nouwelant et al. 2016). Students were particularly over-represented, as were younger-group households, and there were few households with children. Households with members born overseas were also prominent in some parts of this new apartment market, perhaps drawn to a form of housing they experienced in their home countries, or simply greater affordability and accessibility. Yet this represents only one of a number of dominant apartment sub-markets in Australian cities. Randolph and Tice (2013), analysing 2006 ABS census data for apartment-dwellers, found five main sub-market segments in the Melbourne and Sydney higher-density markets: 'battlers' (low-income renters with children), younger economically active people (20–45 years old, singles and couples without children, mix of renting and owning), the higher-income owner occupied apartment 'elite', older retirees, and students. These had distinctive geographies reflecting city-wide property value gradients, but also overlapped in some areas. The eclectic mix of demographic groups that strata properties accommodate presents an intriguing prospect for the new vertical communities being created by urban renewal in our cities. How these new communities evolve over time will be a critical factor in determining the long-term social outcomes of this city-living revolution.

Additionally, there is an explicit assumption behind most of the recent metropolitan strategic plans that promote higher-density residential urban renewal that the greatest need is for non-family and smaller multi-dwelling housing. As a result, these new communities are built with little formal regard for schools and other child-friendly amenities in mind, or indeed of wider community needs (Sherry and Easthope 2016). This is in contrast to greenfield development where the provision for schools and associated 'soft' infrastructure (libraries, health centres, community services, etc.) is hardwired into the planning system. Allocating sufficient land to provide these public services in rapidly densifying neighbourhoods is becoming increasingly difficult due to the escalating land values that the renewal process itself creates.

In fact, what strata-driven urban renewal may be generating is an intriguing polarisation of the city into higher-density neighbourhoods characterised by the mix of households noted above and the remaining low-density suburbs that surround them, where wealthier and/or family households will predominate. Indeed, such a scenario has been explicitly outlined by leading urban designers (Adams 2009). Splitting the city into these two increasingly disparate social worlds defined by built form may herald a new social morphology of our cities and, indeed, a new phase for the Australian city which, hitherto, had been characterised by a much more homogenous low-density urban form across which social divisions were overlain.

Conclusions

We are undoubtedly moving into uncharted waters in Australia's cities. The impact of the escalating drive to higher residential densities through urban-renewal policies and land intensification, with strata title acting as the underpinning property relation, is changing the balance of housing opportunities that urban populations face. While lifestyle changes are undoubtedly emerging as a result of these opportunities, for many the level of choice involved may be tempered by the constraints placed by the cost and availability of appropriate housing options. As we have discussed, strata-based urban renewal is far from the utopian nirvana portrayed by the development industry, neither is it the result of some spontaneous juxtaposition of growing demand from smaller households and the hidden hand of the market moving to meet this demand. Urban design led prospectuses for transport-orientated densification also fail to grasp the social implications of such renewal proposals (Dovey and Woodcock 2014).

The complex matrix of stakeholders involved in delivering strata development adds new layers of potential conflict and stress for strata communities. The lack of understanding on the part of many strata owners of what being a strata owner actually involves is a major concern when those same owners are tasked with planning and decision-making regarding their collectively owned properties. The lack of oversight of building quality and inadequate planning for management and funding of future repair needs is likely to cause ongoing problems for many strata owners for years to come. Mixed tenure and mixed-use development only exacerbates this situation. A market plagued by examples of poor build quality and accountability may well contribute to a growing public scepticism of the attraction of this new higher-density future. The recent legislative response to the challenges of redeveloping existing strata properties by effectively reducing the property rights of strata owners to facilitate urban renewal will also likely add to a more unfavourable public assessment of the merits of strata-led urban renewal. The possibility that our cities will become increasingly socially aligned along lines demarcated by building type, rather than simply by neighbourhood, is another concerning outcome of the contemporary process of urban renewal.

While these outcomes are largely unintended, the realities of strata-based renewal, especially over the long term, need to be much more explicitly appreciated and accounted for in our urban renewal policies if people's experiences of renewal are going to be positive ones over the longer term.

Note

1 For example both Belgium and Hungary introduced similar systems in the early 1920s (Van der Merwe 2015).

References

Adams, R. (2009) 'Transforming Australian cities for a more financially viable and sustainable future: Transportation and urban design', *The Australian Economic Review*, 42(2): 209–216.

Agola, J. and Kashiyani, B. (2015) 'Conceptual study on construction defects and its solution', *International Journal of Advanced Research in Engineering, Science & Management*, 1(6): 1–7.

Australian Bureau of Statistics (2013) 'Census place of usual residence profile, Australia, 2001.0', http://www.abs.gov.au/AUSSTATS/abs@.nsf/Lookup/2001.0Main+Features12011%20Third%20Release?OpenDocument (accessed 9 September 2015).

Australian Government Foreign Investment Review Board (1998–2015) Annual Reports from 1998–1999 to 2014–2015, https://firb.gov.au/about/publication/ (accessed 29 August 2016).

Birrell, B. and McCloskey, D. (2016) *Sydney and Melbourne's Housing Affordability Crisis Report Two: No End in Sight*, Hawthorn, VIC: Australian Population Research Institute.

Blandy, S., Dixon, J. and Dupuis, A. (2006) 'Theorising power relationships in multi-owned residential developments: Unpacking the bundle of rights', *Urban Studies*, 43(13): 2365–2383

Cancer Council NSW (2015) *Smoke-free Apartments NSW*, http://www.cancercouncil.com.au/31948/cancer-prevention/smoking-reduce-risks/going-smoke-free/achieving-smoke-free-housing-an-information-kit-for-strata-title-accommodation-owners-agents-and-tenants/?pp (accessed 13 April 2016).

Cardew, R. (1980) 'Flats in Sydney: the thirty percent solution?', in J. Roe (ed.), *Twentieth Century Sydney : Studies in Urban and Social History*, Sydney: Hale & Iremonger.

Christudason, A. (2007) 'Defects in common property of strata developments in Singapore', *Structural Survey*, 25(3/4): 306–318.

Cooper, B. and Brown, M. (2014) *Dealing With Defects*, Sydney: City Futures Research.

de Silva, N. (2011) 'Promoting the facilities management profession in the project development phase of high-rise buildings in Sri Lanka', *Built Environment – Sri Lanka*, 9–10 (1–2): 37–44

Dovey, K. and Woodcock, I. (eds) (2014) *Intensifying Melbourne: Transit oriented Urban Design for Resilient Urban Futures*, Melbourne: Melbourne School of Design, University of Melbourne.

Dow, A. and Johanson, S. (2015) 'Docklands apartment fire: Hundreds of high-rise towers to be investigated for fire danger', *The Age*, 28 April 2015, http://www.theage.com.au/victoria/docklands-apartment-fire-hundreds-of-highrise-towers-to-be-investigated-for-fire-danger-20150428-1mvc9j.html (accessed 29 August 2016).

Drane, J. (2015) 'Building defects: How can they be avoided? A builder's perspective', paper presented at the Strata and Community Title in Australia for the 21st Century Conference, Gold Coast, 3–4 September 2015.

Dredge, D. and Coiacetto, E. (2011) 'Strata title: Towards a research agenda for informed planning practice', *Planning Practice and Research*, 26(4): 417–433

Easthope, H. (2014) 'Making a rental property home', *Housing Studies*, 29(5): 579–596.

Easthope, H. and Randolph, B. (2016) 'Principal-agent problems in multi-unit developments: The impact of developer actions on the on-going management of strata titled properties', *Environment and Planning A*, 48(9): 1829–1847.

Easthope, H. and Tice, A. (2011) Children in apartments: Implications for the compact city, *Urban Policy and Research*, 29(4): 415–434.

Easthope, H., Randolph, B. and Judd, S. (2009) *Managing Major Repairs in Residential Strata Developments in New South Wales*, Sydney: City Futures Research Centre.

Easthope, H., Randolph, B. and Judd, S. (2012) *Governing the Compact City: The Role and Effectiveness of Strata Management, Final Report*, Sydney: City Futures Research Centre.

Fincher, R. (2004) 'Gender and life course in the narratives of Melbourne's high-rise housing developers', *Australian Geographical Studies*, 42(3): 325–338.

Guilding, C., Ardill, A., Fredline, E. and Warnken, J. (2005) 'An agency theory perspective on the owner/manager relationship in tourism-based condominiums', *Tourism Management*, 26(3): 409–420.

Guilding, C., Bradley, G. and Guilding, J. (2014) 'Examining psychosocial challenges arising in strata titled housing', *Property Management*, 32(5): 386–399.

Han, E. (2016) Insurer rejects storm-damaged Lidcombe apartment building claim because of numerous defects, *The Sydney Morning Herald*, 9 April, http://www.smh.com.au/nsw/insurer-rejects-stormdamaged-lidcombe-apartment-building-claim-because-of-numerous-defects-2060408-go1kz5.html (accessed 29 August 2016).

Johnston, N. and Too, E. (2015) 'Multi-owned properties in Australia: a governance typology of issues and outcomes', *International Journal of Housing Markets and Analysis*, 8(4): 451–470.

Kern, L. (2010) 'Selling the "scary city": Gendering freedom, fear and condominium development in the neoliberal city', *Social & Cultural Geography*, 11(3): 209–230.

Lehrer, U., Keil, R. and Kipfer, S. (2010) 'Reurbanization in Toronto: Condominium boom and social housing revitalization', *disP – The Planning Review*, 46(180): 81–90.

Leshinsky, R. & Mouat, C. (2015) 'Towards better recognising "community" in multi-owned property law and living', *International Journal of Housing Markets and Analysis*, 8(4): 484–501.

Murphy, C. (2011) 'Building control changes: The on-going battle against the leaking building', *Architectural Science Review*, 54(2): 157–163.

Ngai-Ming, Y. and Forrest, R. (2002) 'Property owning democracies? Home owner corporations in Hong Kong', *Housing Studies*, 17(5): 703–720.

Noble-Allgire, A. (2009) 'Notice and opportunity to repair construction defects: An imperfect response to the perfect storm', *Real Property, Trust and Estate Law Journal*, 43(4): 729–796.

NSW Department of Planning (2005) *City of Cities: A Plan for Sydney's Future*, Sydney: NSW Government.

NSW Fair Trading (2012) 'Multi-storey buildings: Home Building Compensation Fund requirements', http://www.fairtrading.nsw.gov.au/ftw/Tradespeople/Home_warranty_insurance/Multi_storey_buildings.page (accessed 13 April 2016).

NSW Fair Trading (2016) 'Frequently asked questions: Home building law changes', http://www.fairtrading.nsw.gov.au/ftw/about_us/legislation/changes_to_legislation/major_changes_to_home_building_laws/frequently_asked_questions_home_building_law_changes.page? (accessed 27 July 2016).

Ong, S. (1997) 'Building defects, warranties and project financing from pre-completion marketing', *Journal of Property Finance*, 8(1): 35–51.

Randolph, B. and Easthope, H. (2014) 'The rise of micro-government: Strata title, reluctant democrats and the new urban vertical polity' in B. Gleeson and B. Beza (eds), *The Public City: Essays in Honour of Paul Mees*, Melbourne: Melbourne University Press.

Randolph, B. and Tice, A. (2013) 'Who lives in higher density housing? A study of spatially discontinuous housing sub-markets in Sydney and Melbourne', *Urban Studies*, 50(13): 2661–2681.

Reid, S. (2015) 'Exploring social interactions and sense of community in multi-owned properties', *International Journal of Housing Markets and Analysis*, 8(4): 436–450.

Rogers, D. (2015) 'The politics of foreign investment in Australian Housing: Chinese investors, translocal sales agents and local resistance', *Housing Studies*, 30(5): 730–748.

Sharam, A., Bryant, L. and Alves, T. (2015) *Making Apartments Affordable: Moving from Speculative to Deliberative Development*, Melbourne: Swinburne University of Technology.

Shaw, W. (2006) 'Sydney's SoHo syndrome? Loft living in the urban city', *Cultural Geographies*, 13: 182–206.

Sherry, C. and Easthope, H. (2016) 'Undersupply of schooling in the gentrified and regenerated inner city', *Cities*, 56: 16–23.

Troy, L., Randolph, B., Crommelin, L., Easthope, H. and Pinnegar, S. (2015) *Renewing the Compact City: Economically Viable and Socially Sustainable Approaches to Urban Development*, Sydney: City Futures Research Centre.

Troy, L., Easthope, H., Randolph, B. and Pinnegar, S. (2017) '"It depends what you mean by the term rights": Strata termination and housing rights', *Housing Studies*, 32(1): 1–16.

van den Nouwelant, R., Crommelin, L., Herath, S. and Randolph, B. (2016) *Housing Affordability, Central City Economic Productivity and the Lower Income Labour Market*, AHURI Final Report No. 261, Melbourne: Australian Housing and Urban Research Institute, http://www.ahuri.edu.au/research/final-reports/261 (accessed 22 August 2016).

Van der Merwe (ed) (2015) *European Condominium Law*, Cambridge: Cambridge University Pres.

Woolcock, G., Gleeson, B. and Randolph, B. (2010) 'Urban research and child-friendly cities: a new Australian outline', *Children's Geographies*, 8(2): 177–192.

Part II
Inner-city regeneration

6 Barangaroo

Machiavellian megaproject or erosion of intent?

Mike Harris

Introduction

The mixed-use megaproject is a seductive form of urban regeneration. When occurring on re-zoned government land they offer opportunities of massive wealth generation and city-scale public benefit. In particular, megaprojects on post-industrial waterfront sites have been occurring commonly around the world for the past three decades. Most often these projects have been seen by governments as opportunities to reimagine these sites as internationally oriented leisure, commercial and residential city expansions (Hoyle 2000; Malone 1996; Smith and Garcia Ferrari 2012).

Examples include Hudson Yards in New York, Toronto Waterfront, Barcelona Universal Forum, Tokyo Waterfront Sub-centre, West Kowloon in Hong Kong, North Harbour in Copenhagen, HafenCity in Hamburg, there is a Docklands in London, Dublin and Melbourne and Sydney has Barangaroo. These projects are far from new so there is a long and rich history of planning and project-management knowledge to draw from. In fact, one might think that after more than 30 years of these projects around the world and in Australia, their planning and delivery to achieve city-scale benefit commensurate with their size, location and government ownership would now be a honed exercise. The evidence suggests the opposite.

Barangaroo, formerly known as East Darling Harbour (EDH), occupies a 22-hectare strip of land along the north-western edge of Sydney's city centre (Figure 6.1). In 2003, the NSW government and Patricks Corporation announced plans to relocate stevedoring operations from the site and a process began to map out a future for the site. Since 2003 to the time of writing,[1] when the bid submissions for the final parcel were under evaluation, Barangaroo has been a highly controversial megaproject. This public controversy stars an international development company who wins the tender for the project, forms a partnership with government and then later takes the same government partner to court, a confrontational ex-prime minister of Australia with a single-minded vision, a popular Sydney mayor who resigns from the project delivery authority board in protest, a billionaire casino owner and developer and a host of famous international architects.

Figure 6.1 "International Towers" at Barangaroo South viewed from Sydney Harbour (source: author)

This chapter first frames Barangaroo within the literature on global megaprojects and second provides an analysis of the course of events of this megaproject from the original design competition onwards. The ambitions, progression and reception of Barangaroo are examined. What can be seen through this analysis is a shifting and deeply divisive delivery process that has been fought out publicly between the developer, high-profile figures, state government agencies, local government, consultants and a range of stakeholder groups.

This analysis draws on the debates played out in the media between 2009–2017, publicly available planning documents, development applications, formal responses to those applications, independent reviews and semi-structured interviews conducted with former members of the Barangaroo Design Advisory Panel, former NSW government architects, planners and councillors at the City of Sydney and design consultants. The chapter concludes with an attempt to unpack the challenges behind the controversies of Barangaroo and thus frame bigger questions about the role of such a megaproject and the processes that produce the outcome. But before we explore the details of Barangaroo we need to first understand the nature of such megaprojects.

What is a megaproject and what do we know about them?

Megaprojects can be defined as large, complex, transformative ventures worth more than $1 billion that are delivered over many years and involve multiple public and private stakeholders (Flyvbjerg 2014). Megaprojects can be transport, recreational, cultural, defence, exploration, commercial or housing oriented. As such they can include bridges, tunnels, motorways, airports, sports stadiums, concert halls, marinas, conference centres, military bases, transport vessels, weapons programs, universities or office and housing complexes. Mixed-use megaprojects are an example that include combinations of these within one project, commonly including commercial, housing, recreational, cultural and transport c omponents.

As Flyvbjerg (2005) points out, the intent of these projects is to change the structure of society, or the city, therefore they are not merely bigger versions of smaller projects that work within existing structures. Megaprojects are a completely different type of project requiring much higher levels of interdisciplinary, interdepartmental and public–private-sector coordination. Simply put, these are complex and high-risk undertakings.

Mixed-use megaprojects on government land have been increasingly occurring around the world over the past three decades. This proliferation is related to common responses to four globally interlinked political and economic restructuring processes. These are city-based international competition (Florida 2002; Moretti 2013b), the mobility and growth of knowledge economies (Montgomery 2007; Moretti 2013a), the shift in global investment from physical to human capital (Sassen, 2001), and the dominance of market-rule ideology and politics (Brenner and Theodore 2005). In the context of Barangaroo, the first process has been particularly important – the development of the plan for the project has been accompanied by a sharp narrative about Sydney as a competitive global city (Clark 2016; Adamson 2004).

Barangaroo is a typical example, but not Sydney's first, or last. In the late 1980s, a few hundred metres south of Barangaroo, the 54-hectare Darling Harbour entertainment, cultural and commercial megaproject was developed on post-industrial waterfront land. After 25 years, Darling Harbour is currently undergoing its second wholesale redevelopment, by the same developer as Barangaroo. One kilometre west of Darling Harbour, the 80-hectare Bays Precinct is in preparatory stages. Nevertheless, true to the seductive narratives of internationally oriented mixed-use megaprojects the state government (NSW Government, 2010) claims Barangaroo is a "once in a 200-year opportunity".

In a review of the literature on 42 mixed-use megaprojects in 20 countries Harris (2017) identified globally common criticisms and characteristics of these projects. This review highlighted the role of mixed-use megaprojects in a perceived or real global environment in which place marketing and the provision of competitive infrastructure is considered imperative to attract increasingly mobile financial and human capital and to grow knowledge-based industries (Beauregard 2005; Florida 2002; Marshall 2003; Moretti 2013b).

The globally common characteristics of mixed-use megaprojects identified in this review include:

- high-end residential and A-grade office space in order to attract (often foreign) investment and subsequent high-profit companies and affluent residents;
- leisure and consumption amenities in order to attract affluent residents and visitors;
- large and striking buildings to symbolise new economic growth and provide high marketing.

The five globally consistent criticisms of mixed-use megaprojects are:

1 introverted governance models that circumvent local planning frameworks, traditional channels of democratic participation and accountability;
2 global economic positioning and marketing towards a mobile elite prevailing over the concern of local issues;
3 physically and socially self-contained, isolated and disconnected from the context of the host city;
4 similar urban form regardless of the host city that encapsulates a narrow definition of urban life and culture;
5 minimal commitment to public benefit or socially just policies arising from a primary focus on profitability.

The evolution of Barangaroo

With some clarity on the nature of megaprojects, this section turns to an analysis of the events that have unfolded since the competition. This analysis focuses on the roles and actions of key players and two controversies in particular: the increasing size of the built form overall and the arrival of a major casino-complex eight years after the original concept plan was approved, necessitating a major reconfiguration of the site plan with dramatically increased building heights and floor space.

The early days

Like many large urban projects in an aspiring "global city", the process for Barangaroo began with an international design competition. When the competition was launched it was promoted by the state premier as an "historic opportunity to return a substantial part of Sydney's foreshore back to the people" (NSW Government 2005a).

In expanding the commercial centre of Sydney,[2] the brief required design proposals that strengthened local communities such as the heritage-listed Millers Point neighbourhood with historic connections to the industrial waterfront and a significant proportion of publicly housed tenants. The new project would integrate with the surrounding neighbourhoods by relatable scale of built form, spatial structure and access. It would express the historical character of the site and surrounds that had evolved over time and retain harbour functions. The housing report commissioned by the state government to inform the Concept Plan found that the "The provision of a mix of housing types, size and price points" was required to achieve social sustainability, particularly focusing on key-worker housing (Randolph 2006). The report considered comparable projects in New York and London which provided 20 per cent and 38 per cent affordable housing and recommended Barangaroo follow suit, stating that "Current NSW affordable housing planning targets ... are below accepted and achieved international standards."

However shortly into the process things began to change, and continue to change, to the present situation of widespread disillusionment and mistrust of the project's planning and political processes (Reinmuth 2012). Despite the optimism after the winning scheme was announced, the process has demonstrated how very different outcomes of a megaproject can emerge over time. Access to decision-making, transparency and accountability have all been heavily criticised by surrounding local governments,[3] popular press, particularly in *The Sydney Morning Herald*, and academia (Johnston and Clegg 2012; Stickells 2010; Walliss 2012; Weller 2010). It is evident the primary aim of Barangaroo represents a particular strand of economic competition targeting headquarters of large globally active commercial tenants (Sydney Harbour Foreshore Authority 2006), while mounting criticism that local social issues have been relegated to private gain, and arguably have been exacerbated (Barlass 2013; City of Sydney 2015). There is an entrenched reluctance in providing affordable housing[4] or other forms of public benefit besides that implicit in public access to new open space. The built form has been steadily becoming taller and bulkier from pressure to satisfy the perceived desires of large commercial tenants (Sussex and Penn 2011). There is a focus on attracting "leading international firms" alongside an absence of genuine engagement with the layers of cultural and spatial history of the site (Burton 2015).

The key actors

The number, range and influence of actors alone has been remarkable. The state-led project has seen seven state premiers (four Labor, three Liberal) in 12 years. The highly popular independent Clover Moore has been mayor of Sydney the entire period, beginning one year before the competition in 2004. She held a position on the Barangaroo Delivery Authority board before resigning under increasing opposition to the project. The City of Sydney has no formal powers to influence the design or planning. Former Australian prime minister Paul Keating entered the process as a member of the Stage 2 competition jury and then as chairman of the Design Excellence Review Panel between 2005 and 2011, with long-held views on what the role of the site should be and what it should look like, that happened to be quite different from the Stage 1 winning scheme (Johnson 2016; Mould 2017; Weirick 2016). Multinational megadeveloper Lendlease has been the sole developer of Barangaroo to date. Casino magnate James Packer entered the process at the application of "Modification 8"[5] in partnership with Lendlease to propose a highly controversial six-star, high-roller casino-hotel complex with luxury apartments. The "world class" project comes replete with world-class (international) architects – a common practice of inter-city competition in which spectacular buildings by famous architects are intended to bring distinctiveness, while often resulting in sameness (Ponzini and Nastasi 2011). US heavyweights Skidmore Owings Merrill and Peter Walker played roles on master planning and public domain design respectively. Fellow globetrotting "starchitects" Richard Rogers, Renzo Piano and Wilkinson Eyre are designing all the major buildings. A sprinkling of local Australian firms is taking care of the smaller-scale stuff.

A steady stream of high-profile conflicts has punctuated the planning and delivery process. In 2010 Clover Moore resigned from the Barangaroo Development Authority claiming the public were being "railroaded" with persistent breaches of transparency and poor public consultation (Spencer, 2010). In 2013 Jan Gehl, the Danish urbanist called in to bring his "human scale" touch to the development, resigned from the project, citing an absence of communication and concerns that the quality of the public domain was declining in favour of increasingly bigger buildings. Gehl claims that despite repeated attempts to participate in the design process for almost two years his firm was not informed about any progress. Consequently the firm submitted a request that their name and logo be removed from the project (Hasham 2013).

In 2011 Clover Moore presented a petition of 11,000 names in the NSW parliament calling for an inquiry into Barangaroo's planning process. In response Paul Keating publicly ridiculed Moore and other critics of Barangaroo as "sandal-wearing, muesli-chewing, bike-riding pedestrians" (Moore 2011a). Keating received a sharp reprimand from the NSW planning minister for his verbal attack who insisted he consider views other than his own in his role as chairman of the Design Excellence Review Panel. In the minister's words "Such comments do not encourage full and robust debate around a development which has the ability to impact the Sydney CBD and NSW's economy" and that the mayor was "well within her rights to voice her concerns with Barangaroo – and those of her community". In a characteristic power-play, Keating handed in his resignation declaring "Let me tell you, I am not to be muzzled" (AAP 2011).

After resigning from the panel, he remained a highly influential figure, persisting in promoting his visions for Barangaroo publicly through the media and privately through meetings with key players in the state government and developer consortiums. His influence was highlighted at the park opening when the North American landscape architect leading the design referred to Keating as the client and explained how they had communicated throughout the process in order to achieve Keating's design vision – four years after Keating resigned from the review panel (McKenny 2015). The same year Keating resigned, the Design Excellence Review Panel was dissolved in response to conflicts of interest of two members. An independent review commissioned by the state government recommended a new Design Review Panel be constituted "with a clearer governance structure, new Terms of Reference, clear protocols and procedures, and better management of conflicts of duty" (Sussex and Penn 2011: 6). However, this new Review Panel was never established (City of Sydney 2013; Mould 2017).

Lendlease has challenged the state government, in court when necessary, over its contractual obligations in providing developer contributions for the public domain, community facilities and affordable housing for key workers. In 2013 their lawyers argued the contract stipulation of a minimum square metre amount for community facilities, such as childcare centres and recreational facilities, was a "drafting error" and was really supposed to be a maximum amount (Hasham 2014). This would mean Lendlease could provide up to that amount, rather than at least, allowing them to develop more profitable commercial and residential floor space instead. In this instance, they were not successful but these challenges, whether won or lost, have been costly to the state government. In 2014 after losing a developer contribution court battle and paying Lendlease's costs, the state government was left with a significant budget shortfall (Hasham, 2014).

In 2014 City of Sydney councillors raised concerns that the state government has a conflict of interest, as it is both the landowner and assessment authority for Barangaroo (City of Sydney 2014). Effectively the state government stands to gain financially if developer applications for density increases are approved.

Amongst the political conflicts, the architects were at it, stirred up by the fact their buildings seemed to be regularly replaced by one another by a series of approvals and disapprovals. The recipient of the first Barangaroo ousting, Philip Thalis, a member of the team who won the inaugural design competition, called the hotel-in-the harbour by Richard Rogers, who was installed in his place, as "boofheads hogging the front row" (Feneley 2013). When Rogers' hotel was dumped and the Crown casino tower was proposed in its place by Wilkinson Eyre, Rogers remarked, "I'd have preferred another opera house," insisting his building would have had a "much stronger relationship" with the public (Feneley 2013). Ed Lippmann of Lippmann Partnership, parting ways with the Rogers team after taking second place in the original competition, quipped of the Crown tower "It's all a bit Las Vegas" (Glasgow 2013). The criticism between architects could be taken as jaded backlashes for losing projects but it also highlights the real disillusionment with a process that seems to reward skilled negotiators outside formal channels of decision-making.

Controversies of a changing project

Whilst there have been many troubling aspects of the development of Barangaroo the following section focuses on two dominant controversies. One is the regularly increasing enlargement of the buildings, to more than double the original floor space. The other is the emergence of a high-rollers casino proposal through a recently introduced unsolicited proposals process.[6]

Scale creep

After Modification 8 of the Approved Concept Plan (ACP), the project has approval for 605,911 m^2 (NSW Planning Assessment Commission 2016). If Modification 9 is approved, total floor space will be 681,008 m^2 (Barangaroo Delivery Authority 2014), more than double the allowable floor space in the competition brief. In 2016 Lendlease won approval to double the floor space and increase the maximum height for the casino-hotel complex by 105 metres from the currently approved 170 metres. This will make it the tallest building in Sydney, excepting the Sydney observation tower. This is a considerable departure from the design brief, which stated building heights should range between 5 and 14 storeys – or roughly 20 to 55 metres tall. What makes the regular, successful applications for height increases even less palatable is that as stated in the Stage 1 Jury report, buildings that significantly broke this height limit were rejected on the grounds they did not deliver "sufficient extra public gain to be justified" (NSW Government 2005b). This is precisely the criticism directed at the recent height and floor space increases. The City of Sydney has questioned how the increase has been justified and suitable planning put in place to deal with the increase stating that there are no indications "any additional measures are being implemented at Barangaroo to offset the significant impact that a 70 per cent increase in GFA (gross floor area) in this location will have on transport infrastructure and services" (City of Sydney 2014). The City further highlighted that under the NSW planning system, if council was developing the site, any increases in floor space would require increases in contributions for public benefit which might include community facilities or affordable housing. They also raised the concern that the changes since the original ACP have occurred with no independent scrutiny or wider participation in traditional democratic processes. This is effectively a criticism that deals are being made behind closed doors between the state government and the developer which effectively acts as a profit-sharing arrangement in which both stand to gain financially.

For each of Lendlease's applications for an increase in commercial floor space, the argument ran that it was imperative in order to "meet the needs of the big financial services organisations that Sydney needed to house, if it was to hold on to its role as the Australian centre for the financial services industry" and that "There is very limited scope to accommodate floor plates of this size within the existing CBD" (Sussex and Penn 2011). Their argument was that there was not enough floor space in Sydney to meet current demand so they should be allowed to build more than was allowed on their site. However, in Modification 8 Lendlease sought to reduce commercial floor space so that residential floor space could be increased. This time the argument was there was too much "commercial GFA in the pipeline across Sydney, including the provision of over 300,000 m^2 in Barangaroo South alone" (Lendlease 2015). The simple market logic implicit in this is that luxury apartments now stood to provide a higher financial return to the developer. The developer was using whatever argument suited the best return on their investment.

The National Trust has been a vocal critic of the increasing scale and changing form, describing the planning process as "contemptuous" and that the significant increase in floor space has been approved "with no discernible improved public benefit" (Robertson 2015). The Trust argues the increasing scale and character of Barangaroo ignores the rich archaeological and maritime history of Sydney's harbours and historic neighbourhoods. They also claim the state government has abandoned their own principle of relating the project's scale to nearby heritage precincts and buildings. The nationally significant Millers Point neighbourhood to the northwest is 2–4 storeys. A few metres south of Barangaroo the Macquarie Bank building was completed in 2009 reaching 11 storeys under the existing built form controls. In comparison the tallest built commercial tower at Barangaroo is 48 storeys (Lendlease 2015), the tallest proposed residential tower is 71 storeys, an increase on the previous modification from 36 metres to 246 metres (NSW Department of Planning and Environment 2016) and the Crown casino complex will also be 71 storeys although taller at 271 metres (Crown Sydney Property 2015), or triple the height of the tallest building in the competition winning scheme (Figure 6.2).

In response to the concern of the visual bulk of the proposed triplet "International Towers", the 2011 independent review of planning processes proposed an independent design review be undertaken on the form of the commercial towers and podia and the effects on the public domain (Sussex and Penn 2011). Admitting the project had been "plagued by a lack of transparency", premier O'Farrell commissioned the review (Moore 2011b). The review found the first tower to be of high-quality design and supported its resolution. However, it found the proposal to replicate this tower as a triplet, with each adjacent tower increasing in height, "creates the effect of an homogenous mass and exacerbates the overall perception of significant visual bulk and a 'wall' of buildings" (Penn et al. 2011: 6). The review's principal recommendation was that each building should be expressed as separately designed towers, potentially taller and thinner to retain approved GFA, and possibly by three different architects. This would, according to the review panel:

> enliven the precinct as a whole at both the city scale where the towers operate as a signal for Barangaroo and the city image; and at the micro

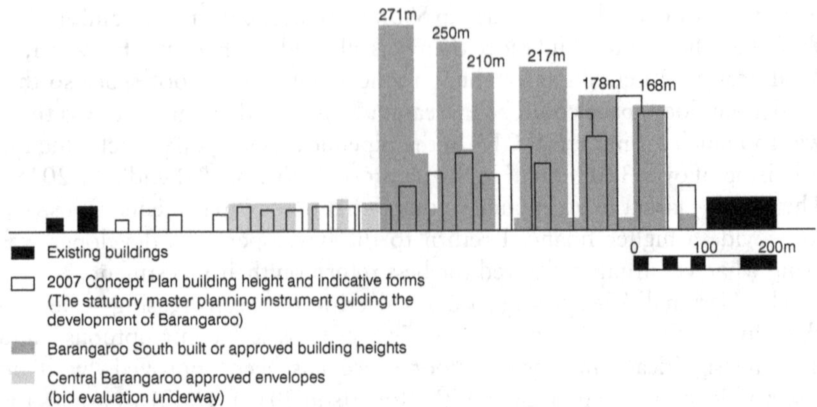

Figure 6.2 Built form comparison of the 2007 Concept Plan (the statutory master planning instrument guiding the development of Barangaroo) with currently built or approved massing, as viewed from the west (source: author)

> level through a finer sense of local character and urban grain. It would also provide greater opportunity to develop diversity in the podia and public domain in an authentic way, which would also assist in integrating the development as a more organic extension of the city.
>
> (Penn et al. 2011: 6)

These qualities were, in fact, precisely what the original winning scheme explicitly attempted to write into planning codes for Barangaroo, what the subsequent developer led scheme iteratively eroded and what the City of Sydney had been arguing for in repeated submissions. Yet despite personally commissioning the review, premier O'Farrell dismissed the findings as "esoteric debate between architects", claiming he was not willing to delay the project (Moore 2011c).

At the other end of the site, Keating's vision of recreating the headland as it was depicted in 1836 became problematic for a range of reasons. First, this was one of the primary sources of conflict between Keating and Hill Thalis/ Paul Berkemeier/Jane Irwin (HTBI), the team that won the competition. The winning scheme proposed the headland park as an expression of all its historical layers whereas Keating demanded it be cleansed of all industrial history and transformed to an idyllic pre-settlement image. The considerable earthworks and modifications to the seawall, as well as the large-scale reconfiguration of site-cut sandstone and complex on-structure planting, executed with admirable technical expertise, that was required to achieve this image pushed the cost to a reported $250 million (Chang 2015). This equates to $41.6 million per hectare. By comparison, Ballast Point Park, across the harbour from Barangaroo and designed by a local firm for the state government on a similar post-industrial headland site and achieving global design acclaim, cost $3.4 million per hectare when it was completed in 2009 (McGregor+Coxall 2010). The Barangaroo headland park is funded by developer contributions from the commercial precinct. It has been speculated that approving increases in floor space in Barangaroo South has been influenced by the need to cover the increasing cost overruns of the headland park (Burton 2015; Weirick 2016).

Then there was the design problem of actually recreating a headland that had not existed for 150 years. While Keating claimed it would be "more representative of any headland as it was before European settlement" (Chang 2015), the design and history community were less convinced. Peter Walker, the "world-class" American landscape architect charged with designing Keating's vision was worried how they would recreate the headland so it didn't "look corny, like Disneyland" (Munro 2010). Locally the idea of recreating a long-lost headland was described as "phoney naturalism" and "nostalgic kitsch" (Weller, 2010). The landscape heritage advisor Keating hired to map out the evolution of headland parks on Sydney Harbour and the Barangaroo headland in particular describes the outcome as a "simplistic image of a green headland" rather than a place to inhabit. He describes the actual landform before settlement possessing drama in its variation of topography and water access, whereas its new form appears "highly mannered, all smoothed out. It's a fake, like Disneyland" (Burton, 2015). The National Trust agreed, producing research showing the proposed design showed almost no resemblance to the original landform (National Trust, 2010) describing it as "an Italian marble mansion claiming to be an authentic recreation of an original colonial cottage" (Munro 2010).

Big buildings are not inherently bad and more open space for an expanding city is undoubtedly a good thing. The criticisms go the heart of the process of how and why decisions are made, who makes them, who is permitted to participate meaningfully in the process, who stands to gain most and how transparent and accountable it all is.

The casino

To understand the casino controversy, we need to first consider the proposal it replaced. As part of Lendlease's 2010 "non-conforming" Modification 4 application, a 150-metre-long pier supporting a 213 metre tall hotel was proposed. The proposed pier was external to the site boundary and prohibited under state planning policy by the 2005 Sydney Harbour Catchment Regional Environment Plan, which expressly prohibits the construction of entertainment and tourist facilities in maritime waters. This has been a longstanding bipartisan position against reclamation of the harbour for commercial purposes (Sussex and Penn 2011).

This proposal to construct a private hotel on reclaimed public waters outside the site boundary of Barangaroo elicited intense criticism. A month before resigning from the Barangaroo Delivery Authority board, Sydney lord mayor Clover Moore declared the proposal unacceptable (Moore and Munro 2010). The National Trust's director criticised the proposal as "privatisation of the harbour" and contrary to the ACP, which required buildings to step down to the water's edge (Moore 2010). The new proposal had one of the tallest towers stepping into the water.

In 2011, a group of 57 prominent architects and planners called for a public inquiry into how the currently approved plan had moved so far from the ACP. They proposed removing the hotel from the harbour, abandoning the ''naturalistic'' headland design and reducing the bulk of the towers (Munro, 2011). Jack Mundey,[7] one of the Stage 1 competition jury members who unanimously selected the scheme by local team HTBI, has called their removal from the project and subsequent design changes by the developer led consortium a "terrible outcome", the exclusion of the City of Sydney a "disgrace" and labelling the process a sign of corruption within the NSW government (Mundey 2010).

Despite broad opposition the minister approved the proposal. The NSW Planning system enabled the minister of planning to be the sole approval authority for Barangaroo and he was free to make a decision to include any area within the provision of that policy, without reference to any process (Sussex and Penn 2011). The proviso the minister required was that the height of the proposed hotel in the harbour was to be reduced from 213 to 170 metres and floor space limited to 33,000sqm.

However public anger intensified, culminating in a legal challenge against the government and Lendlease, triggering an independent review (Cheng 2015). The review found that approving the proposal, while technically legal, "had the effect of undermining confidence in a variety of related laws and decisions, and was therefore not good public policy". Locating the hotel elsewhere on site, it suggested, would be a "demonstration of goodwill" (Sussex and Penn 2011).

Lendlease heeded this advice and in March 2015 presented an alternative hotel proposal. This time it was in partnership with Crown Resorts casino magnate James Packer and included a high-rollers casino with almost half of the floors allocated for luxury apartments (Lendlease 2015). The building would be 105 metres taller than the approved hotel-in-the-harbour, making it the tallest habitable building in Sydney, and be more than double the floor space at 77,500 m^2. To make things more provocative, the hotel-casino-apartment complex was proposed to be located on land designated as public park, educational and cultural space.

Premier O'Farrell described the proposal as "exciting" prior to any assessment (Moore 2012a). Others disagreed. Clover Moore argued the proposal would go against years of planning work, likening it to "plonk(ing) a clumsy Dubai-style hotel with a new mega-casino right in the middle of Barangaroo Central" (Moore 2012b). In their formal submission to the development application the City of Sydney criticised the planning process, lack of public benefit and community engagement, the bulk and scale and the lack of affordable housing (City of Sydney 2015). Mayor Jamie Parker of neighbouring municipality Leichhardt and member for Balmain in the NSW parliament described the process as "Packer's red carpet treatment by the (state) Government." Keating (no longer officially connected with Barangaroo) got involved and told Packer in no uncertain terms that "no major hotel should be built in what is now reserved as public open space" (Nicholls and Moore 2012).

The hotel-in-the-harbour controversy had been replaced with another, more elaborate controversy. It was not just the introduction of a high-rollers casino at Barangaroo and its location in a public park. Again, it was the questionable process it went through. The problem for Packer was that NSW state legislation stipulated that there may be only one casino licence in force at any particular time (NSW Parliament 1992). A NSW government review in 2003 concluded the 'one casino licence' restriction provided a net public benefit by supporting community social standards and was required to effectively meet the objectives of the legislation including "minimising gambling-related harm, protecting local amenity, and ensuring the integrity and proper conduct of market participants" (NSW Government 2003). This licence, awarded through a competitive tender process in 1995, was operating under the Star Entertainment Group less than a kilometre from Barangaroo.

Packer approached the NSW government directly through the "unsolicited proposals" process in which the private sector can approach the government with development proposals rather than respond to a request and participate in a tender process. What exactly transpired has been kept largely secret through "commercial in confidence" procedures but the result was that legislation was changed to allow a second casino licence, and this newly available licence was handed to Packer's Crown Resorts without going to a competitive tender process. A NSW parliament briefing described the Crown proposal as "the most controversial" of all proposals to progress past Stage 1 (Roth 2013: 7). The report highlighted concerns that without a competitive tender process the public do not know if the state is getting the best value for the deal; that the Crown proposal was not sufficiently unique to justify not using a competitive tender process; that the state did not investigate non-gambling alternatives; and, that documents relating to the Crown proposal were not made public until after the announcement to progress to Stage 3 of the process.

City of Sydney councillor John Mant, former planning consultant, lawyer and ICAC Commissioner, has been a vocal critic of the process in which "the most valuable state asset you could think of" was given to Packer in a private deal rather than being put out to public tender (Mant 2015).

The Star Entertainment Group claimed they would have been prepared to bid for the second licence, while suggesting 10 expressions of interest would have been likely based on previous, similarly scaled tenders (Clennell, 2012). Jamie Parker argued for a transparent, international competitive process for a second casino licence, to no avail (Parker 2013). At a dinner address on Sydney Harbour, Packer dismissed the criticisms of improper deal-making, claiming that Crown would have won the tender anyway, before thanking both Labor and Liberal parties for their support of his proposal (Clennell 2012).

By the time the legislative changes in 2012 occurred, two previously senior figures in the NSW Labor party and a former Liberal federal senator for NSW were employed by Packer as either lobbyists or board members of Crown. Clover Moore questioned O'Farrell's impartiality for publicly praising the proposal, days after receiving it and before it could be properly scrutinised (Moore 2012b). Former federal Liberal leader John Hewson questioned the probity of process, raising concerns of Packer's ability to "bulldoze" his application through the approvals process (Clennell 2012).

In 2017, the chairman of the NSW Independent Liquor and Gaming Authority at the time the licence was granted described the Parliamentary scrutiny of the Crown proposal as "inadequate" and "superficial" stating "I don't think there was an appetite for thorough scrutiny. I think there was a wish simply to get the job done in terms of having some basic level of examination and doing the deal" (Wilkinson et al. 2017).

Councillor Mant was not surprised:

> I've worked in other state governments, I've worked in Canberra and Adelaide, and I think I'm qualified to say malevolent old partnerships and agendas and forces are deeply ingrained and very much at play in this state.
>
> (Hackney 2013)

In the Determination Report approving Modification 8, including the Crown proposal, the Planning Assessment Commission expressed "a great deal of sympathy" for the objections raised, however as the NSW parliament had effectively approved the casino and its location with legislative changes to the Casino Control Act, 1992, "the Commission had no power to direct relocation or to change the associated legislation" (NSW Planning Assessment Commission 2016). In other words the casino and its relationship to Barangaroo, and the city, was approved with no reference to the Barangaroo Delivery Authority, no reference to urban design consequences to do with public open space, the waterfront, and built form and certainly no reference to the City of Sydney's planning guidelines. All of this according to former NSW government architect Peter Mould (2017) "fundamentally undermines the whole proposition of urban development on state land and the way it's done."

The significant increase in floor space and the introduction of a casino, in addition to the seeming ease in which legislation changed that made Packer's application first legal, then available to his company exclusively, and the lack of transparency surrounding the process, has created more controversy that the controversial proposal it replaced. The requirement to put forward an alternative proposal to the "hotel in the harbour" was supposed to restore faith in the state government and the planning system. It did the opposite.

Compounding controversies: the casino in the biggest building

The Lendlease/Crown Resorts team argue that more than doubling in size is necessary for the casino complex to achieve the "iconic" or "landmark" status that the hotel on a pier would have achieved. The application argued for the revised urban form with superficial phrases that the tallest building viewed in isolation from other buildings would "create an articulated pinnacle" and "book-end the city's north western edge" (Lendlease 2015). The City of Sydney (2015) rejected these arguments in their formal response listing a range of more tangibly worded assessments such as "reduction in the quantity and quality of public open space", "loss of a consistent public foreshore promenade", "significant overshadowing of the public domain" and "disconnection between the waterfront promenade and parkland." They pointed out that the concept of a monumental tower in isolation at the water's edge was a major departure from both the ACP and the City of Sydney's built-form guidelines.

The more than doubling of size, the application argues, "responds directly to the ambition of the new Crown Hotel proposal to be a world class resort" (Lendlease 2015). The argument is that in order for this building to be "world class" and achieve landmark status, it must more than double in size, become the tallest building in Sydney and stand away from other buildings. In a *Sydney Morning Herald* opinion piece, Mould (2016) pointed out that a building does not become iconic simply because it is big and isolated, but because it resonates with the public imagination and becomes recognised as a symbol of the city, such as the Sydney Opera House and the Harbour Bridge (Mould 2016).

In their development application, and through Packer in the media (Crown Resorts Limited 2013; Lehmann 2013), Crown compared the intended 'iconic' status of the proposed casino complex to that of the Sydney Opera House. Considering this comparison, Sydney's Opera House is indeed isolated, on a peninsula. However, it maintains its 'iconic' status despite being a comparatively small building, more than 200 metres smaller than the Crown proposal. And of course, it is an opera house, not a casino-resort. In recommendation for World Heritage status, UNESCO (2007) described the Sydney Opera House as an outstanding conjunction of architecture, landscape and urban design achieving integrity and authenticity with Sydney Harbour's topography and settlement scale. In contrast the Crown proposal has been criticised by the City of Sydney and the National Trust as significantly compromising the historical and heritage value of the nationally significant Observatory Hill and Sydney Observatory and the state significant Millers Point Heritage Conservation Area. Claiming a mega-casino-resort complex will hold a status similar to the UNESCO World Heritage listed Opera House is an absurd proposition.

The building complex is being promoted by its developer as a future icon of Sydney, primarily on the grounds of its size (Lendlease 2015). However, many have questioned what it will mean for Sydney if its largest and most notable building is a high-rollers casino. The NSW parliament's member for Sydney, Alex Greenwich, said the proposal would "turn Sydney Harbour into a mini-Las Vegas" (Glasgow et al. 2013).

Fitting the building into a different part of the site has, according to the developer "necessitated a reconfiguration of the urban structure" (Lendlease 2015). One of the changes is the relocation of open space from between the waterfront and built form to behind the built form, giving the casino-complex direct promenade frontage. The City argues this compromises the clarity of the public domain hierarchy established in the ACP and adhered to in subsequent iterations, in which the primary public spaces were located along the waterfront. They argue the new park location is "poorly integrated into the street grid of Barangaroo, poorly connected to Barangaroo Point to the north and Darling Harbour to the south, and does not relate to cross links to the city from Hickson Road" (City of Sydney 2015).

Since designing the Star Casino, which he describes as a "disaster", veteran architect Philip Cox has become a strong critic of casino architecture, urban design, culture and social impact (Bleby 2013). Cox laments the idea of the casino in this location, the secrecy of the planning process and the scale and style of the complex, labelling it more suited to the casino mecca of Macau (Glasgow 2013). Joe Agius, the Australian Institute of Architects NSW chapter's president at the time asked "What will be the impact of such a monumental building on the low-scale and people-friendly community facilities proposed for the harbour side of Barangaroo Central?" (Bleby 2013).

At a "Barangaroo Over Development Rally" at Sydney Town Hall in 2010, former NSW planning and environment commissioner and NSW government architect Peter Webber argued "The built form of our city symbolises our values ... the more prominent buildings should represent the best of our cultural, civic, spiritual values and aspirations" (Webber 2010).

The City of Sydney has been particularly concerned about the casino at Barangaroo: "Is this the right site for it? If it was an innovation centre that might be different but a casino? That's not innovative, that's a gaming facility. Do we want to be known as a city for that? They are comparing it to the Opera House! That's a cultural facility, this is a high-end casino." (Coffey 2015).

These questions are important as from the conception of Barangaroo the project has been championed by the state government, as these types of projects around the world invariably are, as a new identity of Sydney.

The outcome and the consistent criticisms

Each step of each controversy outlined in this chapter contributes to bigger questions of what this megaproject is all about. Considering the exceptional opportunities presented by a large, government owned site along the water's edge of the city centre, important questions about the role of this megaproject need to be answered. How is it addressing the particular challenges the city faces such as housing and transport? How does it contribute to a sense of identity and meaning for the city and its people? Perhaps most importantly has the process been fair, transparent and without any doubt of the public benefit commensurate with the scale of the project?

Analysing the project's progression reveals a disjuncture between the project's early framing, delivery process and outcomes to date. Besides expanding the city's commercial floor space and reserving 50 per cent of the site for open space, other principles have been either eroded over time to a point where they are no longer viable or abandoned outright.

When it comes to presenting Sydney as a "centre of the Asia-Pacific" and being a "centre for corporations" Barangaroo is meeting its principles. Its tall, bulky and glassy buildings present a formidable commercial district rising from the water's edge. It is tenanted by big internationally active tenants. The centrepiece will be the high-rollers casino-resort complex with luxury apartments occupying the tallest and most dominant habitable building in Sydney.

The inability of the project to deal with social goals and broader public benefit is a failing of Barangaroo when considered against its global peers. Many of the criticised mixed-use megaprojects in Harris's review (2017) offer a redeeming public benefit aspect. Most commonly this equates to a commitment to providing a large portion of the residential component to affordable housing, ranging between 20–60 per cent of residential floor space, or leveraging the value created through the site's re-zoning to fund infrastructural deficits of the city. In comparison, despite explicit knowledge and advice of global precedent, Barangaroo's affordable housing requirements were exceptionally modest and strategic coordination with other state assets and deficiencies at the city scale has been absent. From a public benefit point of view the provision of new open space and promenade could be considered Barangaroo's one redeeming feature. Although in the context of comparable waterfront projects around the world these provisions would be considered standard.

In light of all that has transpired, we might be left wondering cynically if the early optimistic start of Barangaroo was a "bait and switch" strategy where the public is promised one thing, only to be delivered something else. Or was there genuine intent to achieve the things that have not been achieved? After conducting the largest global survey on megaprojects to date, Flyvbjerg (2005: 18) was left with no doubt of the "Machiavellian formula for project approval, even if it means misleading parliaments, the public and the media about the costs and benefits of projects." However, common threads that emerged during the interviews suggests a pattern within the governance and delivery of Australian mixed-use megaprojects likened more to an accumulative erosion than a deliberate pre-planned strategy (Mould, 2017). They begin with good and strong ambitions with a mandate clearly in the public interest before two eroding processes begin to happen. First, due to the pressure the delivery authority is placed under to deliver their project (and their project only) they become more inward-looking. This means they stop thinking about anything that is not directly related to their delivery objectives. They stop thinking about the interface with the existing city and they stop dealing with the local government and existing planning regulations, which is legislatively incentivised through state government "exceptionality" policies. Second, as delivery authorities are there to develop, they start acting like developers and increasingly focus on profitability. While making more money for the government could be argued to be in the public interest, the pursuit of profit over other goals undermines not only the early project principles but the ability to realise more strategic and demonstrable public benefit related to the project and its role in the city. These patterns compound with the lack of accountable and transparent targets and monitoring frameworks related to the early principles.

Changes along the way are not necessarily bad. Large projects with long timeframes need to have a degree of flexibility to allow for unforeseen changes. The question is for what purpose, how and with whose input? These questions are normally examined through planning frameworks that assess developments and any changes with due process and outcomes that can be demonstrated as the best outcome for the sites, and in the case of megaprojects for broader public benefit related to the particular social and infrastructural challenges facing the city. It is the planning process that has been most acutely criticised. Barangaroo has been described as representative and a cause of the deep mistrust the people of NSW have with the planning system (Harris et al. 2014; Reinmuth 2012). These criticisms of due process have been reported frequently in the media and expressed in interviews with the City of Sydney, former NSW government architects, former members of design review panels and project consultants, who articulate a process that is seen to be geared towards satisfying the interests of project-based private development over strategic coordination, where deals are done behind closed doors, where the government relinquishes control of large extremely valuable sites to a single developer and in doing so lose their ability to enforce good urban o--utcomes.

The analysis in this chapter indicates that Barangaroo aligns with all of the *five globally consistent criticisms*. First, the governance approach is inward looking, existing planning frameworks have been spectacularly contravened and normal channels of participation and accountability for an urban development project have been bypassed. Second, Barangaroo is marketed as a "world class" internationally oriented high-end office and luxury residential offering competing in a global economy, which fails to address local concerns and issues. Third, in departing from the city's built-form guidelines and historic development patterns, Barangaroo has been decontextualised and sits as a spatial anomaly in the city. Fourth, the international architects have produced a series of international buildings and spaces – a packaged image offering a narrow definition of urbanity that lacks idiosyncrasies and a layering of old and new. Fifth, the delivery authority has exhibited a persistent focus on profitability over socially just policies.

In the case of Barangaroo we should not be surprised. This is a mixed-use megaproject excised from the established planning controls of the city and delivered by a separate authority engaging a single developer. This has engendered a profit-sharing condition between the developer and the government landowner who is also the approval authority. Planning principles are seemingly interchangeable and non-accountable. Participation outside the delivery authority and developer is highly restricted. Recommendations from independent reviews have been ignored by the same government agency that commissioned them and significant protest from community and local government have had no impact. "Unsolicited outcomes" resulting in major structural and land-use changes have been encouraged.

If Australian cities are to maximise the opportunities from megaprojects, a change in approach is needed. Project principles need to balance flexibility with accountability. They need to appreciate that conditions will change over the course of the project's long timeframe. They also need to protect their own integrity by preventing the intent of the goals, arrived at through public debate and discussion, being eroded over time. A clear set of principles that filter down to measurable targets need to provide the flexibility to deal with unforeseen changes while providing assurance that the whole process is 'on track' and in accordance with agreed values.

Given the deep interconnections between Australian governments and the property sector in Australia, well demonstrated by Kate Shaw's analysis of Melbourne's megaprojects (Chapter 7), it is hard to be optimistic about the likelihood of this change. Citizens will need to ask in a very loud voice when the next megaproject is being planned – "Does this project provide what the city needs?"

Notes

1 June 2017.

2 In a strategic planning sense expanding the commercial core of Sydney was an important element of Barangaroo given the constrained nature of Sydney's CBD and the increasing prominence of residential buildings in the CBD.

3 Both the nearby councils – the City of Sydney and Leichhardt Council – have been long-standing critics of the planning approvals process, access to decision-making, lack of public benefit and the increasing scale of the development.

4 A requirement for 2.3 per cent key-worker housing on-site and 0.7 percent off-site currently is in place. Project scoping documents cited global benchmarking at 20–38 per cent. An independent review commissioned by the state minister for planning in 2011 recommended doubling the then provision of 2.3 per cent. The City of Sydney has been arguing for at least 10 per cent, preferably 20 per cent, since project inception.

5 Since the approval of the Concept Plan by the NSW Minister of Planning in 2007, there has been a number of modifications to the original plan. The

casino proposal was part of Modification 8 which was approved in June 2016. The latest modification – Modification 9 – is currently being considered by the Minister (as of June 2017).

6 The 'unsolicited proposals' process is a NSW government process designed to encourage non-government sector participants to approach government with innovative infrastructure or service-delivery solutions, where the government has not requested a proposal and the proponent is uniquely placed to provide a value-for-money solution. Details of the process are provided in NSW Government (2014).

7 Jack Mundey was a trade-union official who had led the fight to preserve historic parts of Sydney in the 1970s. In later life, he held a number of government appointments concerned with heritage and heritage protection.

References

AAP (2011) 'Keating resigns over Barangaroo spray', *Sydney Morning Herald*, 6 May, http://news.smh.com.au/breaking-news-national/keating-resigns-over-barangaroo-spray-20110506-1ec89.html (accessed 10 July 2017).

Adamson, M. (2004) *Global Cities*, Oxford: Oxford University Press.

Barangaroo Delivery Authority (2014) *Modification to Barangaroo Concept Plan: Central Barangaroo and Headland Park. Preliminary Environmental Assessment Report and Request for Director-General's Requirements*, https://majorprojects.accelo.com/public/a3596f4fb2a1286eb19bfdb7b362b534/Central Barangaroo CP Mod Final Request for DGRs 060913.pdf (accessed 12 January 2017).

Barlass, T. (2013) 'Barangaroo a plague on all their houses', *The Sydney Morning Herald*, 25 August, http://www.smh.com.au/nsw/barangaroo-a-plague-on-all-their-houses-20130824-2sied.html (accessed 10 July 2017).

Beauregard, R. (2005) 'The textures of property markets: Downtown housing and office conversions in New York City', *Urban Studies*, 42(13): 2431–2445.

Bleby, M. (2013) 'Philip Cox: Barangaroo "a stupid urban resolution" and Echo's Star casino "my worst building by far"', *Financial Review*, 12 September, http://www.brw.com.au/p/professions/philip_worst_barangaroo_stupid_urban_gmVZXINVvfm2ubLbXrlxdI (accessed 10 July 2017).

Brenner, N. and Theodore, N. (2005) 'Neoliberalism and the urban condition', *City*, 9(1): 101–107.

Chang, O. (2015) 'Sydney's new $250 million park, Barangaroo Reserve, is open to the public', *Business Insider Australia*, 23 August, https://www.businessinsider.com.au/sydneys-new-250-million-park-barangaroo-reserve-is-open-to-the-public-2015-8 (accessed 10 July 2017).

Cheng, L. (2015) 'Lendlease's Barangaroo South sparks backlash', *Architecture Australia*, 20 May, http://architectureau.com/articles/barangaroo-south-slammed/ (accessed 12 January 2017).

City of Sydney (2013) *Barangaroo Update, February 18: Transport, Heritage and Planning Sub-committee*, Sydney: City of Sydney.

City of Sydney (2014) *Barangaroo Update, May 6: Transport, Heritage and Planning Sub-committee*, Sydney: City of Sydney.

City of Sydney (2015) *Submission to the NSW Department of Planning in Response to Barangaroo Concept Plan Modification 8 and State Environment Planning Policy Amendment (Barangaroo) 2015* (MP06_0162 MOD 8), Sydney: City of Sydney.

Clark, G. (2016) *Global Cities: A Short History*, Washington, DC: Brookings Institution Press.

Clennell, A. (2012) 'James Packer's Barangaroo casino is high-roller heaven', *The Australian*, 26 October, http://www.theaustralian.com.au/news/james-packers-barangaroo-casino-is-high-roller-heaven/story-e6frg6n6-1226503489924 (accessed 10 July 2017).

Crown Resorts Limited. (2013) *Crown Sydney Hotel Resort: Volume 1A – Project Submission.* http://www.crownresorts.com.au/CrownResorts/files/5b/5bd0ec73-b1fb-416a-a49a-238f5612d5d6.pdf (accessed 2 October 2017).

Crown Sydney Property (2015) *State Significant Development Application SSD 15_6957; Crown Sydney Hotel Resort*, Report prepared by JBA. Sydney: Crown Sydney Property.

Feneley, R. (2013) 'Barangaroo blow but top architects let the Crown rule', *Sydney Morning Herald*, 2 November, http://www.smh.com.au/nsw/barangaroo-blow-but-top-architects-let-the-crown-rule-20131101-2wsdb.html (accessed 10 July 2017).

Florida, R. (2002) *The Rise of the Creative Class: and How It's Transforming Work, Leisure, Community, and Everyday Life*, New York: Basis Books.

Flyvbjerg, B. (2005) 'Machiavellian megaprojects', *Antipode*, 37(1): 18–22.

Flyvbjerg, B. (2014) 'What you should know about megaprojects and why: An overview', *Project Management Journal*, 45(2): 6–19.

Glasgow, W. (2013) 'Cox and co denounce Echo, Packer casino "secrecy"', *Australian Financial Review*, 25 June, http://www.afr.com/business/cox--co-denounce-echo-packer-casino-secrecy-20130624-jhixx (accessed 10 July 2017).

Glasgow, W., Carapiet-Fanous, L. and Hutchinson, S. (2013) 'Dubai? Macau? No, it's Las Vegas on Barangaroo', *Australian Financial Review*, 17 May, http://origin-www.afrsmartinvestor.com.au/p/national/dubai_macau_no_it_las_vegas_on_barangaroo_I4llRR1Hy6X0DVZwjdgI5I (accessed 10 July 2017).

Hackney, P. (2013)' Urban impressario: a profile of Cr John Mant', *Altmedia*, 28 February, http://www.altmedia.net.au/urban-impressario-a-profile-of-cr-john-mant/71086 (accessed 10 July 2017).

Harris, M. (2017) 'Competitive precinct projects: The five consistent criticisms of "global" mixed-use megaprojects', *Project Management Journal*, 48(6): 76–92.

Harris, M., Phibbs, P. and Simpson, R. (2014) *Megaprojects: A Global Review and an Outline of Planning Principles*, Festival of Urbanism, University of Sydney. https://www.leichhardt.nsw.gov.au/ArticleDocuments/153/mega_projects_planning_principles.pdf.aspx (accessed 2 October 2017).

Hasham, N. (2013) 'Designer's Barangaroo bombshell', *Sydney Morning Herald*, 5 October, http://www.smh.com.au/nsw/designers-barangaroo-bombshell-20131004-2uzri.html (accessed 10 July 2017).

Hasham, N. (2014) 'Lendlease squeezes Barangaroo', *Sydney Morning Herald*, 10 June.

Hoyle, B. (2000) 'Global and local change on the port-city waterfront', *Geographical Review*, 90(3): 395–417.

Johnston, J. and Clegg, S. (2012) 'Legitimate sovereignty and contested authority in public management organization and disorganization: Barangaroo and the grand strategic vision for Sydney as a globalizing city', *Journal of Change Management*, 12(3): 279–299.

Lehmann, J. (2013) 'Barangaroo rising – the three-way race to build Sydney's next big thing', *The Daily Telegraph*, 3 May, http://www.dailytelegraph.com.au/

realestate/barangaroo-rising-the-three-way-race-to-build-sydneys-next-big-thing/story-fncv6y42-1226634233082 (accessed 10 July 2017).

Lendlease (2015). *Environmental Assessment Report: Concept Plan Modification 8 and Major Development SEPP, State and Regional Development SEPP and Sydney Harbour SREP Amendments*, Report prepared by JBA. Sydney. Sydney: Lendlease.

Malone, P. (ed) (1996) *City, Capital, and Water*, London: Routledge.

Marshall, R. (2003) *Emerging Urbanity: Global Urban Projects in the Asia Pacific Rim*, London: Spon Press.

McGregor+Coxall (2010).'Ballast Point Park', *Landezine*, 23 November. http://www.landezine.com/index.php/2010/11/ballast-point-park-by-mcgregorcoxall-landscape-architecture/ (accessed 2 October 2017).

McKenny, L. (2015) 'Barangaroo's headland park takes shape', *The Sydney Morning Herald*, 9 April, http://www.smh.com.au/nsw/barangaroos-headland-park-takes-shape-20150407-1mg2cs.html (accessed 10 July 2017).

Montgomery, J. (2007) *The New Wealth of Cities: City Dynamics and the Fifth Wave*, Aldershot: Ashgate.

Moore, M. (2010) 'Critics and backers face off over Barangaroo tower', *Sydney Morning Herald*, 25 February. http://www.smh.com.au/nsw/critics-and-backers-face-off-over-barangaroo-tower-20100224-p3lz.html (accessed 10 July 2017).

Moore, M. (2011a) 'Keating lashes "muesli-chewer" Barangaroo inquiry', *The Sydney Morning Herald*, 6 May, http://www.smh.com.au/nsw/keating-lashes-mueslichewer-barangaroo-inquiry-20110505-1each.html (accessed 10 July 2017).

Moore, M. (2011b) 'Premier plans on goodwill in Barangaroo shift', *Sydney Morning Herald*, 9 August, http://www.smh.com.au/national/premier-plans-on-goodwill-in-barangaroo-shift-20110808-1ij89.html (accessed 10 July 2017).

Moore, M. (2011c) 'Long wait over, work starts on harbour', *Sydney Morning Herald*, 26 October, http://www.smh.com.au/nsw/long-wait-over-work-starts-on-harbour-20111025-1mi4t.html?deviceType=text (accessed 10 July 2017).

Moore, M. (2012a) 'O'Farrell applauds Packer proposal for a second casino', *Sydney Morning Herald*, 27 February, http://www.smh.com.au/nsw/ofarrell-applauds-packer-proposal-for-a-second-casino-20120226-1twlk.html (accessed 10 July 2017).

Moore, M. (2012b) 'Not so fast with that Dubai-style tower, says mayor', *Sydney Morning Herald*, 28 February, http://www.smh.com.au/nsw/not-so-fast-with-that-dubaistyle-tower-says-mayor-20120227-1tyyj.html (accessed 10 July 2017).

Moore, M. and Munro, K. (2010) 'Moore declares hotel shadow too intrusive', *Sydney Morning Herald*, 19 August, http://www.smh.com.au/nsw/moore-declares-hotel-shadow-too-intrusive-20100818-12f8m.html (accessed 10 July 2017).

Moretti, E. (2013a) 'Are cities the new growth escalator?', World Bank's Sixth Urban Research and Knowledge Symposium, 8–10 October, Barcelona.

Moretti, E. (2013b) *The New Geography of Jobs*, Boston, MA: Mariner Books.

Mould, P. (2016) 'Barangaroo is more insult than icon', *The Sydney Morning Herald*, 2 May, http://www.smh.com.au/comment/barangaroo-is-more-insult-than-icon-20160501-gojak9.html (accessed 10 July 2017).

Mundey, J. (2010) 'Barangaroo Development Sydney: Ep 20 – Jack Mundey environmentalist speaks out against Lendlease, Sydney', https://www.youtube.com/watch?v=Y1B0Fsxqv90 (accessed 12 January 2017).

Munro, K. (2010) 'Bringing life back to Barangaroo', *The Sydney Morning Herald*, 11 December, http://www.smh.com.au/nsw/bringing-life-back-to-barangaroo-20101210-18swx.html (accessed 10 July 2017).

Munro, K. (2011) 'Hotel moved from harbour in new Barangaroo vision', *Sydney Morning Herald*, 19 February, http://www.smh.com.au/nsw/hotel-moved-from-habour-in-new-barangaroo-vision-20110218-1azmc.html (accessed 10 July 2017).

National Trust. (2010) 'Barangaroo headland is nothing but a Lendlease landfill', Australasian Special Events. http://www.specialevents.com.au/barangaroo-headland-is-nothing-but-a-lend-lease-landfill/ (accessed 2 October 2017).

Nicholls, S. and Moore, M. (2012).'Keating warns Packer over casino site plan', *Sydney Morning Herald*, 29 February, http://www.smh.com.au/nsw/keating-warns-packer-over-casino-site-plan-20120228-1u13i.html#ixzz3zGAYGylm (accessed 10 July 2017).

NSW Department of Planning and Environment (2016) *Major Projects Assessment: Barangaroo South Residential Building R4A*, Sydney: Department of Planning and Environment.

NSW Government (2003) *Review of the NSW Casino Control Act 1992*, Sydney: NSW Government.

NSW Government (2005a) *East Darling Harbour, Sydney: Urban Design Competition Brief*, Sydney: NSW Government.

NSW Government (2005b) *East Darling Harbour Stage 1 Jury Report*, Sydney: NSW Government.

NSW Government (2010) *Budget 2010–11 Planning; Barangaroo Delivered at No Cost to Community*, Sydney: NSW Government.

NSW Parliament (1992) Casino Control Act 1992, section 4A(1)(,b) NSW Parliament.

NSW Planning Assessment Commission (2016) *Determination of Section 75W Modification Application for the Barangaroo Concept Plan, Hickson Road, Barangaroo (MP06_0162 MOD)*, Sydney: Planning Assessment Commission.

Parker, J. (2013) Addressing the Casino Control Amendment, Sydney: NSW Parliament. Retrieved from http://www.jamieparker.org.au/addressing-the-casino-control-amendment-barangaroo-casino/ (accessed 12 January 2017).

Penn, S., Mould, P.and Brown, R. (2011) *Barangaroo South – Commercial Precinct: Report of the DRP 48 Design Review Panel*, Sydney.

Ponzini, D. and Nastasi, M. (2011) *Starchitecture: Scenes, Actors and Spectacles in Contemporary Cities*, Turin: Umberto Allemandi.

Randolph, B. (2006) *East Darling Harbour and the Intermediate Housing Market*, Sydney: UNSW.

Reinmuth, G. (2012) 'Barangaroo: the loss of trust?', *The Conversation*, 20 November, http://www.theconversation.com/barangaroo-the-loss-of-trust-10676 (accessed 10 July 2017).

Robertson, J. (2015) 'Barangaroo "usurps" Sydney Harbour: The National Trust says "Sydney deserves better"', *The Sydney Morning Herald*, 7 September, http://www.smh.com.au/nsw/barangaroo-usurps-sydney-harbour-the-national-trust-says-sydney-deserves-better-20150907-gjgz7e.html (accessed 10 July 2017).

Roth, J. (2013) *Unsolicited Proposals*, Sydney: NSW Parliament. Retrieved from https://www.parliament.nsw.gov.au/researchpapers/Documents/unsolicited-proposals/unsolicited proposals.pdf (accessed 12 January 2017).

Sassen, S. (2001) *The Global City: New York, London, Tokyo*, Princeton, NJ: Princeton University Press.

Smith, H. and Garcia Ferrari, M. S. (eds.) (2012) *Waterfront Regeneration: Experiences in City-Building*, London: Routledge.

Spencer, A. (2010) 'Lord mayor resigns from Barangaroo Authority', *ABC*, 22 September, http://www.abc.net.au/local/stories/2010/09/22/3018469.htm (accessed 10 July 2017).

Stickells, L. (2010) 'Barangaroo: Instant urbanism – just add water', *Architecture Australia*, 99(3): 47–51.

Sussex, M., and Penn, S. (2011) *Barangaroo Review*, Sydney, http://www.barangaroo.com/media/43967/barangaroo%20review%20final%20report%2031%20july%202011%20compressed.pdf (accessed 12 January 2017).

Sydney Harbour Foreshore Authority (2006) *East Darling Harbour State Significant Site Proposal, Concept Plan and Environmental Assessment*, Sydney: Sydney Harbour Foreshore Authority.

UNESCO (2007) *Advisory Body Evaluation of Sydney Opera House*, http://whc.unesco.org/en/decisions/1329 (accessed 12 January 2017).

Walliss, J. (2012) 'The politics of aesthetics : expanding the critique of Headland Park, Sydney', *Journal of Landscape Architecture*, 7(2): 6–13.

Webber, P. (2010) 'Barangaroo development Sydney: Ep 17 Prof Peter Webber, international architect. Sydney', https://www.youtube.com/watch?v=SThNFSiQfMU.

Weller, R. (2010) 'Barangaroo', *Landscape Architecture Australia*, 12: 17–18.

Wilkinson, M., Cronau, P. and Davies, A. (2017) 'Crown's local casino licences under scrutiny if China hands down conviction, former regulator boss warns', *Four Corners ABC*, 6 March, http://www.abc.net.au/news/2017-03-06/crown-resorts-local-operations-under-scrutiny-four-corners/8326428 (accessed 10 July 2017).

Interviews

Burton, C. (2015) Heritage advisor to Paul Keating, CAB, interview with M. Harris, Sydney, 13 May.

Coffey, D. (2015) City of Sydney urban design specialist, Barangaroo, interview with M. Harris, Sydney, October 28.

Johnson, C. (2016). Member of Stage 1 and Stage 2 East Darling Harbour Design Competition and Barangaroo Design Excellence Review Panel, interview with M. Harris, Sydney, October 16.

Mant, J. (2015) City of Sydney councillor, former Commissioner for the New South Wales Government Independent Commission Against Corruption, planning and governance expert, interview with M. Harris, Sydney, September 2.

Mould, P. (2017) Former NSW government architect, interview with M. Harris, Sydney, April 17.

Weirick, J. (2016) Member of Barangaroo Design Review Panel, interview with M. Harris, Sydney, 12 August.

7 Murky waters

The politics of Melbourne's waterfront regeneration projects

Kate Shaw

Introduction

This chapter examines the redevelopment of Melbourne's waterfront in the context of the city's recent history of strategic planning. It makes the case that despite endless participatory visions and revisions of metropolitan and local plans, none of the key decisions for the inner-city's three largest regeneration projects in the last few decades accorded with anything previously agreed. While such megaprojects are conceived in a framework of increasing economic competitiveness (Lehrer and Laidley 2008; Oakley and Johnson 2013) the argument here is that these developments were shaped less by inter-city competition than by the increasingly toxic battle between the two centrist political groupings that dominate the Victorian government to create a legacy that might propel one or the other beyond one or two electoral terms. Notwithstanding the similar objectives of both parties and periodic commitments to transparent processes and fresh faces and approaches, their tactics led to constant questioning of in whose interests the decisions were being made and with what accountability. That the competing political ideologies have far more in common than differences is evidenced by the same key individuals continuing to circulate in the murky realm of government-appointed planning and development corporations, producing and re-producing the same sets of assumptions and aspirations, regardless of the increasingly disappointing outcomes on the ground.

Melbourne was founded in 1835 by white settlers in what was named the state of Victoria in 1851. It became a prosperous city during the gold rush, contributing many fine public buildings to the Commonwealth. Victoria sustained simultaneously a powerful squattocracy, which amassed its fortune from mining claims, pastoral occupation and land deals, and labour movement, which worked the mines and pastures, built the buildings, transported the produce, and prepared it for export at Australia's busiest docks. Three political parties emerged after Australian federation: the Liberal (Tory) party, the Country party (predecessor to the contemporary National party) and the Australian Labor party (ALP). In Victoria in 1947 the Liberal/Country parties formed a coalition which has persisted ever since, alternating in government in monotonous regularity with the ALP in a mutual quest to exclude any alternative political arrangement.

The waterfront projects in question are Southbank, begun in the early 1980s; Docklands, in the early 1990s; and Fishermans Bend, the announcement for which was made in February 2011. Each was the largest urban development project in Victoria at the time, and each was the subject of a political declaration prior to its articulation in any coherent planning strategy. All have the status of continuing projects at least into the 2020s with cleared sites yet to be redeveloped, although already Southbank has transformed Melbourne's skyline with five of the city's 30 tallest buildings, and the new-build at Docklands covers two-thirds of its total land area of 150 hectares. They are significant not just because of their size – the three sites considered here account for nearly 600 hectares of inner-city land, three times the area of the historic city centre – but because the opportunities presented by them for building what the city needs in the late twentieth and early twenty-first centuries are greater than they ever will be again. The ways in which these opportunities have been dealt with over the last three decades illustrate a malaise in the state's governance. Australian capital cities have an affordable housing crisis. Income inequalities are based on access to property. Metropolitan Melbourne is profoundly divided, socio-economically and spatially. While successive state governments compete with each other over central regeneration and infrastructure projects, the outer suburbs continue to sprawl with inadequate planning and funding. It is the inner and middle local governments that maintain Melbourne's justified (if overblown) reputation for liveability, sometimes with support from, but mostly in open tension with the state.

Southbank is the name given to the one kilometre of north-facing Yarra River frontage directly opposite the central city grid (Figure 7.1). For more than a century it was low-lying land accommodating temporary and industrial uses mainly associated with port and railway infrastructure. Docklands is further west and just north of the Yarra, and from the 1890s was the largest port in Australia. Fishermans Bend was the colloquial name for the land on the curve of the Yarra River which, after winding through the eastern suburbs and past Southbank and Docklands, sweeps around Port Melbourne to join the waters of Port Phillip Bay (Figure 7.1). The land on the bend was set aside for heavy industry in 1935; before that, from the mid-1850s, it was a European shack settlement for fishermen and itinerant dock workers (Victorian Places 2015).

Before all this, all was wetlands and inhabited by clans of the Boon Wurrung (known as the coastal tribe) who were members of the Kulin nation, long since swept aside (Oakley and Johnson 2013). They hunted and fished and lived here for many tens of thousands of years before British settlers dispossessed them of their lands and systemically annihilated them. The Boon Wurrung called the river Birrarrung, meaning place of mists and shadows.

In the late twentieth century, as Australian inner-city manufacturing declined and imports increased, and container ships got bigger and

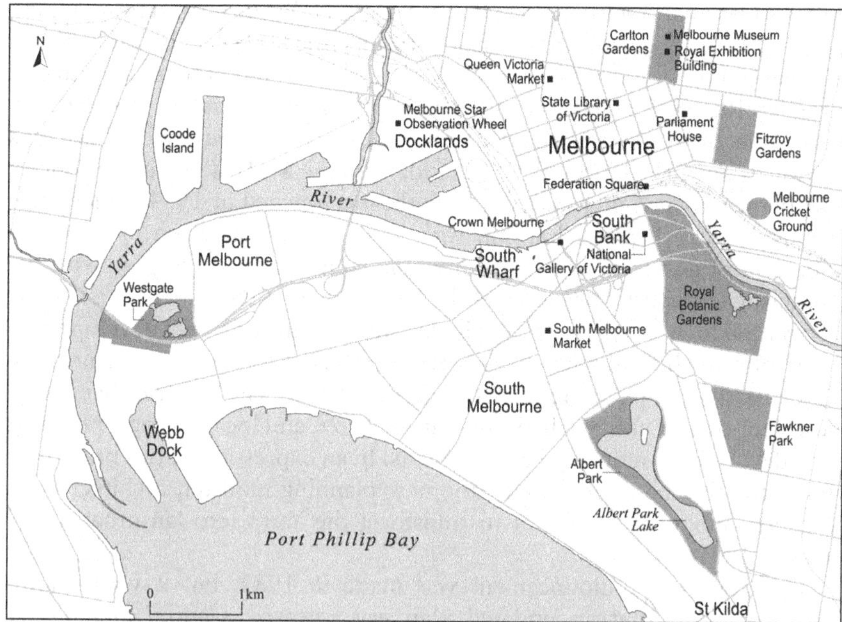

Figure 7.1 The Yarra River flows through Melbourne, past Southbank, Docklands and around Port Melbourne before merging with Port Phillip Bay (source: author, adapted from Google Maps)

technologies changed accordingly, the shipping manager, the Port of Melbourne Authority, began to relocate the working docks downstream to the mouth of the Yarra. As in many cities Melbourne's river was polluted, and residential, retail and entertainment uses were oriented with their backs to the waterways. Even given the riverbank's potential value with its deindustrialisation, the city's approach to the water was slow. It was a campaign in the 1970s by one of the daily newspapers, *The Age*, in concert with civic leaders and collected real-estate interests, to clean up and save the Yarra that drew collective consciousness back to the river. That consciousness consisted of more than the people of Melbourne, who were similarly cautious. The key players in the transformation of the inner city were those who controlled the timing and location of private finance. This is the story of the strategic embrace of Melbourne's waterfront by property developers and investors, and the lengths successive Victorian governments went to, once they found their thirst, for more.

Southbank, from 1982

The development of metropolitan Melbourne was subject to a series of comprehensive plans from 1954 under the firm authority of the Melbourne Metropolitan Board of Works (MMBW). It was also subject to conservative

Liberal/National government from 1955 to 1982. While little in the planning system changed during that period of "incremental growth" and district centres policy (McLoughlin 1992: 71), in the 1970s a remarkably liberal premier Rupert Hamer saw the potential for an extended arts precinct just south of the Yarra, next to the already existing national gallery. His government worked through that decade with the MMBW to build a concert hall (which opened in 1982) and state theatre (opened in 1984). They were edging toward the river but still with their backs to it when in 1982 the Liberals lost government to a fresh young Labor party. In keeping with the optimistic left politics of the time, the Labor government under John Cain Jr committed to public participation in decision-making on urban development. Most of the MMBW's functions were transferred to the more democratic state ministry for planning and environment. Water and sewerage went to a new state-owned utility called Melbourne Water, and local strategic planning was 'devolved' to municipal governments. In an expression of this brave new world for the shape of the city, the new planning minister, architect Evan Walker, announced his intent to transform the Yarra into 'an urban river' (Dovey 2005: 35).

The minister's announcement was made in 1982, but it wasn't until two years later that an updated plan was released. *Central Melbourne: Framework for the Future* (Victorian Government 1984) detailed the planned redevelopment of Southbank and four other large state land-holdings (the Latrobe Central and Flagstaff stations, the Jolimont railway yards, and Station Pier in Port Melbourne). Southbank was in the local government area of South Melbourne but the state government gave itself responsibility for projects of 'state significance', theoretically working in collaboration with local government. The government and South Melbourne council did collaborate to link the established arts precinct with the river, to lever the state's considerable land holdings "in an entrepreneurial way ... to initiate change and redevelopment" (Victorian Ministry for Planning and Environment 1986: 123). Land was consolidated for public infrastructure, and subdivisions sold to the private sector 'at considerable profit to the public purse' (Dovey 2005: 37). Plans for a wide riverside promenade were prepared, and design guidelines specifying public access, 6 to 8 storey height limits stepping down to the river, active street-fronts and high permeability (Victorian Ministry for Planning and Environment 1986).

In 1987, when global stock markets crashed and Melbourne's property market became more volatile, the state–local collaboration became more tense. Differences over the design guidelines began to appear as potential land values increased. A Major Projects Unit (MPU) was created within the state government to form a buffer between local and state, to 'separate the urban development process from the constraints of democratic governance' (Dovey 2005: 37). According to Dovey, the MPU was also intended to mediate between public and private interests as 'the government was engaged in a major wealth creation activity and the private sector wanted more of the action' (2005:37). Evan Walker became minister for major projects, with a junior minister taking over the planning and environment portfolio. The MPU maintained an 'arm's length' relationship to other state departments and local councils, and began negotiating directly with the private sector.

The promenade was publicly funded and built in 1990 with an award-winning design by local architects, DCM. The first building to be approved was Southgate, a medium-rise shopping and dining complex, to which was attached a 22-level hotel. This was a speculative venture, begun in 1990 by Jennings builders and financed by the Australian Guarantee Company (AGC), a private-sector property financier. It was sold on completion in 1992 to a subsidiary of the Westpac banking corporation with management from Centro Property Group, a large, regional shopping centre manager, since delisted. The shopping centre is currently owned by Deutsche Bank. The hotel was managed initially by Sheraton, and is currently in the hands of Langham, another high-end international corporate hotel chain. The Southgate complex was followed by 26-storey twin office towers for *The Herald* and *Weekly Times*, which in addition to their highly derided contribution to the skyline, completely block through-access on the ground. By the time these were completed the recession of the early 1990s had hit, and market interest at Southbank had dampened.

Southbank consists now of a sunny and pleasant public riverfront promenade which is very popular with Melburnians and visitors. Lining the promenade is a wall of ordinary medium to high-rise buildings too obviously designed to deliver profits to their investors and developers. Behind them, a snarling traffic sewer is interspersed with residential towers only properly accessed by car (see Dovey 2005; Lucas 2016). This difficult hinterland was not intentional, of course. The first two decades of Southbank's redevelopment are chronicled at length in Dovey (2005) but suffice it to say here that progress in the late 1980s was piecemeal. The project suffered from the state's vacillation between participatory processes and local governance on the one hand, and practices that were commercially favourable to the private sector on the other. The traditional Labor-left values of democratic governance, conservation and social justice were under attack from the Labor right, which was coming to embrace neoliberal capitalism, deregulation and privatisation as the way to strong economic performance. The uncertain economic climate compounded the lack of activity at Southbank, and the government's efforts at a solution in the MPU only exacerbated the tensions. The unit came to be seen as a barrier to the democratic processes the Labor government once stood for, and a funnel for private finance into the built environment with executive capacity to vary planning and design frameworks as necessary.

Premier John Cain, whose Keynesian approach had catalysed Southbank, but in the looming recession had disastrous results elsewhere, resigned. He was replaced by Joan Kirner, whose appointment as Victoria's first woman premier when the ship was already sinking was unavoidably symbolic. Her only hope was for rescue by a flagship project that would bring untold riches into the state, requiring abandonment of her personal ideology to wave it down.

Docklands, from 1989

The relocation of shipping activity through the 1980s from Docklands to Webb Dock released 44 hectares of water and over 150ha of industrially zoned, publicly owned, under-valued and under-utilised land just west of the city centre (Figure 7.1). The metropolitan strategy dated back to 1981, and the MMBW planning scheme was in transition to local government-prepared schemes under the new 1987 Planning and Environment Act. The docks appeared on metropolitan maps as a blank area, under the opaque control of the Port of Melbourne Authority. The *Central Melbourne* plan (Victorian Government 1984) and its metropolitan update *Shaping Melbourne's Future* (Victorian Government 1987) made no mention of Docklands. But in 1989 the Major Projects Unit appointed a Docklands taskforce to consider alternative uses for the area, with representation from state departments, Melbourne city council and the community and private sectors. A strategy was prepared that included heritage and height controls, public transport links, green open space, a range of housing types and a levy on developers to pay for public housing (Docklands Task Force 1991). The tensions within the Cain–Kirner Labor government in its final years were on full display however, with proposals also entertained ranging from a state park and wetlands to a Japanese government-led 'multi-function polis' (MFP). An Olympic Village was also in the mix – part of a 1996 games bid led by former Melbourne lord mayor and Liberal party treasurer at the time, Ron Walker. In mid-1992, on the recommendation of the MFP and Olympic Village advocates, the entire land holding was transferred from the Port of Melbourne Authority to an independent development corporation, the Docklands Authority. The authority was not only at arm's length from state departments and local councils, but governed by a board of corporate professionals with no community representation.

Not much happened at the docks in the early 1990s other than the diversion of the MFP concept to Adelaide (where it was never built) and the failure of Melbourne's Olympics bid (the games went to Atlanta). While Premier Kirner publicly courted the Liberals' natural allies in a desperate bid for investment, 'smooth-talking US casino salesmen' arrived in Melbourne, according to Costello and Millar (2000:167), 'spinning wonderful stories of economic rejuvenation and jobs. They found recession-struck governments vulnerable and receptive'. The introduction of gaming machines into Victoria was hotly debated in the media, but Labor needed some investment at Docklands, anything. In April 1992 a statewide plan, *Shaping Victoria's Future* (Victorian Government 1992) pronounced hopefully:

> The Docklands Authority has been established to develop the area between Spencer Street and the waterfront. This will greatly extend central Melbourne's access to the Yarra and will be the culmination of the Government's strategy of making the river part of the city ... It will involve greening the waterfront and opening it up for public recreational use. The development will include residential and recreational components and could house a casino, elements of a multi-function polis, higher education and research facilities and a teleport to provide cheap around the clock data transmission to and from the rest of the world.
>
> (Victorian Government 1992: 46)

Later, Kirner acknowledged she was 'disturbed by the impact of commercial gambling on working-class women and on Asian communities' in Costello and Millar (2000:167) who slammed the 'failure of public policy and governance' in this and other Australian states at the time (p. 170). The strategy was unsuccessful anyway: by the October 1992 election the lack of activity combined with the government's internal tensions and economic woes to deliver a landslide win to a radically neoliberal coalition led by Liberal Jeff Kennett. From this point on, until Labor returned to power in 1999, decisive action from the unequivocal Coalition government – the slogan for which was 'Victoria: on the move' – transformed the hinterland of Southbank and the entirety of Melbourne Docklands with road extensions, site layouts, podium footprints and building heights that flew in the face of all the urban design strategies, and were structurally irredeemable.

The road to lasting legacy: first take local government

Local governments in Victoria are creations of the state and have no constitutional recognition, but they do have responsibility for implementing state planning policy and developing local plans within constraints. They are perceived to be closer and more accountable to the electorate, and as councillors tend not to be professional politicians and are often drawn from the local business or resident base, their capacity for galvanising opposition to state intervention is considerable. The growing tendency to exclude local councils and communities from state decision-making in the late 1980s and early 1990s led to unrest and difficult relations that were compounded by the widening gulf between the left and right within Labor. When the Kennett government came into power in late 1992 it commenced a radical program of reforms. Among these were the sacking of all local councillors, forced amalgamations, and the redrawing of municipal boundaries amid much protest. The entire southern length of the Yarra, from South Yarra in the east to Port Melbourne in the west, was excised from the Cities of Prahran, South Melbourne and Port Melbourne respectively, and claimed for the City of Melbourne (Figure 7.2). The new city map included newly designated

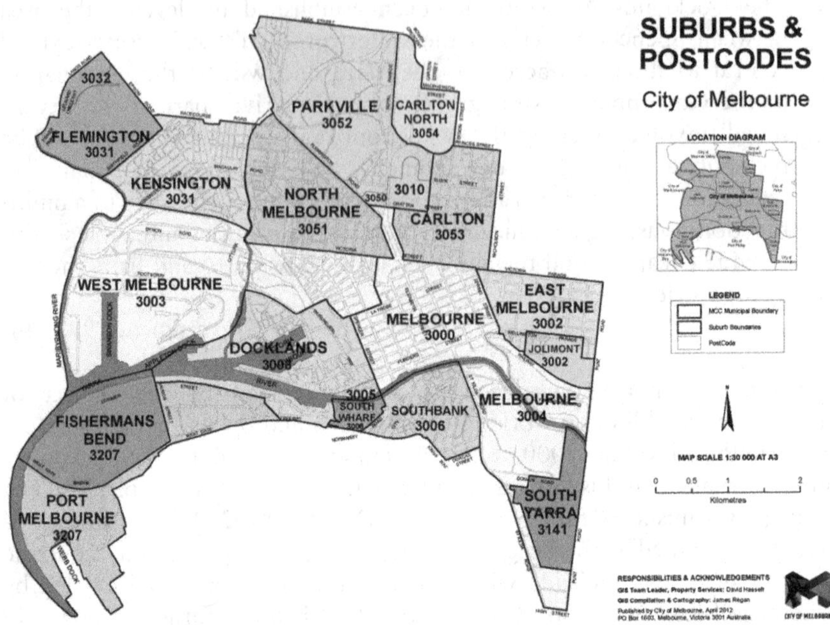

Figure 7.2 Designated districts of the City of Melbourne after council amalgamations in 1994 (source: City of Melbourne)

districts of Southbank, South Wharf, Fishermans Bend and Port Melbourne. The remnants of South Melbourne and Port Melbourne were merged with the City of St Kilda further south to become the City of Port Phillip.

Overseen by Kennett's minister for planning and local government, Rob Maclellan, these reforms were accompanied by amendments to the Local Government Act and the Planning and Environment Act that reduced council powers and consolidated state powers. This was followed by a wholesale privatisation of state-owned land and utilities and compulsory competitive tendering of council services. The amalgamations included not only the redrawing of boundaries but a restructuring of the voting system to advance state strategic and commercial interests. The transfer of the southern banks of the Yarra from the working class Port Melbourne and conservative South Melbourne and Prahran councils to the central City of Melbourne came with a voting structure that virtually assured a business-dominated capital city council, and compliance with the neoliberal trajectory of the state. The exclusion of the City of Melbourne from all decisions on Southbank and Docklands was meekly accepted by newly elected councillors in the restructured local governments some years later, under the watchful eyes of council CEOs chosen personally by government-appointed interim commissioners.

Then create special acts of parliament and development corporations

Having dismissed local councils from the political arena, the next move was to bring in corporate business. *Agenda 21*, more of a 'to do' list than a strategic plan, listed nine 'major civic projects' for central Melbourne, in the process rebranding the MPU the Office of Major Projects (OMP) but keeping all its basic functions including a senior minister. Top of the list was a Melbourne Casino for Southbank. Premier Kennett's friend Ron Walker, who by then was chairman of the Melbourne Major Events Company (appointed by Kirner) and the Grand Prix Corporation (appointed by Kennett), was interested in building a casino but not at the undeveloped Docklands, where the risk was too high. Southbank seemed more appropriate, as much of the infrastructure was already in place. What remained of the planning strategy for Southbank was abandoned and decisions for the rest of its development were driven mainly by the requirements of the casino, which would be developed by Walker and his friend, property developer and racehorse owner Lloyd Williams. The government assured the public that funding for the casino (at an estimated cost of $500 million) would be 'totally privately sourced with *no* public sector guarantee or underwriting' (Victorian Government 1993: 2, emphasis in original). A special Casino (Management Agreement) Act 1993 was introduced that exempted it from normal planning and licensing laws. In addition, the act made the owners' obligations 'negotiable', allowing the government to waive requirements and provide concessions at a cost to taxpayers of millions of dollars per year (*Australian Financial Review* 1998). These conditions continue to apply to Crown's current owner, James Packer (owner also of the casino at Barangaroo on Sydney's waterfront – Chapter 6) who in 2014 secured an extension to 2050 of the Melbourne licence with support from both the Victorian Coalition and Labor (ABC News 2014). A freeway traversing the Southbank hinterland leads directly into Crown's private underground car park before continuing on under the complex and up again across the river. The casino consumes and renders impermeable half of Southbank's river frontage with bars, restaurants and shops intersected by grand entrances into the complex, while the rear is devoted to the car park and service access. The residential and office buildings and streets beyond, which make up the Southbank hinterland, are haphazardly distributed through the network of roads that converge behind the complex.

At Docklands, the Kennett government retained the Docklands Authority but 'refocused' its mission from 'manag[ing] development of Melbourne Docklands on behalf of the Victorian State Government' to 'undertaking development in partnership with the private sector' (Docklands Authority 2000: 72; 2002: 10). Already with a board and CEO with no electoral accountability, the authority was distanced further from the rest of government and public input by the introduction of 'commercial-in-confidence' provisions that protected all financial negotiations and transactions from public scrutiny. In 1994 the government released a 'capital city policy', *Creating Prosperity*, which only briefly mentioned Docklands to say that the area would be 'progressively opened up from 1995' (Victorian Government 1994: 47). The policy added that with the agreement of Melbourne city council, the minister for planning would be the responsible authority for all decisions on Docklands and the Port of Melbourne (1994: 59).

The Kennett government often reiterated the virtues of giving the private sector free rein. In keeping with this commitment the Docklands Authority divided the entire 200ha of Docklands (including water) into seven precincts with no planning controls, flexible design guidelines and no infrastructure plan, stressing that Docklands would be developed 'at no cost to the public purse' (Docklands Authority 1995: npn). Developers were invited to bid for whichever precinct they wanted and for whatever they wanted to build, including infrastructure: roads, decontamination, water and sewerage, they would be responsible for the lot. There was no response.

In late 1995 a new CEO was brought in, John Tabart, who came from 15 years with the developer Lendlease and in the 1980s was its managing director. Tabart kept the precinct divisions, but made it clear that developers would be able to negotiate with the government for infrastructure funding. In 1996 the tender process was relaunched and Docklands' fate was sealed. In the following years the government discreetly picked up the costs of infrastructure provision and decontamination works, along with the less discreet stimulatory interventions of a football stadium, roads, bridges and a major freeway, while still allowing developers to build whatever they deemed to constitute highest and best economic use. The main condition was that construction start before the end of the decade (and the 1999 state election). Lendlease was allocated the prized central precinct of Victoria Harbour. Court documents from a challenge years later by Lendlease and a second key Docklands developer, MAB Corporation, against the state's revenue office (which collects taxes on transfers of land) revealed that some of these land sales occurred at prices less than 1 per cent of their value at the time (Johanson 2012).

The lengths the Kennett government went to in order to catalyse economic activity was unprecedented in Victoria's history. The publicly owned docks, wharves and water were privatised with no planning leverage, no value capture, no specifications for community amenities or public facilities or social services. The developers were effectively paid to build there. They were not required or even encouraged to work together, with the result that there is little connection between the precincts and no centre to the whole place. Even within the precincts the connections between buildings are poor, consisting mostly of high-rise buildings on large podiums that channel the strong winds into narrow tunnels and cast much of the northern public realm into shade. The criticisms from social equity, affordable housing and tenants' rights advocates were fierce, and united in their condemnation of the uniformly high-cost office/residential/retail product that provides little variation in employment, housing and street-life. Commentary from the planning, urban design and architecture disciplines (eventually) was at one that Docklands lacked a 'heart'; that it was a display of 'corporate homogeneity, lacking in grit, just the usual, run-of-the-mill, second-rate modernism' (Leon Van Schaik, then Dean of the RMIT School of Architecture and Design, quoted in Millar 2006: 1).

The two decades of development of Melbourne Docklands are detailed in Shaw (2013), but a summary from an *Age* columnist after her ride on Docklands' Star Observation Wheel (a giant ferris wheel modelled on the London Eye) which, plagued as it was with structural problems and derided as a white elephant, became something of a metaphor for the entire project, seemed to encapsulate the public response:

> I'm not alone in being unimpressed with Docklands, but it's when you see it from above that its true hideousness is laid bare; a bleak, depressing monument to missed opportunity, where greedy developers were allowed to run rampant without any thought for the people who would actually live there.
>
> (Dux 2014)

In 1999, after two terms, the Kennett government lost office to Labor. Kennett attributed his loss to the private sector being too slow to come to the party, and to the inability and lack of imagination in the electorate to appreciate what he was trying to do. The figures showed that in fact he lost in the middle and outer suburbs and regions long neglected by his inner-city agenda. For the inner-city voters the issue was more of how he went about it. Under premier Steve Bracks and planning minister John Thwaites the new government committed to more consultative governance, and immediately released a 'state planning agenda', *A Sensible Balance* (Victorian Government 1999).

The Bracks government's signature was third-way – public–private partnerships, collaborations and consensus building. A comprehensive metropolitan plan was prepared, *Melbourne 2030* (Victorian Department of Infrastructure 2002) which made no mention of the industrial land in Port Melbourne or Fishermans Bend. It did identify areas for regeneration however, emphasising the role of local councils in urban consolidation (albeit controversially and with insufficient resourcing). Development at Southbank and Docklands continued apace, with all the contracts entered into under the Kennett government carried through. Throughout the 2000s Labor introduced a series of themes for Docklands – ecologically sustainable, creative, 'a place for everyone' (McCracken 2005: 113); the government's slogan for the decade was 'Victoria: the place to be' – though none of the basic arrangements changed. The themes came to be used as narratives for the redevelopment as the effects of the Kennett government's stimulatory interventions began to be felt, and private-sector interest built momentum, and Docklands took shape precisely as the authority had designed it.

In 2005 John Tabart was able to triumphantly declare the market value of Docklands at $10 billion. Riding the wave, premiers Bracks then Brumby, and their erstwhile (and short-lived) ministers for planning Thwaites, Delahunty, Hulls and Madden respectively, made efforts to connect Southbank and Docklands at their south-west meeting point. With assistance from the small but sophisticated urban design division of Melbourne city council, more modest, detailed, finer grained and lower scale development was delivered. Two 'keyworker' affordable housing projects at Docklands were initiated, and responsibility for the developed parts of Docklands was finally in 2010 handed over to the council. The level of institutional and public support for this move suggested that the critiques of Docklands' governance to that point were widely shared. The city council, despite its limited powers, has relatively transparent decision-making processes. It immediately began remedial efforts at Docklands, including a 'community hub' with a library and childcare centre on the small amount of remaining public land, and began to look for ways of humanising the in-between spaces. The City of Port Phillip, in the spirit of the more democratic mood of the 2000s, began a substantial strategic planning process for the parts of South and Port Melbourne that remained within its boundaries.

The opaque processes within the development corporations persisted, however. The Docklands Authority was merged in 2003 with the state's public land agency to become VicUrban, though still with a politically insulated structure, and Tabart became CEO of the larger body. The act of parliament that created VicUrban specified that the functions of the new authority 'are to be carried out on a commercial basis' (Victorian Government 2003: 7). The Office of Major Projects remained, and in 2007 a Growth Areas Authority was added as another 'arms-length' planning body, with Peter Seamer as CEO. Seamer came from a career of local government CEO appointments, notably with the regional City of Greater Bendigo 1994–96 when he oversaw the amalgamation of six municipalities into one. His CV says he was:

> building a strong economic development capacity, creating sound long term budgets, lifting service levels and extending the capital works budget while reducing staffing and rates significantly.
>
> (Seamer 2016)

Local governments continued to be sidelined in new and strategic decision-making, and while they were given a role in implementing Melbourne 2030 received neither the powers nor resources to effectively do so. It became clear later in the decade that not that much had changed: transparency and public participation on Melbourne's regeneration projects were as remote as ever. The Southbank hinterland had become a forest of towers, and the narrative for the undeveloped parts of Docklands returned to Kennett's: full funding from the private sector with no state outlay notwithstanding that the detailed figures had never been made publicly available. The Growth Areas Authority didn't seem to do much at all, other than the CEO looking reassuring and making the right public pronouncement from time to time.

In late 2010, after three terms, Labor lost office. Although it had inherited a legacy project, it hadn't sufficiently redeemed it; in making a clear break from the 'can-do' style of the Kennett government, it hadn't done enough. To distinguish the new Coalition government, premier Ted Baillieu and planning minister Matthew Guy announced the preparation of a new metropolitan plan. But they had a problem: with publications like the glossy promotional *Waterfront Spectacular* (Keeney 2005) in the public domain, the economic success of Docklands, if nothing else, was being attributed to the Labor government. The Coalition set to creating a new urban renewal authority, which took over from VicUrban in 2011. Trading as Places Victoria (PV), this body maintained most of VicUrban's functions, especially its commercial basis. Its first annual report explained that:

> the business has experienced a fundamental transformation as Places Victoria refocuses its operations to better align with the Government's policy of prioritising urban renewal in defined precincts in metropolitan Melbourne.
>
> (Places Victoria 2012: 4)

The second annual report clarified, this meant bringing to 'a development-ready state urban renewal sites which are not suited to the private development industry'; that is, 'the business will now operate alongside the private development industry, not in competition' (Places Victoria 2013: 3).

Tabart had left VicUrban on the crest of the wave in 2006 for Dubai. VicUrban's long-standing chief financial officer, Sam Sangster, became acting CEO of Places Victoria. The new organisation's first chairman, friend of premier Baillieu and hand-picked by him without a formal selection process, was sitting Melbourne city councillor and former Liberal party vice-president Peter Clarke. Prior to his political career Clarke had been an executive director of the Property Council of Australia (Victoria), a development industry lobby group, and a senior executive of Lendlease. The Growth Areas Authority under Peter Seamer was retained.

Muddying the waters: public and private sector churn

Having watched the same well-connected individuals travel from one taxpayer-funded position to another across multiple political regimes, the commentariat didn't pull its punches:

> Melbourne City Councillor, Peter Clarke, has called it a day and has given notice of his intention to quit the Council having served just over one and a half terms. Peter Clarke has accepted a generous offer from

> his good friend and Premier of Victoria, Ted Baillieu, to head up the Urban Renewal Authority. A position that comes with twice his salary as a Councillor, his own office, secretary and plenty of opportunity for international travel. Why would you hang around a powerless and non influential City Council when you can enjoy the trappings and benefits of a plum position that has little accountability or oversight? Clarke's pending resignation, which takes effect in three weeks time, will cause grief for the City Council and the Victorian Electoral Commission (VEC).
> (van der Craats 2011)

Clarke was also a director of a listed company, Australian Property Custodian Holdings (APCH). Just 14 months into his appointment to Places Victoria, Clarke had to stand down while the Australian Securities and Investments Commission (ASIC) investigated claims that he and other directors of Australian Property Custodian Holdings had improperly used their positions to approve a $33 million payment to a managing director (itself not contested) (Butler 2012). Despite admitting he had given false evidence to ASIC, Clarke denied any wrongdoing and was exonerated on appeal (Butler and Desloires 2014). He continued to draw his Places Victoria chairman's salary pending the resolution of the investigation (Rolfe 2012).

The relationships between VicUrban, the Docklands developers, the state department for planning and Melbourne city council had been remarkably smooth during the 2000s. John Tabart had worked closely with his former colleagues at Lendlease, and with Lyndsay Neilson, state secretary of the department responsible for planning from 2000 to 2006, and with David Pitchford, the CEO from 2003 to 2007 of Melbourne city council, to oversee what became the period of Docklands' most rapid growth. In 2006, when the project's shape was set in concrete, Tabart left VicUrban for Sama Dubai Real Estate. Neilson left for Dubai soon afterwards. They were followed in 2007 by Pitchford who, on the basis of his expertise in waterfront regeneration, was appointed to lead the Palm Jumeirah Dubai development. He remained there until 2013 when he returned to Australia to head up the NSW state authority Urban Growth, once again joining Tabart who by then was established as CEO of the Barangaroo Delivery Authority, whose developer, of course, is Lendlease.

Notwithstanding the likely kinds of policy transfer in these instances, there were glimpses of institutional learning back home. At a meeting of Melbourne city council's Docklands committee, Sangster, now Places Victoria CEO, demurred: 'Everyone in this room will acknowledge it hasn't worked, for various reasons' (cited in Perkins and Dowling 2011). But as he introduced Melbourne Docklands' 'D2' – the second decade – a concept in apparent denial of those critical years in the 1990s, rationalisation quickly followed: 'The first decade has very much been about putting in those large blocks of infrastructure and some of the large buildings, and the second decade is much more about the people' (Perkins and Dowling 2011). Even if he or anyone else in the room did believe that the 'large blocks of infrastructure' of 'D1' had not rendered the objectives of 'D2' fanciful, Sangster didn't carry that burden long. He resigned two months later, to be replaced after a year of chaos and staff cutbacks at Places Victoria by one much more accustomed to the requirements of the job: Peter Seamer. Movements such as these are fairly routine in Australia, and tend not to attract attention. They are rationalised as bringing much needed private-sector experience to government and vice versa, and as maintaining corporate memory, which they do, and they ensure that nothing really changes.

And then there was Fishermans Bend

The Coalition must have felt robbed by the narrative of economic success at Docklands under Labor. In February 2011, just three months into the new government's term, Matthew Guy, minister for planning, announced that a new 'high-rise density suburb at Fishermans Bend' would be the 'Baillieu Government's legacy' (McMahon and Wright 2011: npn). At first there was some confusion. Did the minister mean the district of the City of Melbourne? This was still an important industrial hub, with General Motors Holden, Toyota, Boeing, the Commonwealth Aircraft Corporation and Aeronautical Research Laboratory, and the Defence Science and Technology Organisation all operating there. The matter went quiet until late June the following year, when Guy suddenly declared 250 hectares of both the Melbourne and Port Phillip local government areas new precincts of an expanded central city (Figure 7.3). The minister continued to call this area Fishermans Bend but, he had to clarify, it did not include the designated district Fishermans Bend – rather, the land to the east and south, mainly in the City of Port Phillip. This was news to the City of Port Phillip, which had completed the strategic plan for continuing industrial and complementary uses there. An ABC radio announcer observed that most of the land marked for re-zoning was privately owned, unlike much of the land in the proper Fishermans Bend. Yes, the minister had to agree, the landowners would have to be 'incentivised' to take part.

Four days later the 250 ha area was re-zoned from industrial to 'capital city' zone – a mixed use zone that, on paper, immediately tripled land values. The Coalition's metropolitan strategy was still in preparation. Up until that point, Fishermans Bend had not been identified as a priority while Docklands remained unfinished and several other regeneration areas in the City of Melbourne's west and north were in their early stages. How would this huge new project – the largest of its kind anywhere in Australia – be managed? If Places Victoria were restricted to operating alongside private developers rather than 'in competition', who would develop it, and what principles would guide the private sector? There was no strategic plan, no infrastructure plan, no provision for open space. There were no requirements for affordable housing, schools, community facilities or public services, no

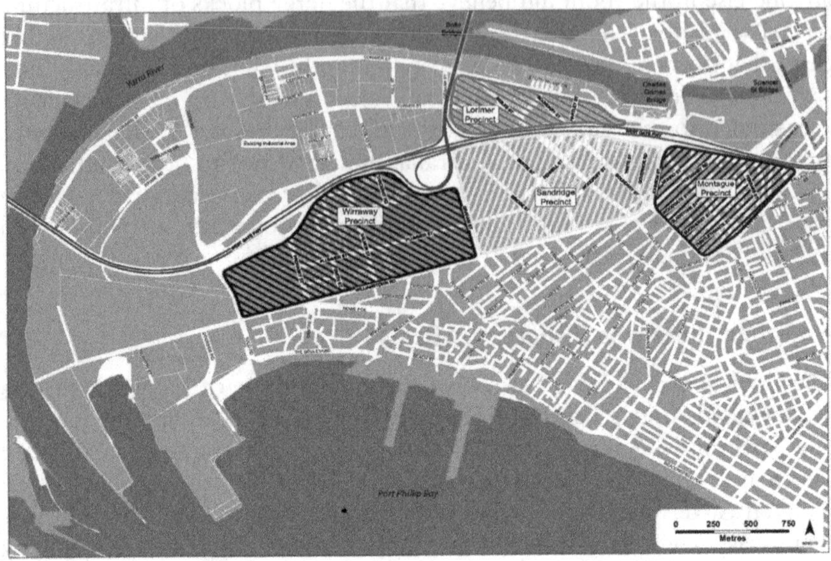

Figure 7.3 Fishermans Bend Renewal Area as announced in 2012 (source: Victorian Department of Planning and Community Development)

environmental or heritage protections, no mechanisms in place for value capture or leverage.

A ministerial advisory committee had been appointed in May 2012 to direct the metropolitan strategy. The committee was instructed in July to include both Fishermans Bend and another massive new infrastructure project north of the city centre – a freeway and road tunnel known as the East–West Link – which had the effect of corroding many of the basic planning principles the committee had brought to the table. After 12 months further work and a stand-off with the minister's officers, in which key sections of the committee's draft were rejected, the majority of committee members resigned. Three months later, in October 2013, a 'Plan Melbourne' discussion paper was released with Fishermans Bend and East–West Link central planks of the metropolitan and central city planning strategies.

In the outcry after the re-zoning, the minister had insisted a strategic plan would be prepared to guide development at Fishermans Bend. Places Victoria was assigned to the task, and a year later, in July 2013, it submitted a comprehensive report to government. The report made urgent recommendations, including to institute a public acquisition overlay, capture the value uplift, build public transport, exact development levies on private land-holders for affordable housing, community infrastructure and open space, introduce an inclusionary zoning overlay (which would specify a percentage of affordable units in each new development) along with height controls and mandatory high environmental standards. The minister rejected all the recommendations and the report was never made public. Three months later, in September 2013, an alternative 'draft vision' was released for consultation. This version offered no implementation mechanisms for the vague indications it did contain, but it did display the comforting smile of Peter Seamer.

The following year, in February 2014, Places Victoria was removed as urban renewal authority for Fishermans Bend. Its mission was revised once again to become the 'Victorian Government's property development agency'. A newly created Metropolitan Planning Authority (MPA) took over 'strategic planning functions' at Fishermans Bend, with Seamer at the helm – his CV added that before moving to the MPA:

> Peter undertook a seven-month part-time role as CEO of Places Victoria. Under Peter's direction significant changes were made in staffing and strategy along with a write down of over $190m. As a result of Peter's stewardship PV moved forward as a successful property development agency for the State.
>
> (Seamer 2016)

The MPA prepared a Fishermans Bend Strategic Framework Plan in July 2014, which declared the intent of accommodating 80,000 residents and providing commercial space for 40,000 jobs, and included broad planning and design guidelines. Of the 146 guidelines in the document, 'only 34 are mandatory and much of the area did not even include suggested height limits' (Fishermans Bend advisory committee 2015: 11).

Later in 2014, contents of the first Places Victoria report were leaked to journalists at *The Age* who relayed that more than $2 million had been spent commissioning thorough background reports on matters from land acquisitions to soil conditions. *The Age* also reported that advice had been provided to the minister as early as February 2012 – four months before the re-zoning – urging him to purchase key strategic sites as soon as possible. The advice, from senior officers at Places Victoria just before the appointment of Seamer, listed ten important sites and noted that land prices at Fishermans Bend had already doubled since the preliminary announcement in 2011. The advice was ignored.

It was later revealed in a series of ministerial briefing notes from departmental advisors that the boundaries were drawn up as early as March 2011; that is, the politicians and advisors knew precisely the area for re-zoning 16 months in advance, but this was kept quiet, not for the purposes of strategic government acquisitions. The briefs did initially advise the government to purchase key sites, provide services, and require affordable housing from developers, but these recommendations fell away in the lead up to the July 2012 re-zoning. A final brief on 27 June 2012 (Victorian Department of Planning and Community Development 2012) made no mention of any of these issues. It advised that the Melbourne and Port Phillip councils not be formally consulted ('to allow greater flexibility for decisions' (p.8)), that public consultation 'is not a preferred process as delays will cause uncertainty' (p.7), and noted that the brief contained 'the refinements and changes as discussed with your office' on various meeting dates with the minister. That is, the departmental heads changed their advice, presumably at the behest of the minister, dropped their exhortations for government planning and intervention, and supported Guy's preference for *ad hoc*, laissez-faire development of the entire area.

The metropolitan strategy had been denounced by the original committee and broader planning and design professions as a 'retrospective plan' (like so many before it) that attempted to accommodate in a coherent framework disconnected decisions that had already been made. But the Fishermans Bend re-zoning and subsequent strategic framework plan were met with a kind of paralysis. The MPA plan dismissed Places Victoria's early recommendation for a tram route right into the regeneration area, downgrading it to a 'possible' route, which Guy later 'relegated to "long term" status' (Millar, Vedelago and Lucas 2014: npn).

In the late 2000s under the Labor government, plans had started for an expansion of the inner-city underground train system (another project that did not extend to the middle or outer suburbs where such work was much more necessary). The Melbourne Metro, as it was known, had been in departmental preparation for five years when minister Guy binned it and proposed an alternative route that skirted the eastern edge of the newly-designated Fishermans Bend – the part of the precinct already best serviced by public transport. Internal disruptions within the Coalition led to a leadership change, with Baillieu being replaced in 2013 by Denis Napthine. Guy introduced a development levy of $15,900 per dwelling at Fishermans Bend, but it was dismissed as too little to cover infrastructure costs with the increased land values, and too late. A well-regarded economist and planner observed that:

> these properties will now take longer to develop, as owners wait for housing, office and retail prices to rise enough to absorb the unexpected levies. This is opposite to the effect that government might have been looking for when it re-zoned the land to clear the way for early investment.
>
> (Marcus Spiller, cited in Millar et al 2014: npn)

In the three years following the re-zoning, despite the chaos at Fishermans Bend, development applications for 68 office and residential buildings were made for the area, 46 of which were between 20 and 64 storeys. *The Age* revealed that senior Liberal party members were the recipients of massive potential and realised windfalls from land sales. In particular, the Liberal party's honorary federal treasurer at the time, Andrew Burnes, who is one of the party's most generous donors, bought a key site in March 2012, just months before the re-zoning. *The Age* reported that he 'could pocket tens of millions of dollars from a 40-storey apartment tower' if the application were approved (Millar, Vedelago and Schneider 2015: npn). Other beneficiaries listed in the same report included 'former Liberal party activist and on-going financial supporter' Harry Stamoulis; the Master Builders Association, 'traditional Liberal allies'; *BRW* richlister[2] John Higgins, Liberal party donor; and the directors of MAB Corporation, the Docklands developer that had gone to court to contest the government's stamp duties. All had bought land in the newly-designated Fishermans Bend before or immediately after the re-zoning. *The Age* calculated the potential and realised profits to be anything from three to many ten times the original outlays.

Sites began to be sold and resold at massive profits with no works occurring at all, while development applications poured into the MPA. Minister Guy approved 12 randomly-located high-rise towers at Fishermans Bend before any controls were in place. One of these belonged to another major donor to the Liberal party, Bill McNee and his development consortium Maxvic Holdings, which bought a site in the area in 2013 for $10 million. After planning approval was received for nearly 1300 apartments across four towers up to 51 levels tall, it was sold on for $41 million without a sod being turned (Johanson 2017).

Place of mists and shadows

Is this corruption? Not of the brown paper bag variety, probably, but in process, certainly. The extent to which all governments in Victoria are beholden to private-sector interests, especially through political donations, is almost too deep to fathom. In November 2014, just weeks before the state election, the Coalition government appointed a new chairman to Places Victoria: card-carrying Liberal Tony de Domenico, from the development industry lobby the Urban Development Industry Association. The government also signed a contract with the preferred developer of the East–West Link – a consortium headed by Lendlease – amid heated controversy about the merits of the project. The Labor opposition seized on the point of difference, declaring it would cancel the project and calling for the issue to be decided at election. Just days before the polls opened, the Coalition locked in a deal that would generously compensate the developer if the contracts were rescinded. The Napthine government became the first in Victoria for half a century to serve only one term. The in-coming Labor government under premier Daniel Andrews had no choice but to deliver on its election promise, resulting in a payout to the consortium of around $1 billion.

In a forlorn attempt to make things right, the applications at Fishermans Bend were transferred from the MPA to the new minister for planning, Dick Wynne. The strategic plan was 'recast', and an alternative decision-making process established. A Fishermans Bend taskforce would be composed of representatives of Places Victoria, the MPA, the Department of Environment, Land, Water and Planning, and the Cities of Melbourne and Port Phillip. It would work with a ministerial advisory committee (MAC), and report to a Fishermans Bend coordination group made up of officers from all the relevant state and council departments. The local government officers were sworn to confidentiality, including not informing their city councillors of the discussion. The MAC and the coordination group had undefined lines of communication and reference to each other, and reported separately to the minister (Victorian Government 2015). The Fishermans Bend recast vision promised to engage with 'the community to develop a blueprint for Fishermans Bend that will transform it into a place for everyone' (Places Victoria 2015: npn). In the meantime, minister Wynne approved another 13 towers ranging from 11 to 49 storeys in height, saying his hands were tied. The impossibility of this governance structure had many supporters of consultative processes at a loss as to how future decisions would be made.

Minister Guy's handling of Fishermans Bend was universally condemned by the planning, design and political professions, with the MAC declaring it 'unprecedented in the developed world in the 21st century' (Fishermans Bend advisory committee 2015: 19). The committee report noted that already 17,600 apartments were proposed or approved:

> The number and scale of the development approvals is alarming, but more alarming is that development is proceeding without reliable mechanisms to ensure that contamination will be addressed, that planned new roads can be built, that connected open space can be secured and that infrastructure will be delivered in time, or at all.
>
> (Fishermans Bend advisory committee 2015: 5)

But business continued as usual. The Andrews government changed the name of the Metropolitan Planning Authority to the Victorian Planning Authority: the new CEO was Peter Seamer. It retained de Domenico as chairman of Places Victoria. Controls were introduced that would 'encourage' developers of larger residential projects to allocate six per cent of apartments for social housing, with no requirements or incentives for them to do so. The Fishermans Bend regeneration area was expanded to include the City of Melbourne's designated industrial precinct, increasing the entire project area to 450ha. Matthew Guy was elected by the Coalition to lead the opposition.

Places Victoria maintained responsibility for the completion of Docklands, although the remaining land had already been allocated to private developers. Despite serial 'place-making' workshops and community consultation sessions as the reformed development authority attempted to reform its public perception, Place Victoria had no capacity to do anything beyond implementing the contracts. Bright young urban-planning graduates were engaged to make the unfinished parts of Docklands 'more about the people', but even had they been willing to commit their employer to public transport, diverse employment opportunities, affordable housing and community facilities, they were fatally constrained by Places Victoria's terms of operation.

The City of Melbourne is similarly hamstrung in the developed parts of Docklands. Making the council's task even more difficult is the legal contracts between the established developers and VicUrban/PV, which specify the price paid for the land, the expected developer contributions, the timing of development stages and penalties for failure to comply, along with other specifications and agreements, were at the close of 2016 still not available to the council. The City of Melbourne as responsible authority is not able to assess whether the landowners in its new patch are meeting their original obligations, and Places Victoria appears in no mind to pass this information on.

Nevertheless, the same faces linger. In 2014 John Tabart left the Barangaroo Delivery Authority and, after a short stint advising the Victorian government on urban renewal in the inner west, returned to the private sector as CEO of Australian Education City (AEC). In early 2016 AEC was named preferred developer of a new, $25 billion, 600 ha regeneration project in Melbourne's inner west.

Most recently, perhaps to round this story off, James Packer won approval from the Andrews Labor government to build a $1.75 billion, 90-storey addition to Crown casino (Willingham and Lucas 2017). The glamorous hotel, once completed, will be 25 metres higher than Melbourne's existing tallest building, the Eureka Tower, also at Southbank. The government's support for the project saved Crown $61 million in fee exemptions (Lucas 2017).

Melbourne's waterfront regeneration projects – Southbank, Docklands, and now Fishermans Bend – have transformed the city more comprehensively than any other single project. With strategic planning processes scrambling to catch up, the key decisions for all of them have been made with no reference to anything previously agreed or understood. It goes largely without public protest or even comment in this age of neoliberal governance and pervasive inter-city competition that local plans are deemed provincial and so readily dismissed. More alarming in Victoria perhaps is that the race to return the now passionate embrace of the private sector is conducted not between state governments but within them – between two political establishments that have few checks and know no limits in their quest to out-do the other. All the major decisions for Melbourne are now made by private interests with no regard for the city other than the extent of the financial yields on their investment. These continue to be mediated by government-appointed middlemen who appear to express no doubt regarding the merits on the ground of their economic calculations.

There are many people in Melbourne who long for, if not the wetlands long gone, at least some humanity where they feel a part of their city – in its collective decision-making processes as well as its sense of place. But the mists are deeper than they've ever been, and the shadows are much, much longer.

Conclusion

The inter-city competitive impulse to attract economic activity and jobs from neoliberal urbanisation is clear in this story of the development of Melbourne's three key waterfront projects. What is startling here is the length to which each of the dominant political parties is prepared to go to satisfy this imperative. Even when private financier and developer interest in the city is clear and the market is ready and able to be regulated, the Liberal and Labor parties that have long governed the state of Victoria compete further, not for investment from elsewhere, but for dominance against the other. This has resulted in a continual loosening of standards and increasingly poor outcomes for the people of the city and state. The opacity of decision-making processes in Australian governance is almost unknown in similarly privileged nations, and this will only change with an expansion of political possiblities. Until then, a strengthening of the powers of Victoria's Independent Broad-based Anti-Corruption Commission (IBAC) to the level of NSW's Independent Commission Against Corruption (ICAC) may go some way to providing redress.

Notes

1 This research occurred with the support on an Australian Research Council Future Fellowship.
2 *Business Review Weekly* (*BRW*) list of Australia's wealthiest individuals.

References

ABC News (2014) 'Labor backs Crown Casino license extension to 2050 despite "serious misgivings"', 17 September, http://www.abc.net.au/news/2014-09-16/labor-backs-casino-extension-despite-serious-misgivings/5748298 (accessed 4 December 2016).

Australian Financial Review (1998)' Exasperating Jeff Kennett', Editorial, 7 July.

Butler, B. (2012) 'Urban renewal chief facing corporate ban', *Sydney Morning Herald*, 28 August, http://www.theage.com.au/victoria/urban-renewal-chief-facing-corporate-ban-20120827-24wqg.html (accessed 4 December 2016).

Butler, B. and Desloires, V. (2014) 'Prime Trust director Peter Clarke gave false evidence', *Sydney Morning Herald*, 28 July, http://www.smh.com.au/business/prime-trust-director-peter-clarke-gave-false-evidence-20140727-3cnkx.html, (accessed 4 December 2016).

Costello, T. and Millar, R. (2000) *Wanna Bet? Winners and Losers in Gambling's Luck Myth*, Sydney: Allen and Unwin.

Docklands Authority (1995) *Annual Report 1994–1995*, Melbourne: Victorian Government.

Docklands Authority (2000) *Melbourne Docklands 2000+*, Melbourne: Victorian Government.

Docklands Task Force (1991) *Melbourne Docklands: Draft Strategy for Redevelopment*, Melbourne: Victorian Government.

Dovey, K. (2005) *Fluid City: Transforming Melbourne's Urban Waterfront*, Sydney: UNSW Press.

Dux, M. (2014) 'Bird's eye view of dismal Docklands', *Sydney Morning Herald/The Age*, 3 October.

Fishermans Bend Advisory Committee (2015) Report 1, October. Melbourne: Fishermans Bend Advisory Committee.

Johanson, S. (2012) 'Docklands: Kennett was almost giving the land away', *The Age*, 3 March.

Johanson, S. (2017) 'Developer McNee makes a $30 million killing on South Melbourne apartments site', *The Age*, 24 January.

Keeney, J. (2005) *Waterfront Spectacular— Creating Melbourne Docklands: The People's Waterfront*, Roseville, NSW: ETNCOM.

Lucas, C. (2016) 'World's most liveable city should be ashamed', *Sydney Morning Herald/The Age*, 7 July .

Lucas, C. (2017) 'Crown saves millions under special deal with Labor for Melbourne's biggest skyscraper' , *The Age*, 16 March.

Lehrer, U. and Laidley, J. (2008) 'Old mega-projects newly packaged? Waterfront redevelopment in Toronto', *International Journal of Urban and Regional Research* 32(4): 786–803.

McCracken, D. (2005) 'Living Melbourne', in J. Keeney (ed) *Waterfront Spectacular— Creating Melbourne Docklands: The People's Waterfront*, Roseville, NSW: ETNCOM.

McLoughlin, J.B. (1992) *Shaping Melbourne's Future? Town Planning, the State and Civil Society*, Cambridge: Cambridge University Press.

McMahon, S. and Wright, A. (2011) 'Richmond train station, Fishermans Bend set to be transformed', *Herald Sun*, 18 February.

Millar, R. (2006) 'Waterfront spectacular or wasted opportunity?', *The Age*, 17 June.

Millar, R., Vedelago, C. and Lucas C. (2014) 'Fishermans Bend: anatomy of a high-rise ghetto', *The Age*, 2 November.

Millar, R., Vedelago, C. and Schneider, B. (2015) 'Liberals profit at Fishermans Bend', *The Age*, 1 November.

Oakley, S. and Johnson, L. (2013) 'Place-taking and place-making in waterfront renewal, Australia', *Urban Studies*, 50(2): 341–355.

Perkins, M. and Dowling, J. (2011) 'Down on the Docks', *The Age*, 28 September.

Places Victoria (2012) *Annual Report 2011–2012*, Melbourne: Places Victoria.

Places Victoria (2013) *Annual Report 2012–2013*, Melbourne: Places Victoria.

Places Victoria (2015) 'Central Melbourne's key growth area: Fishermans Bend', http://www.places.vic.gov.au/precincts-and-development/fishermans-bend, (accessed 4 December 2016).

Rolfe, P. (2012) 'Places Victoria chairman Peter Clarke keeps Government salary', *Herald Sun*, 26 September.

Seamer, P. (2016) Curriculum Vitae, https://www.linkedin.com/in/peter-seamer-ba527340/ (accessed 4 January 2017).

Shaw, K. (2013) 'Docklands Dreamings: Illusions of sustainability in the Melbourne docks redevelopment', *Urban Studies*, 50(11): 2158–2177.

van der Craats, A. (2011) 'Clarke calls it quits as he takes on the design of greener pastures', Melbourne City Council: holding them to account, http://melbournecitycouncil.blogspot.com.au/2011/06/clarks-calls-it-quits-as-he-takes-on.html, (accessed 4 December 2016).

Victorian Department of Infrastructure (2002) *Melbourne 2030*, Melbourne: Victorian State Government.

Victorian Department of Planning and Community Development (2012) *Ministerial Briefing Note*, 27 June.

Victorian Government (1984) *Central Melbourne: Framework for the Future. Land Use and Development Strategy no.* 6, December.

Victorian Government (1987) *Shaping Melbourne's Future: The Government's Metropolitan Policy*, August.

Victorian Government (1992) *Shaping Victoria's Future: A Place to Live. Urban Development 1992–2031*, April.

Victorian Government (1993) *Agenda 21*, Office of Major Projects.

Victorian Government (1994) *Creating Prosperity, Victoria's Capital City Policy*, Office of the Premier, November.

Victorian Government (1999) *State Planning Agenda, A Sensible Balance*, Minister for Planning, December.

Victorian Government (2003) Urban Renewal Authority Act.

Victorian Government (2015) *Fishermans Bend Advisory Committee Report 1*, Melbourne.

Victorian Ministry for Planning and Environment (1986) *Central Melbourne, Southbank: A Development Strategy, Technical Papers,* May.

Victorian Places (2015) 'Port Melbourne. Monash University and The University of Queensland', http://www.victorianplaces.com.au/port-melbourne (accessed 16 September 2016).

Willingham, R. and Lucas, C. (2017) 'Crown casino wins approval for 90-storey tower at Southbank', *The Age*, 9 February.

8 Growth plans and growing pains in Brisbane's sunbelt metropolis

Phil Heywood

Introduction

This chapter traces the changing intentions, scope and impacts of urban regeneration in Greater Brisbane over the last three decades within their contexts of evolving socio-economic conditions, urban forms and political controls. It is divided into three parts. The first part examines changing growth patterns and traces the evolution of the urban regeneration that has transformed the city over the last 30 years from an overgrown country town to an emerging sunbelt metropolis with "World City" aspirations. The second part illustrates these general trends by focusing on two major recent renewal schemes in Brisbane's city centre, located on either bank of the Brisbane River, one now approved for implementation over the next decade, the other appearing to have collapsed in the face of vigorous public and political opposition. A final section draws conclusions about changing planning assumptions, trends, opportunities and options.

Following Roberts and Sykes (2002), the scope of urban regeneration is taken to encompass not only economic regeneration and funding but also physical, environmental, social and community factors. As well as housing, employment, and education, these also include sustainability. Regeneration is seen as "an outcome of the interplay between these many sources of influence and, more importantly ... a response to the opportunities and challenges which are presented by urban degeneration in a particular place at a specific moment in time" (Roberts and Sykes 2002, p. 9) – in this case Brisbane City in the opening decades of the twenty-first century .

Changing growth patterns

With a current population of over 1.1 million, and an area of more than 1300 square kilometres, Brisbane City is by far the largest and most populous local government in Australia (BCC 2014a). Accommodating more than half the total metropolitan population (Figure 8.1), its urban governance, development and renewal have become highly centralised in City Hall.

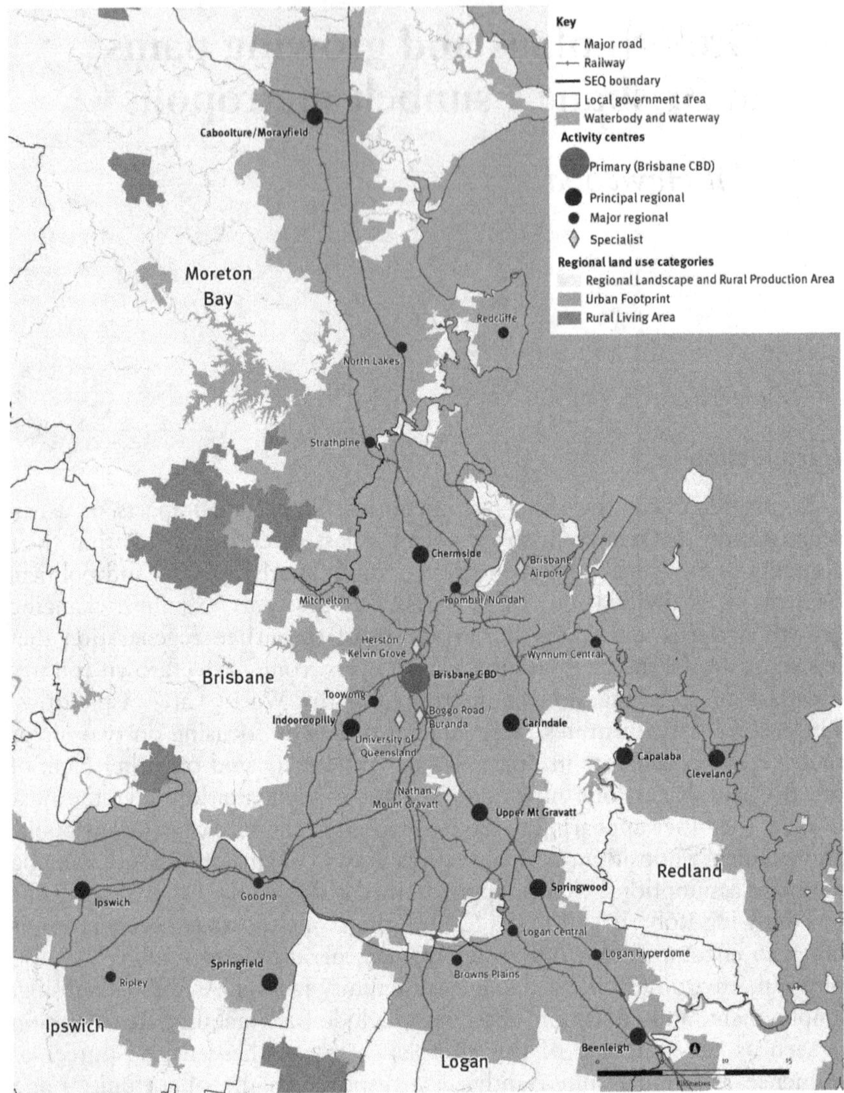

Key:

NPP-001 Retail precinct

NPP-002 Quay Street precinct: city fringe location

NPP-003 Queen's Wharf precinct: Priority Development Area to facilitate an integrated resort development

NPP-004 River precinct: public space, access to the riverfront, and improved Riverwalk infrastructure.

NPP-005 Howard Smith Wharves precinct: parklands, revitalised heritage buildings dining, retail, hotel and event facilities.

Figure 8.1 Greater Brisbane, local government boundaries and sub-regional activity centres, South East Queensland Regional Plan, 2009–2031 (source: Queensland Department of Infrastructure, Local Govenrment and Planning 2009)

This concentration of political power has been amplified by the lack of any political representation at more local scales, with the city's smallest electoral units being the council's twenty-six electoral wards, each with an average of over 50,000 people, which is more than that of many entire council areas elsewhere in the state. This concentration of control is further reinforced by the system, unique in Queensland, of electing the lord mayor "at large" from all voters, rather than from among elected councillors. These characteristics have combined to produce Australia's nearest approach to a presidential system. As a result, successive lord mayors have often been larger-than-life figures, from the swashbuckling Clem Jones (1961–1975), and the eloquent and elegant Sallyanne Atkinson (1984–1992), to the highly effective Jim Soorley (1992–2003) and the pugnacious Campbell Newman (2004–11). Throughout the past 50 years, political control has alternated between the Australian Labor Party (ALP) and the Liberal National Party (LNP), formerly the Liberal Party (LP) (Table 8.1).

Impacts of differing political agendas

It can be seen that over the 40-years from 1975–2015, Labor and Liberal/National city administrations enjoyed power for very similar periods of 20 years each. Although individual regimes had their own distinctive styles, there has also been a discernable contrast between the social and distributional policies of Labor regimes, and the more developmental and growth-promoting ones of the Liberals. Urban regeneration under Labor has been largely directed to public services such as the sewering of the middle and outer suburbs of the 1960s and 1970s and the introduction of dedicated track busways of the 1990s. This contrasts with the emphases of 1980s Liberal administrations on modernising and growth promotion polices and of successor Liberal National Party (LNP) ones, since 2004, on largely privatised intensive inner-city redevelopment supported by tunnel and bridge building.

For instance, while the incoming ALP Soorley administration of 1992 maintained the developmental policies of its Liberal predecessors, it paid increased attention to the middle and outer suburbs. The Commonwealth's *Building Better Cities* program supported both urban renewal in the inner northeastern suburbs and clustered residential development along the southern corridor rail extension to Beenleigh (Neilsen 2008), at the same time that "Gateway" schemes were improving port access in Sydney, Melbourne and Perth (Chapters 6, 7 and 9). Exclusive track busways were opened to serve south and southeastern suburbs. Community development projects were implemented for the outer suburbs of Inala and Carole Park. Urban form considerations also began to be taken seriously. By the time of the 2001 City Plan, Brisbane's city centre had come to be seen as only one focus of a polycentric metropolis (Chapter 2) (Figure 8.1).

Table 8.1 Brisbane City lord mayors, Queensland state premiers and key urban regeneration policies, 1961–2016

<table>
<tr><th colspan="2">Brisbane City</th><th colspan="2">State of Queensland</th></tr>
<tr><th>Mayor</th><th>Operative urban regeneration policies</th><th>Premier</th><th></th></tr>
<tr><td rowspan="3">Clem Jones (ALP, 1961–1975)</td><td rowspan="3">Widespread urban improvement and densification, including sewering new and existing residential suburbs, through a combination of city's public works program and developer contributions via infrastructure charges.</td><td>George Nicklin (CP/ Lib, 1957–1968)</td><td rowspan="3">Successive conservative governments encouraged incremental growth</td></tr>
<tr><td>Charles Pizzey (CP/ Lib, 1968)</td></tr>
<tr><td>Gordon Chalk (1968)</td></tr>
<tr><td>Bryan Walsh (ALP, 1975–1976)</td><td>Consolidation of Jones' improvement schemes for 1982 Commonwealth Games.</td><td rowspan="3">Johannes Bjelke Petersen (CP/ Lib, 1968–1987)</td><td rowspan="7">Support of conservative governments for economic and physical growth</td></tr>
<tr><td>Frank Sleeman (ALP, 1976–1982)</td><td>Decentralised residential and recreational facilities, associated with 1982 Commonwealth Games venues in middle suburban.</td></tr>
<tr><td>Roy Harvey (ALP, 1982–1985)</td><td>Emphasis on road-building through inner suburbs and amplification of infrastructure charging codes to manage outer area development.</td></tr>
<tr><td rowspan="2">Sallyanne Atkinson (Lib, 1985–1991)</td><td rowspan="2">Balance between central area and inner-city partnership schemes in Queen Street Mall and South Bank Gardens, combined with conservation of creek corridors and outer green belt.</td><td>Michael Ahern, (Nat/ CP/Lib, 1987–1989)</td></tr>
<tr><td>Russell Cooper (Nat/ CP/Lib, 1989)</td></tr>
<tr><td rowspan="3">Jim Soorley (ALP, 1991–2003)</td><td rowspan="3">Continued balancing of growth between inner-city and central area urban renewal, and development around four regional centres served by new radial busways, with local area plans (LAPs) to coordinate suburban development. Support for community development programmes of state government in outer suburbs.</td><td>Wayne Goss (ALP, 1989–1996</td></tr>
<tr><td>Robert Borbidge (Nat/ Lib, 1996–1998)</td></tr>
<tr><td>Peter Beattie, (ALP, 1998–2003)</td><td>State-city collaboration to promote balanced growth</td></tr>
</table>

<table>
<tr><th colspan="2">Brisbane City</th><th colspan="2">State of Queensland</th></tr>
<tr><th>Mayor</th><th>Operative urban regeneration policies</th><th>Premier</th><th></th></tr>
<tr><td>Tim Quinn (ALP, 2003–2004)</td><td>Support for inner-city redevelopment for residential and recreational uses, as well as wider scale planning of first South East Queensland Regional Plan.</td><td>Peter Beattie (ALP, 2003–2004)</td><td></td></tr>
<tr><td rowspan="2">Campbell Newman (Lib/LNP, 2004–2011)</td><td rowspan="2">Channelling of residential and commercial development into city centre and inner suburbs, supported by privatised tunnel construction for car travel with limited change in outer areas. Move away from policies of radial corridors development, served by public transport.</td><td>Peter Beattie (ALP, 2004–2007)</td><td rowspan="3">City development policies held in check by ALP state govt.</td></tr>
<tr><td>Anna Bligh (ALP, 2007–2012)</td></tr>
<tr><td rowspan="2">Graham Quirk (LNP, 2011–2016)</td><td rowspan="2">Collaboration with state government to develop Brisbane as a "world city" including ambitious riverside urban regeneration schemes of Queens Wharf and Kurilpa Point, and continued intensive high-rise residential development throughout inner suburbs, with limited intensification outer suburbs.</td><td>Campbell Newman (LNP, 2012–2015)</td></tr>
<tr><td>Anna Palaszczuk (ALP, 2015–)</td><td>Emergence of strong city centre urban regeneration and "world city" policies</td></tr>
</table>

Source: Author

ALP - Australian Labor Party; CP = Country Party; Lib = Liberal; LNP = Liberal National Party; Nat = National

Central area redevelopment

Meanwhile high-rise redevelopment of the city centre for commerce and administration was clustered within the city's central peninsula, bounded on three sides by the central city river meander, and on the fourth by the historic slopes of Spring Hill with its heritage of workers' cottages and small-scale mixed uses (Figure 8.2).

Within this core, new offices and restaurants grew on space vacated by disused wharves and docks, reuniting the city centre with the riverside. Along the Town Reach, Harry Seidler's strongly stated triangular Riverfront Place of 1986 set the standard for new office development. Soon after, the well-loved South Bank Gardens were developed to occupy the site of

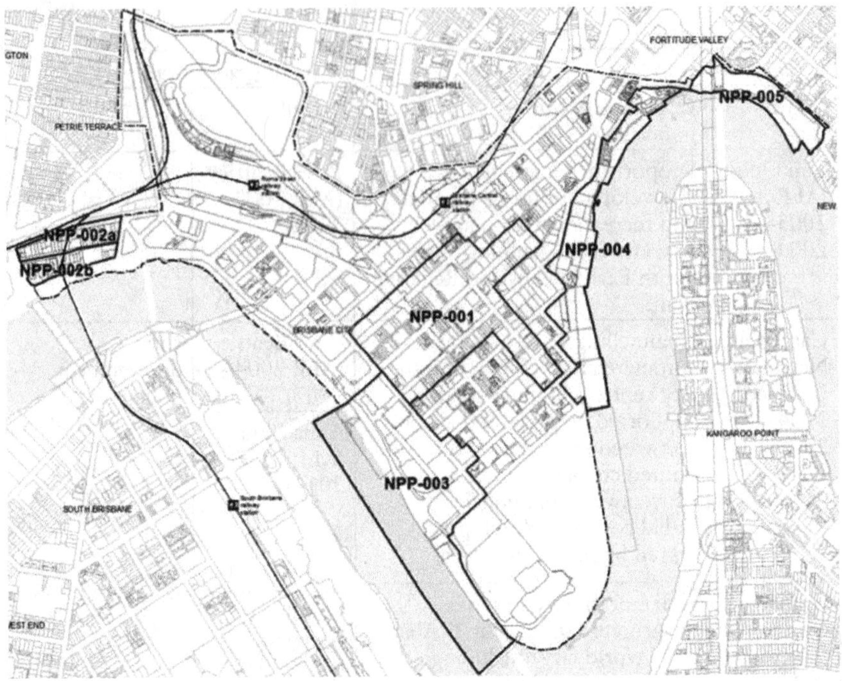

Figure 8.2 Draft city centre neighbourhood plan boundary, 2015 (source: BCC City Centre Neighbourhood Plan 2015)

the 1988 Bicentennial Expo, providing much needed riverside recreational space (Heywood 2011: 47). High-rise offices clustered along the Town Reach, and new apartments, hotels and business premises replaced the old warehouses and car parks along the spine of the peninsula. Following construction of the Queen Street Mall, the city centre, for long in decline as a shopping location, recovered and, for a time, increased its share of the city's total retail revenue.

A three-decade demographic boom, fed by high rates of interstate and overseas migration, prompted and sustained these developments. Greater Brisbane was gaining more than 50,000 people a year through migration, increasing by the first decade of the new century to 60,000–80,000, many of whom lived or worked in Brisbane City itself (Urban Renewal Brisbane 2011). New office employment from the mining boom of the 2000s also contributed. The incoming LNP city administration of 2004 sought to further leverage investment and jobs by promoting the city's lifestyle and climate. Using his engineering training, lord mayor Newman quickly introduced ambitious redevelopment plans. By 2006 he was publicising his "TransApex" proposals to speed middle and outer suburban traffic to work and play opportunities around the metropolis with a triangle of new toll tunnels bypassing the city centre (Moore and Stephens 2014).

These strategies had the electoral advantage of preserving the traditional low-density outer suburban environments of the ruling party's support base at the same time as channeling large-scale investment to ambitious redevelopment schemes in the inner city. However, not all of the three strands of inner city bypass routes, outer suburban conservation and inner city redevelopment were compatible. While the TransApex and outer suburban conservation policies were congruent and highly popular with the LNP's natural political base among the established homeowners of the outer suburbs, neither of them assissted the major new investment proposals for the city centre. Although TransApex appeals to the party's outer suburban constituencies, by promoting traffic flow and reducing city centre gridlock, it does nothing to encourage accelerated employment, tourist and residential growth in the city centre.

As in Sydney where the cross-city toll tunnel suffers from low and fluctuating usage (Kitney, 2013), operational and financial difficulties have also resulted. The first TransApex scheme, the Clem Jones toll tunnel, merely acted as an extra eastern bypass for the city centre, replicating the very efficient (and toll free) Gateway Bridge Freeway. Originally budgeted at $2 billion, it never made a profit and its owners, River City Motorways went into receivership in February 2011 (Brannelly 2011). BrisConnections, responsible for the second scheme, the Airport Link, followed suit in January 2012. Due to underestimated costs and overestimated revenue, the constructors, Leighton Holdings, announced a pre-tax loss of $430 million on the project. Adele Ferguson, writing in the *Sydney Morning Herald* reported:

> In the case of Airport Link, the project was originally forecast to produce $407 million pre-tax profit. It is now expected to report a $430 million loss, representing a $560 million deterioration since November
>
> (Ferguson 2011)

Unlike the previous administration's busways, the TransApex roads and tunnels do little to buttress the competitive position of the city centre relative to the regional shopping and employment centres, and may be exacerbating the current lack of demand for city centre office space. The third TransApex scheme, the Legacy Freeway and Tunnel joins the Western Freeway from Ipswich and the western suburbs to the Bruce Highway which serves the northern suburbs and the Sunshine Coast. While this useful road should assist the economies of both the south-western and northern cities and suburbs, it may further accelerate the relative decline of the city centre as an employment focus by compounding the dispersing effects of rapid transport and electronic communication. It is therefore not surprising that recent city centre redevelopment proposals have switched from office to residential apartments, hotel accommodation and casino–tourism schemes.

The vision of a new world city

By the second decade of the century, construction of toll tunnels had exhausted its attraction and regeneration strategies started to concentrate on projects to re-brand Brisbane as a "new world city" aiming to capitalise on riverside central area sites made available by changing land uses and industrial relocation. Enshrined in the City Centre Master Plan (CCMP) and River's Edge Strategy (BCC 2014b; 2014c), these schemes bear some spatial similarity to the proposals being developed by state governments to the south for the central areas of their capital cities. In Sydney, the larger-scale and more integrated The Bays Precinct (UrbanGrowth NSW 2015) covers 95 hectares and eight related harbourside action areas. In Melbourne, the even more extensive Fishermans Bend Urban Renewal Area proposes to redevelop 455 hectares (4.5 km^2) of industrial and dockland areas, to accommodate 80,000 extra residents and 60,000 new jobs (State of Victoria Department of Environment, Land, Water and Planning, 2015; Chapter 7). By comparison, Brisbane's CCMP and accompanying River's Edge Strategy are more concerned with "place branding" than land-use planning, proposing visionary possibilities to fulfil four chosen themes of "place", "play", "connect" and "enable", expressed in a collection of 16 individual schemes, whose major commonality is proximity to the banks of the Brisbane River.

Current urban regeneration schemes

The North Bank/Queens Wharf proposals

Part Two now examines in more detail two of the most significant and contentious of these proposals, the first of which has been approved by the city council and state government and is currently under construction, while the second remains in limbo, inviting concerned parties to "watch this space", with the state government requesting that the city council return to the drawing board, and the city council declaring that the ball is now in the state government's court.

The site of both these schemes is the semi-derelict space around and under the elevated Riverside Expressway next to the city centre. Cleared of its former riverside uses, North Bank and Queens Wharf currently support little beyond a riverside bikeway, and some tour boat jetties and ferry terminals, which form a belt of semi-derelict land under the freeway, fronting the riverside, and separating the city centre's government and heritage buildings along William and George Streets from their original riverside setting. In 2002 the State Government and Brisbane City Council agreed to invite private developers to redevelop this North Bank area for city centre uses. In 2008, the selected construction firm, Multiplex, eventually presented a proposal containing eight skyscrapers for office space and residential uses, with the highest being 46 storeys, all overlooking or straddling the freeway, and extending as far as 70 metres into the Brisbane River. There was also a pedestrian bridge, open spaces, riverfront promenade and an Olympic-sized public swimming pool. Pronounced public concerns about flooding, design and access, and opposition from architects and the Planning Institute led the state government to initiate an "Enquiry by Design", with an independent chairperson and an interstate facilitator to help participants explore different scenarios, all including large areas of office space, new hotels, residential apartments and tourist facilities.

These proposals generated widespread public opposition, concerned over loss of riverside access, narrowing of the river and associated increase in dangers of flooding (Skinner 2008). In addition, the Coordinator General's Department discovered that the required construction over the expressway would involve prohibitive insurance costs. Another issue acknowledged, but then ignored by both Multiplex and the Enquiry by Design, concerned the alternative possibility of placing the Riverside Freeway underground for the 750 metres from the Captain Cook Bridge to the William Jolly Bridge, and defraying those costs by the value of the resulting building sites, tourist, entertainment and amenity areas linking the city centre to an enhanced river frontage. Although featured in the City Council's 2006 City Centre Master Plan, this approach was not considered, probably because the state government had committed itself to zero financial liability, though in the long term such a scheme might well have proved very economically and environmentally beneficial, also creating a complementary metropolitan central space opposite the South Bank Gardens. Skinner (2008) described public concerns over the proposed scheme:

> Opposition has come from the public as well as Queensland's peak body of town planners, a major group of architects, and a ... Queensland University lecturer ... There is currently much debate regarding the value of the project, given the extensive ecological, visual, physical and infrastructure consequences involved in building so far out over the river.
>
> (Skinner 2008)

As a result, the Minister for Infrastructure and Planning scrapped the scheme, leaving the City Council to pick up the pieces and consider possible successor proposals, avoiding such unintended consequences.

Meanwhile, an incoming LNP state government, led by Newman, proclaimed its own urban regeneration intentions by proposing to redevelop the large William Street car park opposite the Parliamentary Annexe for a major new building of privately financed government offices on its own land, overlooking the former North Bank site. The building provides 75,000 square metres, constituting the largest commercial office building in Queensland, and one of the largest in Australia. Its massive elliptical shape rising to more than 50 storeys already dominates the city centre skyline and introduces a contrasting powerful oval geometry amongst the prevalent rectilinear building shapes and street structures of the city centre.

Costing more than $1 billion to construct, 1 William Street accommodates 5,000 public servants and now replaces the Executive Building in George Street, freeing that site for redevelopment to accommodate a new high-rise hotel which is part of the proposed Queens Wharf scheme discussed below (Queensland Treasury and Trade 2013).

In 2013, soon after launching this scheme and more than ten years after the first abortive attempt to redevelop North Bank, the state government invited applicants to tender for rights to redevelop the surrounding nine hectare area, extending along William and George Streets for several hundred metres and down to the river frontage, in a scheme now renamed "Queen's Wharf". Under the overriding powers of the state's Integrated Resort Development Act, residents and businesses were invited to present their ideas to revitalise the area to create:

> a high quality, mixed use destination ... providing tourism, leisure and entertainment facilities that appeal to as broad a demographic as possible, including the international market ... [and] deliver broad direct and indirect benefits to the economy and community. [Proposals] might also include casinos and offer gaming activities integrated with other service offerings to go beyond the traditional stand-alone casino.
>
> (Queensland Treasury 2013: 1)

These state government intentions were reflected in Brisbane City Council's 2014 City Centre Master Plan (CCMP), including the scheme as a key element in the city's strategy of being "Open for Business" (BCC 2014b: 19). Queens Wharf was listed as one of the master plan's first three projects, along with Howard Smith Wharves and the City Reach, among the sixteen schemes in its implementation plan (BCC 2014b: 102–106).

However, the site is not an easy one. Riverside access is impeded by the expressway, which, though elevated, severely restricts waterfront construction throughout its length. Extensive development into the river has already been proposed, challenged and rejected. Intensive administrative office space for the 5,000 workers in 1 William Street already occupies part of the site. Over 20,000 staff and students, who flock daily to and from the Gardens Point campus of the Queensland University of Technology abutting the site to the immediate east, constitute the city centre's largest single concentration of daily commuters. In addition, much of the west of the designated area is already occupied by heritage buildings, which are amongst the most significant in the state.

Of the two finally selected proposals the "Crown-Greenland" consortium representing Packer Associates and Chinese interests concentrated on hotel and gaming facilities to attract an international clientele; while the Echo consortium, composed of the proprietors of Sydney's troubled Star Casino and major operators of Macau's casinos, tempered this with ingenious pathways, recreation and environmental features linking the site to its riverside and city centre context. In July 2015 the state government announced that Echo's "Destination Brisbane" proposals had been successful, citing its greater sensitivity to the city centre context, provision of public amenities for other city centre users and the advantage of the consortium's intention to turn the existing Treasury Casino (which it already owns and operates), into a specialist luxury shopping centre. This avoided the unfortunate outcome of "bookending" the state's main parliamentary and administrative street with gambling casinos separated by only 400 metres of heritage buildings and government offices.

Echo's CEO, Matt Bekier, in a newspaper interview of 2015 extolls the scheme's recreational provisions of a mangrove walk, bikeway and kayak hire facility, as contributing to linking the new resort to the city and the river (Passmore 2015: 44). As approved, the scheme will include a six-star integrated resort to be located in a signature 'arc' building, topped by a sky deck with restaurants and bars giving all round river and skyline views, from an elevated sky deck said to have been inspired by the Queensland veranda which Bekier anticipated would become "the beating heart of entertainment in the CBD" (Passmore 2015: 44). The integrated resort would also feature ballrooms, function spaces, a rooftop cinema, restaurants and bars, and a new hotel and hospitality school (Johnson 2015). Between the casino and George Street, three buildings of 30–50 storeys are proposed to accommodate five hotels, fifty restaurants and bars, a cinema and "I2 football fields" of public event space (Johnson 2015). Passmore (2015) summarised the conundrum posed by the development in qustion as "punters' palace or people's playground?"

The artist impressions and video "walk throughs", of the proposal depict a tactically astute scheme, which has been designed to acknowledge and integrate with its context. The giant 16-storey arc building which forms its centrepiece both echoes the elliptical shape of the massive bulk of 1 William Street and acts as an intermediary between that building's 45 storeys and the more human scale of the historic Treasury, Lands Office and other heritage buildings of William Street. Riverside bike and pedestrian paths would provide for increased pedestrian and cycle movement along the riverside. Links with surrounding educational, administrative, shopping and tourist uses both through the development and to the riverside would be significantly enhanced though the likely results have been criticised as cosmetic and inadequate.

Criticisms focus more on the strategic intention of the whole scheme than on these tactical aspects of design, internal arrangements and links. They point to problems of appropriate land uses and activities, scale, cultural context, aesthetics and symbolism, and impacts on the flood regime of the Brisbane River and its banks. Richard Kirk, president of the Queensland Chapter of the Architects Institute, voices a strong concern that the proposed location of the very large project would diminish the cultural and physical significance of the adjacent Government Precinct:

> The Government Precinct is an unsuitable choice for the development of such an overwhelmingly large size and scale. A world size, contemporary casino project demands a much larger grid, or block size than that available in ... the relatively small block size of Brisbane urban core.
>
> (Kirk 2015: 1)

He also questions whether:

> the anticipated qualities of the public spaces. ... would be compromised by the environmental impact of the Riverside freeway that runs the length of the development.
>
> (Kirk 2015: 1)

Kirk concludes that a less sensitive site adjacent to the Roma Street Transit Centre and Gardens would be more suitable.

The aesthetics, symbolism and economics as outlined in the government's original call for expressions of interests are themselves highly questionable. Packaging key sites in the historic and cultural core of the state capital to attract overseas gamblers proclaims an opportunistic approach to planning that is unlikely to yield good long-term results, either in sustainable economic development or long-term international reputation. Economic prospects for such developments have also declined sharply since the scheme was first mooted. The rate of growth of the Chinese economy and Australia's mining boom of the century's opening decade are both trending sharply downwards, causing both international and internal tourism to be facing diminished and insecure returns that offer a brittle basis for a sustainable urban economy.

The proposed large extensions of the river shoreline are also puzzling. Even where occupied by mangrove habitats and nature conservation, they can only reduce the capacity of the river basin to accept and dissipate floodwater. Intensive modeling is needed to predict flood effects. However, were the shoreline extension to be removed, much of the new public space would also go. There are thus serious physical, social, aesthetic, cultural, economic and safety concerns, with possibly damaging unintended long-term effects. Cost–benefit analyses are needed to compare this scheme with more socially and economically balanced ones, including re-routing underground the currently elevated Riverside Freeway.

The Kurilpa Riverfront Renewal

The most recent scheme, extending for nearly a kilometre from the Go Between Bridge along the opposite bank of the Brisbane River, has attracted equal attention and controversy. Next to the large cultural complex of the Queensland Performing Arts Centre, Art Gallery and State Library and South Bank Gardens, this 25-hectare area of Kurilpa Point had long been occupied by industrial uses of food processing, glass manufacture and concrete batching and was considered ripe for redevelopment. Starting in 2013, the BCC planners collaborated with state government colleagues to consider redeveloping this riverside land.

Initial consultation with planners and developers included a meeting with the West End Community Association (WECA). However, these discussions were not extended to include the community at large, and subsequent consultations and workshops were confined to professional and developer groups. In mid-2014, the council and state governments announced the launch of the proposed Draft Kurilpa Master Plan at a breakfast with the Brisbane Development Association (BDA) in the Queensland Convention Centre, to be followed by a three-week period of public consultation (BDA 2014).

The scheme that was unveiled brought gasps from the assembled audience of government officials and developers, who might have been thought to have become inured to high-density proposals. The proposal was to accommodate over 11,000 new residents, amounting to nearly 6,000 new dwellings on the 25-hectare site, in 54 apartment blocks, ranging from 12 to 40 storeys, as well as 8,000 office jobs, with a total of 2.3 hectares of public open space or less than 10 per cent of the overall site. Little attention was paid to urban design and the scheme contained no reference to traffic generation, land-based public transport, or education or health facilities (BCC 2014d).

Widespread opposition quickly gave rise to a new community group, which became known as the Kurilpa Futures Campaign Group (KFCG), based in the neighbouring suburbs of West End, Highgate Hill and South Brisbane. Combining local residents, political activists, policy and design professionals and academics, the group also included local elected representatives, councillor Helen Abrahams and the deputy leader of the state ALP opposition, Jackie Trad. In September 2014, KFCG (which has since become simply Kurilpa Futures) joined WECA in organising a public meeting, attended by over 300 people, at which it was resolved to oppose the proposed development. A Saturday morning "Community Conversation" gathered reactions and concerns from over a hundred local residents and a candidates forum, held in February 2015, preceding the 2015 state election, brought together three of the four parties (ALP, Greens, and the local "People Decide" candidate) who all committed themselves to scrapping the draft master plan. A vacant chair indicated the absence of the candidate of the then ruling LNP who had declined the invitation to attend.

The election produced a surprise result: the gain of more than thirty seats by the ALP, resulting in a majority depending on the speaker's casting vote. Jackie Trad was appointed deputy premier and, importantly, as minister of planning. She lost no time in making good her promise to reject the draft KMP, and the BCC responded by placing responsibility for managing future development of the site on the state government. The existing owners had set very high values upon their land (one proposing a price of $100 million for their eight hectare site) and indicated that they would in the meantime be content to continue their current profitable and conveniently located operations.

Concerns that politics, like nature, abhors a vacuum, led the KFCG to gather a large team of over 50 professional design policy and environmental professionals to join local residents and citywide interested parties in an "Ideas Forum" to develop an alternative vision and set of policy proposals for the site. On 17 May 2015, more than a hundred community participants joined the 56 volunteer facilitators, scribes, resources persons and visualisers in a cumulative half-day program, with outcomes being recorded on butchers paper and later synthesised to produce a vision and polices which were posted for review and amendment on the group's website. Once these additions had been integrated, the final outcome was presented to the lord mayor and the state minister for infrastructure and planning in the first week of August 2015 (KFCG 2015b).

Resulting proposals were comprehensive. Emphasising mixed uses, context sensitivity and links to the surrounding cultural, recreational and creative industries, they integrated connectivity, open space, housing, education, community services, tourism, Aboriginal culture and art, retail, specialist and convenience shopping, water sensitive and energy conserving design with governance and funding. A comprehensive alternative vision incorporated a wide mix of productive and residential activities into a green space matrix, including a new Kurilpa Nature Park accommodating retention basins to gather and dissipate water during flood events and an active pathway network (KFCG 2015a).

The Ideas Forum scheme also proposed that responsibility for future development be put into the hands of a mixed public–private development corporation, possibly reporting to Economic Development Queensland (EDQ), the government's land development arm, with a preference for the extension of the powers and authority of the neighbouring South Bank Corporation to manage the extended area.

Meanwhile the site continues to function as an industrial cluster, producing and distributing dairy products, glass containers and batched concrete throughout the inner city. The city council may have been fortunate in avoiding the consequences of a hurriedly conceived and under-researched scheme, based on inadequate consultation of both local communities and existing users. Construction of so many expensive new residential apartments might have foundered in a saturated and international investment market (Smart Property Investment 2016). The result might have been a serious case of highly visible inner-city blight that would have done nothing for the city's image, brand or economy.

Conclusions: planning assumptions, trends, opportunities and options

The neoliberal planning assumption underlying both these schemes is that once inherently benevolent operations of markets are licensed to add value, they will produce streams of social benefits. They find their international and interstate parallels and models in even larger and more ambitious schemes in the UK and elsewhere. The rapid growth of London's Docklands, for instance, throughout the 1980s and 1990s radically adapted principles of classical land economics going back to the nineteenth century. Viewed across the distance of one-and-a-half centuries, these had come to seem to many to be dynamic and ultimately productive, but it is worthwhile noting that in the detailed analysis of Parry Lewis's classic economic study *Building Cycles and Britain's Growth* (1965) they emerged as having been economically wasteful. More immediate views in the literature of the time depict them as also humanly destructive and socially devastating and economically wasteful and socially devastating (Bales 1994; Dickens 1994; Forster 1954; Rowntree 1901). Michael Heseltine's 1980 establishment of London's London Docklands Development Corporation (LDDC) learned from these social disasters to create an overarching public authority to manage development, coordinate infrastructure and balance activities in the redevelopment of these extensive and semi-derelict docklands (LDDC 2009). By contrast, Brisbane City Council's approach has been to package, brand and promote the financially "best and highest" central area land uses and densities to provide the pospect of irresistibly high yields to potential developers. Only political choice can decide whether such contentious policies will prevail or be revised and replaced, and this in turn will depend on unpredictable questions of future political control of both the Brisbane City Council and the Queensland State Government, inextricably yoked together as decision takers for these key sites.

Elsewhere in Australia, in Sydney's The Bays Precinct and Melbourne's Fishermans Bend Urban Redevelopment Authority, there is a more considered mix of public interest and private-profit concerns (UrbanGrowth NSW 2015; State of Victoria Department of Environment, Land, Water and Planning 2015). The Victorian State Government and the councils of the City of Melbourne and Port Philip are balancing public and private; commercial and recreational; and residential and social objectives across the five precincts that make up Fishermans Bend (although Shaw challenges this position in Chapter 7). Although overall density levels are high, densities are lower than the gross figure of 800 persons per hectare of the abortive BCC draft Kurilpa Master Plan, and more care is being given to providing social, physical and environmental infrastructure.

There are also strong parallels between Brisbane's Queens Wharf proposal and James Packer's Barangaroo Crown Casino scheme in Sydney (Chapter 6). The proposal implicitly aligns the public interest with the attraction of international "high rollers" to pursue semi-quarantined gambling in the key central location of the state's capital. With 75 floors (including mezzanines) and a height of 271 metres, the casino will become the city's tallest habitable building. The City of Sydney Council aimed to inject public interest concerns into the Barangaroo scheme by appealing to the NSW Planning Assessment Commission to reduce the height and bulk of the proposed building and relocate it further back from the foreshore (McKenny 2016). The Commission eventually attempted to balance public and private interests, maintaining the height and location of the tower, but prescribing an extension of the public domain by requiring a 30 metre "unencumbered public concourse area" and an extension of Hickson Park on land reclaimed from the harbour. This significant though marginal redress was won by a coalition of state government and city council, which, though only partially successful, does recognise that private-sector pursuit of the "best and highest use of land" for itself can and should be required to frame development proposals within the public interest (Cheng 2016). By contrast, Brisbane City Council has recently removed all height control limits in the city centre – previously capped at 280 metres- to promote maximum market attraction for commercial tourist developers (Atfield 2015).

Both the Queens Wharf and the Kurilpa Waterfront schemes emerge as place-branding initiatives, rather than land-use plans. Aiming to deliver on perceived once-in-a-lifetime opportunities, they involve little public or professional consultation. As a result, many of the proposals lack expert review and are receiving hostile receptions from unprepared public opinion. This aligns with the conclusion of Greenop and Darchen (2016, 391) that Brisbane's city planning practice "needs to incorporate resident buy-in and a more ethical approach, without which place branding becomes merely a chimera, lacking an authentic place identity to inform the brand". They go on to recommend two sets of actions:

1 Incorporate resident – centred processes in the construction of Brisbane's urban strategy
2 Include the suburbs in the construction of this urbanity by planning strategies that consider the centre and the suburbs. Only through acknowledging the contemporary political, social and cultural identifies that are already present in Brisbane, and their roots in its historical past, can the city hope to re-brand itself and come to terms with the haunting spectre of its history.

(Greenop and Darchen 2016: 391–392)

In seeking to re-brand Brisbane as a "new world city" the council is thus focusing on image promotion of these central areas, largely ignoring the development potential of the rest of the city, or the possibilities of more equally distributed polycentric development. Strongly orientated towards place re-branding, this approach pays inadequate attention to equally important concerns of habitability, spatial integration, social capital, economic justice or environmental sustainability.

In Brisbane, as elsewhere, forces of urban and technological change, particularly the opportunities to redevelop superseded dock and central area industrial sites, will continue to influence development in city centres and inner suburbs. The issue is not whether such changes in metropolitan activities and land uses are likely to occur – in mixed-economy market democracies like Australia's they appear to be both inevitable and potentially beneficial – but how the central city *should* change, in whose interests, and what forces or combination of agencies will be empowered to control these changes. Brisbane's experiment of the past decade with market-driven planning and urban regeneration has proved contentious and politically polarising, resulting in more failures than successes. As in Sydney, tunnels and bridges built by private enterprise for private vehicles have failed financially, economically and functionally, having caused severe and continuing dislocation. They remain chronically underused. Riverside schemes such as North Bank and Kurilpa Riverside Renewal have had to be scrapped by embarrassed state governments. Now, Queens Wharf seeks to hitch its market-driven wagon to the wandering star of international tourism and gaming. Such schemes are a far cry from the successful collaboration of the Atkinson and Soorley administrations of the 1980s and 1990s when a mix of public-interest objectives and private-sector investment produced such developments as the city's South Bank Gardens. These earlier initiatives have played a significant long-term role in attracting knowledge and creative industry workers, as well as tourists. By contrast, current city centre luxury apartment and gaming industry resorts may prove far less sustainable or beneficial.

References

Atfield, C. (2015) 'Brisbane City Council negotiates with aviation authorities to raise CBD heights', *Sydney Morning Herald*, 30 August http://www.smh.com.au/business/aviation/brisbane-city-council-negotiates-with-aviation-authorities-to-raise-cbd-heights (accessed 28 June 2016).

Bales, K. (1994), *Early Innovations in Social Research: The Poverty Survey of Charles Booth*. London: London School of Economics and Political Science. Department of Social Policy and Administration.

Brannelly, L.(2011) 'Clem7 owners River City Motorways go into receivership owing $1.3bn', *The Courier Mail*, 24 February. http://www.couriermail.com.au/business/clem7-owners-rivercity-motorways-go-into-receivership/story (accessed 10 November 2015).

BCC (Brisbane City Council) (2014a) 'City Plan', http://eplan.brisbane.qld.gov.au/ (accessed 1 November 2015).

BCC (Brisbane City Council) (2014b), 'City Centre Master Plan' http://www.brisbane.qld.gov.au/sites/default/files/20140417-City%20Centre%20Master%20Plan-full%20document.pdf (accessed 5 November 2015).

BCC (Brisbane City Council) (2014c) 'River's Edge Strategy', https://www.brisbane.qld.gov.au/planning-building/planning-guidelines-tools/neighbourhood-planning/neighbourhood-plans-other-local-planning-projects/rivers-edge-strategy (accessed 28 June 2016).

BCC (Brisbane City Council) (2014d) 'Draft Kurilpa Master Plan' http://www.brisbane.qld.gov.au/sites/default/files/20140821_-_draft_kurilpa_master_plan.pdf (accessed 10 November 2015).

BCC (Brisbane City Council) (2015) 'Kurilpa Riverfront Renewal' http://www.brisbane.qld.gov.au/planning-building/planning-guidelines-tools/neighbourhood-planning/neighbourhood-plans-other-local-planning-projects/kurilpa-riverfront-renewal (accessed 11 November 2015).

BDA (Brisbane Development Association) (2014) 'Kurilpa Riverfront Renewal: Draft masterplan launch event' http://www.bda.org.au/events/2014/kurilpa-riverfront-renewal-draft-masterplan-launch-event (accessed 11 November 2015).

Cheng, L. (2016) 'Approval of Crown's Barangaroo tower "disappointing", Institute says', *Architecture Australia,* 29 June http://architectureau.com/articles/approval-of-crown-barangaroo-tower-disappointing-institute-says (accessed 29 June 2016).

Cole, J. (1984) *Shaping a City: Greater Brisbane, 1925–1985*, Brisbane: William Brooks.

Dickens, C. (1994) *Hard Times,* Peterborough (Canada): Broadview Press.

Ferguson, A. (2011) 'Heavyweight Leighton on the way to being turned into a feather duster', *Sydney Morning Herald*, 12 April http://www.smh.com.au/business/heavyweight-leighton-on-the-way-to-being-turned-into-a-feather-duster-20110411-1db0q.html#ixzz4Cqc46aUL (accessed 28 December 2016).

Forster, E. M. (1954) *Howard's End,* London: Vintage.

Greenop, K. and Darchen, S. (2016) 'Identifying "place" in place branding: core and periphery in Brisbane's "New World City"', *GeoJournal*, 81(3):379–394.

Heywood, P. (2011) *Community Planning: Integrating Social and Physical Environments*, Oxford: Blackwell-Wiley.

Johnson, N. (2015) 'Giant "arc" casino resort by Cottee Parker Architects wins Queens Wharf Brisbane design battle'. *Architecture & Design* http://www.architectureanddesign.com.au/news/giant-arc-casino-resort-by-cottee-parker-architect (accessed 10 November 2015).

Kirk, R. (2015) 'Brisbane casino will diminish Government Precinct' http://architectureau.com/articles/brisbane-casino-will-diminish-government-precinct/ (accessed 10 November 2015).

Kitney, D. (2013) 'Tunnel on road to deeper trouble', *The Australian Business Review*, 13 October http://www.theaustralian.com.au/business/companies/tunnel-on-road-to-deeper-trouble/story-fn91v9q3-1226739906815 (accessed 14 June 2013).

KFCG (Kurilpa Futures Campaign Group) (2015a) 'Ideas Forum outcome's, http://kurilpafutures.org/events/ideas-forum/outputs-ideas-forum/ (accessed 11 November 2015).

KFCG (Kurilpa Futures Campaign Group) (2015b) 'Kurilpa Futures Campaign Group and the Political Process', 3 August 2015 https://kurilpafutures.files.wordpress.com/2015/05/kfcgandthepoliticalprocess.pdf (accessed 11 November 2015).

LDDC (London Docklands Development Corporation) (2009) 'About LDDC' http://www.lddc-history.org.uk/lddcachieve/ (accessed 28 June 2016).

McKenny, L. (2016) 'Reject James Packer's casino tower: City of Sydney', *Sydney Morning Herald*, 28 April. http://www.smh.com.au/nsw/reject-james-packers-casino-tower-city-of-sydney-20160428-gogtq1.html (accessed, 28 June 2016).

Moore, T. and Stephens, K. (2014) 'TransApex is dead – bring on the River's Edge', *Brisbane Times,* 19 June http://www.brisbanetimes.com.au/queensland/transapex-is-dead--bring-on-the-rivers-edge-20140618-zsdvv.html (accessed 11 November 2015).

Neilson, L. (2008) 'The "Building Better Cities'"program 1991–96: A nation-building initiative of the Commonwealth Government', in J. Butcher (ed.) *Australia under Construction: Nation-building – Past, Present and Future*, Canberra: ANU Press.

Parry Lewis, J. (1965) *Building Cycles and Britain's Growth*, London: Macmillan.

Passmore. D. (2015) 'River city takes big gamble', *Courier Mail*, 25 July, p.44.

Queensland Government (2005) *South East Queensland Regional Plan*, 2005–2026, Brisbane, Department of Local Government and Planning.

Queensland Treasury and Trade (2013) '1 William Street, Revitalising Brisbane city's riverside' https://www.treasury.qld.gov.au/projects-infrastructure/projects/1-william-st/index.php (accessed 10 November 2015).

Roberts, P. and Sykes, H. (eds) (2002) *Urban Regeneration, A Handbook*, London: Sage.

Rowntree, S. (1901) *Poverty: A Study of Town Life*, London: Macmillan.

Skinner, P. (2008) 'Some sensible ideas for the northern bank of the Brisbane River', http://media01.couriermail.com.au/multimedia/2008/06/080617_northbank/PS_%20Sensible_NB2.pdf (accessed 2 October 2017).

Smart Property Investment (2016) 'Bank issues dire warning to investors for 2016' http://www.smartpropertyinvestment.com.au/news/14964-bank-issues-dire-warning-to-investors-for-2016 (accessed 6 February 2016).

State of Victoria Department of Environment, Land, Water and Planning (2015) 'Fishermans Bend Urban Renewal' http://www.mpa.vic.gov.au/wp-content/uploads/2014/07/FBURA-Recast-Fact-Sheet-April-2015.pdf (accessed 28 June 2016).

Urban Growth NSW (2015) 'The Bays Precinct Plan: Transformation Plan, Sydney' http://thebayssydney.com.au/wp-content/uploads/2015/10/TheBaysPrecinct_Transformation_Plan_FULL_Oct15_WEB.pdf (accessed 27 June 2016

Urban Renewal Brisbane (2011) 'Urban Renewal Brisbane –Our first 20 years', http://www.planning.org.au/documents/item/3245 (accessed 10 November 2015).

9 Transforming Perth

The evolution of urban regeneration in Perth, Western Australia, 1990–2016

Paul J. Maginn

Introduction

Australian cities are in a perpetual state of transformation. Much of this transformation occurs in a discrete and incremental manner and often takes the form of individual residential sub-divisions on fairly large blocks (approx. 800–1,200m^2) within middle-ring suburbs. In addition, refurbishments and renovations to residential and small-scale retail, commercial and industrial premises also contribute to the transformation of the city. Whilst these improvements may be spread across urban and suburban areas and go largely unnoticed, their cumulative effects in economic, physical and aesthetic terms, arguably have a profound impact on cities. More visible forms of transformation in the form of large-scale state-, private- and/or joint-venture-sponsored urban regeneration projects are evidenced by the cranes, high-rise apartment blocks, commercial buildings, bars, cafes and retail outlets found in CBD areas or along prime waterfront locations such as Barangaroo and Central Park in Sydney (Chapters 6 and 10); Southbank and Docklands in Melbourne (Dovey 2004; Chapter 7); Southbank in Brisbane (Ganis et al. 2014; Chapter 8); Port Adelaide in Adelaide (Oakley 2009); and Elizabeth Quay in Perth (Bolleter 2014).

In overall terms, policy efforts to transform the Perth CBD, as well as a number of key inner- and outer-suburban locations, has been a facet of the Western Australian (WA) state government's policy agenda since the early 1980s when the East Perth Urban Renewal Project was initially proposed by the then Labor state government (Byrne and Houston 2005). However it was not until the early 1990s following the commencement of the Building Better Cities Program (BBCP) (Neilson 2008) and the establishment of the East Perth Redevelopment Authority (EPRA) in 1991 that urban regeneration proper really began (Crawford 2003). The redevelopment authority model borrowed significantly from the urban development corporation approach (e.g. London Docklands Development Corporation) adopted in the UK during the early 1980s by Margaret Thatcher. That is to say, the EPRA and subsequent redevelopment authorities were all-powerful entities in that they acted as both planning regulator and developer of the land under their control. Urban regeneration within Perth since the 1990s has essentially been premised on a partnership-based approach between the state government and private-sector developers, and underpinned by an entrepreneurial governance ethos (Hall and Hubbard 1996) that uses civic boosterism in order to 'promote the name and the positive image of the city' (Short et al. 1993:208) on the pathway to being recognised as a globalising city:

> By 2031, Perth and Peel people will have created a *world class* liveable city: green, vibrant, more compact and accessible with a unique sense of place.
>
> (WAPC 2010:2)

> Working together, the [Metropolitan Redevelopment Authority] and its partners revitalise communities and build a distinctive sense of place that is transforming the city and redefining key areas of metropolitan Perth for future generations.
>
> (MRA 2014:6)

This chapter, then, is primarily concerned with examining the evolution of the urban regeneration landscape within Perth, Western Australia, by showcasing a number of key projects that have been completed or are in the process of being completed, and the governance arrangements underpinning them. The chapter is structured into several parts. First, an overview of what constitutes urban regeneration is presented as a precursor to an outline sketch of the evolution of urban regeneration policy within Australia and the UK. Basically, this shows that there are similarities in the broad policy paradigms and trajectories across both nations. This is exemplified in the next section which highlights the emergence of the Building Better Cities Program (BBCP) and the adoption of a place-based, property-development approach to urban regeneration in Perth that gave birth to a new governance structure – i.e. a redevelopment authority – that had both planning and development powers. The third section highlights that the redevelopment authority model proved to be a bi-partisan political and policy success and is reflected in the proliferation of redevelopment authorities set up during the 1990s/2000s. The penultimate section highlights the next phase (i.e. consolidation and streamlining) in the evolution of urban regeneration governance in Perth as part of the state government's wider planning reform agenda. Finally, in the conclusion some tentative comments about the future direction of urban regeneration policy are noted.

Getting to grips with urban regeneration

Tallon (2012) notes that urban regeneration is a subset of urban policy. Furthermore, he states that 'at the most general level, regeneration has come to be associated with any development that is taking place in cities and towns' (Tallon 2012:4). This suggests that urban regeneration is an activity undertaken by both the public and private sector, either independently or collaboratively. In terms of urban regeneration *policy*, this has generally been seen as the domain of the state. In the UK, the chief policy architect and funder of urban regeneration programs has been central government. However as Ball and Maginn (2005) have noted, urban-regeneration policy has always essentially been a *partnership* with different state and private-sector actors involved in varying ways and degrees depending on the prevailing ideology and policy emphasis of the government of the day. In general, a partnership approach has been seen as a force for positive change; although this is not to deny concerns about the power dynamics within urban partnerships between central- and local-government, property developers, other private organisations as well as the 'the local community' and voluntary sector organisations (Hastings 1996).

In simple terms, urban regeneration policy is concerned with three fundamental issues:

> i *people* – '...enhancing [their] skills capacities and aspirations to enable them to participate in and benefit from opportunities';
> ii *business* – '...[improving] economic competitiveness in terms of business performance, to create more local jobs and prosperity'; and
> iii *place* – '...improve the general appeal of a place [to attract more people and business].'
>
> (Tallon 2013:5)

The policy priority given to these three issues has ebbed and flowed over time with left-leaning governments often championing 'people and place', whereas right-leaning governments have tended to privilege 'business and place'. Relatedly, there have also tended to be differences in the governance arrangements of regeneration programmes. Between 1950–1970 Australian and British governments tended to be relatively more interventionist and actively involved in urban regeneration programmes which were largely welfarist in character. Since the rise of neo-conservativism and neo-liberalism in the US under Ronald Reagan, and in the UK under Margaret Thatcher in the late 1970s, governments from the right and left have increasingly adopted more competitive and market-based structures and mechanisms in urban regeneration programs. In other words, the governance approach to regeneration shifted from being premised on urban managerialism to urban entrepreneurialism (Harvey 1989; Gleeson and Low 2000).

Policy paradigms in urban regeneration

As Roberts (2000) has identified, British urban regeneration policy has undergone a number of broad paradigms – *reconstruction* (1950s); *revitalisation* (1960s); *renewal* (1970s); *redevelopment* (1980s); *regeneration*

(1990s) and *renaissance* (2000s) (also see Maginn 2004; Leary and McCarthy 2013). In summary, over time the policy emphasis has shifted from being concerned primarily with the renewal of public housing estates and alleviation of poverty to broader large-scale property redevelopment and economic renewal. Environmental sustainability emerged as a fashionable policy concern during the late 1980s and into the 1990s and by the late 1990s social exclusion came to replace policy concerns about traditional notions of poverty and social disadvantage (Imrie and Raco 2003). A 'new' paradigm – *recovery* (2010s) – that resonates with the entrepreneurial, laissez-faire economic development approach of the 1980s but emphasises the need for austerity in public-policy expenditure, emerged in the wake of the global financial crisis in 2008 and can be added to the urban-regeneration paradigm list. Peck (2012) has described this approach as 'austerity urbanism', whilst Pugalis and McGuinness (2013:339) argue that the UK Conservative–Liberal Democrat coalition government's policy position signalled that the 'last rites were effectively read for holistic notions of area-based regeneration in England'. Thus far, Australia has managed to evade the *austerity-recovery* paradigm given the fact that the national economy has endured continuous economic growth for the last 25 years. That said, there is a recognition by the federal and state governments that, if Australia and its major metropolitan regions are to remain competitive in a global economy, there is a need to invest in nation-building projects. Hence, as Steele and Legacy (2017) note, there has been an 'infrastructure turn' in urban policy.

Within an Australia context, the Commonwealth government's involvement in urban-regeneration policy has been somewhat sporadic as this policy has been the domain of state governments. Nevertheless, there are traces of the paradigmatic shifts experienced in the UK also clearly evident in the trajectory of Australian policy thereby pointing to the existence of international policy transfer (Cook 2008; McCann and Ward 2013). Orchard (1999a) notes that national government involvement in urban policy has ebbed and flowed since the 1940s following the establishment of the Ministry of Post-War Reconstruction (1943). There was considerable expansion in the housing stock with some 750,000 homes built between 1945–1970 (Sandercock 1977). Notably, approximately 310,000 of these new homes were built and managed by the various state housing commissions. The growth in housing supply was also augmented by a period of extended economic growth, high employment and social mobility that contributed to mass suburbanisation and the physical expansion of metropolitan regions. But as Dodson (2016:26) notes, '[b]y the 1970s the post-World War II suburban model was showing strains from under-servicing and a spatial mismatch in housing and employment location'.

Orchard (1999b) notes that with the emergence of the *(sub)urban problem* a 'wave of reformist zeal' emerged in the late 1960s and early 1970s. Following the election of the Labor Party in 1972, the then prime minister, Gough Whitlam, who had been advocating since the mid-1960s that Australia needed to tackle the '(sub)urban problem', established the Department of Urban and Regional Development (DURD). This was the Commonwealth's first *official* foray into *urban policy*. DURD's policy remit was largely about physical regeneration in that it sought to develop suburban centres, conserve and improve older inner-city housing stock for low-income families, improve services in suburban areas, and connect suburban homes built during the 1950s and 1960s to the sewerage system. Dodson (2016) characterised Whitlam's overall policy paradigm as being focused on '*suburban remediation*' (p.27).

The incumbent Labor government lost the 1975 federal election to the conservative Liberal/National Party coalition which quickly set about abandoning its involvement in urban policy and disbanded the DURD. It would take another 15 years and the election of a Labor government before urban policy was back on the national policy agenda in the form of the Building Better Cities Program (BBCP). However, in this intervening period there was a notable shift in political ideology towards neo-liberalism. Simultaneously, there had also been an 'environmental turn' as nations began to question the issue of sustainable development as the impacts of globalisation began to be increasingly recognised. The 1983 election victory saw the new Labor government 'set about defining an approach ... which embraced free market rationalist ideas in many areas of economic policy' (Orchard 1999b:205). Whilst this tilt towards economic rationalism was very much evident under Paul Keating, who replaced Bob Hawke as prime minister, there was still a recognition of the need to address the perennial issues of the '(sub)urban problem' – social inequality, locational disadvantage and accessibility to affordable housing.

Badcock (1993) notes the BBCP was as much an instrument of crisis management as it was a genuine commitment to urban policy. Despite the sceptical tone of Badcock's assessment of why and how the BBCP emerged, Neilson (2008: 83–84) proclaimed that it can be 'credited with leading the revival of inner cities [and] broke new ground with the style of intervention undertaken [...] [and] also created new forms of intergovernmental agreements built around *outcomes*'. In many respects the BBCP's approach to regeneration resembled the site-specific property-led approach adopted in the UK during the 1980s where land was annexed from local councils, urban-development corporations were set up and charged with the responsibility for both planning and development as well as attracting private investment, and where the central government provided core funding for land assembly and infrastructure provision. The BBCP has had an enduring legacy within the urban-development industry (Neilson 2008) especially within Western Australia where the state government established a series of redevelopment authorities – Subiaco, Midland and Armadale – across Perth during the 1990s and 2000s following the establishment of the East Perth Redevelopment Authority (EPRA) in 1992 as part of the BBCP (Crawford 2003).

East Perth: From contamination to gentrification

Neilson (2008:94) states that the aim of the BBCP was to 'promote improvements in the *efficiency*, *equity* and *sustainability* of Australian cities and to increase their capacity to meet a range of social, economic and environmental objectives'. In order to obtain funding from the Commonwealth government, state governments were encouraged to submit bids for area strategies that had to conform to specific criteria and, more specifically, policy outcomes. That said, an Australian National Audit Office (1996: 3) report noted that whilst:

> The program has been well managed, particularly in recognising and controlling financial risk to the Commonwealth [...] baseline information on urban factors at which the programme was directed had not been established, and there was very little measurement of change or improvement in these factors.

A total of 26 area strategies were finally approved across the states and territories. In WA, four projects were supported with three of these in the Perth metropolitan region – East Perth, Stirling and Fremantle – and one in regional WA (Bunbury). However, only the East Perth project stands out as memorable. This is arguably due to the profound transformation of the area, under the auspices of the East Perth Redevelopment Authority (EPRA), from a contaminated industrial wasteland to an 'upmarket inner city suburb [where] many of the architect-designed homes and apartments have expansive river views' (Gregory 2008:91). The EPRA was established under state legislation – East Perth Redevelopment Act 1991 – and had the following prescribed key functions:

a. to plan, undertake, promote and coordinate the development of land in the redevelopment area; and
b. for that purpose – (i) to prepare and keep under review a redevelopment scheme for that area; and (ii) to control development in that area.

In governance terms, the minister for planning had overall responsibility for the EPRA whilst a seven-person executive board was responsible for its strategic management. Most of the board members were from the private sector and appointed by the minister but it also included two local councillors from the City of Perth, the local government area that East Perth was located within. The minority status of local councillors plus the removal of the East Perth area from the City of Perth was a deliberate decision designed to ensure that the regeneration of East Perth was not prone to cumbersome bureaucratic processes or politicised decision-making at the local authority level. The day-to-day operations of EPRA fell to a small team of professionals headed by a chief executive officer and four directors whose operational portfolios have changed relatively little since the early 2000s and the creation of the Metropolitan Redevelopment Authority (MRA) in 2011 which subsumed the EPRA (EPRA 2003; MRA 2012). When initially conceived, the EPRA was charged with regenerating an area known as Claisebrook Village, however over time its geographical remit and portfolio expanded to include a number of other distinct project sites within the East Perth area – East Perth Power Station and Riverside – and the adjacent suburb of Northbridge – New Northbridge, Perth Cultural Centre and Perth City Link (Table 9.1).

Neilson (2008) notes the EPRA was heralded as a major urban regeneration success story on account of its efficiency and effectiveness in transforming a once industrial and contaminated site into a popular residential and well-designed neighbourhood. It is difficult to disagree with this basic assessment given that, for example:

i the number of residents increased from 100 to 2,571 (between 1992–2011);
ii the number of businesses with more than 50 staff increased from 103 to 401 in 2004 before falling to 257 in 2011;
iii the number of students increased from 1,100 to a maximum of 3,500 in 2010, falling to 2,788 in 2011; and
iv the number of visitors to public facilities increased from 2.06m in 2002 to a peak of 4.01m in 2007, falling to 1.97m in 2010 and subsequently increasing to 2.63m in 2011.

(EPRA 2003; 2011)

Despite these improvements Crawford (2003) expressed concerns about the success of the regeneration of East Perth in terms of the EPRA meeting social justice objectives – providing public and other affordable housing for low income households – that formed part of the BBCP's overarching policy remit:

> Upon the announcement of the East Perth Project in 1990 it became apparent that the original plan to include a substantial affordable housing component had changed, with the emphasis resting more on the redevelopment's appeal to the private market and its potential to reinvigorate the inner city.
>
> (Crawford (2003: 89)

EPRA had set a target of 10–15 per cent social/affordable housing to be built in the East Perth area. It is difficult to discern, however, whether or not this target was realised. Annual reports from EPRA, including the final report in 2010/11, provide no clear indication of how many affordable housing units were built over the lifetime of the project. Moreover, there has never been any kind of independent systematic policy evaluation on whether or not the EPRA met its policy objectives or the wider social, economic and environmental impacts of the regeneration of East Perth. Data from the ABS Census would seem to suggest affordable housing targets may not have been reached. That is, whilst the number of public housing authority dwellings in East Perth increased by 45 units from 144 to 189 (+31.2 per cent), the overall share of affordable housing in the suburb fell from 7.2 per cent to 5.3 per cent between 2001–2011. Conversely, the number of dwellings owned outright or with a mortgage increased by 857 from 487 to 1,344 (+176 per cent), increasing its share of total housing from 24.5% to 37.8 per cent over the same period. In other words, social housing comprised only 5 per cent of total new housing (n = 902).

Further evidence of the 'successfulness' of the property-led regeneration approach to East Perth is reflected in median house prices. Data from REIWA (2017) shows that the median house price for the area peaked in 2015 at $1.37m – more than two and a half times the Perth metropolitan region median house price of $545,000. The median price for apartments/units had peaked in 2013 and stood at $619,500 compared to $440,000 for the Perth metropolitan region. Despite initial political and planning concerns during the early phases of regeneration in East Perth about the market appetite for higher density housing and apartment-style living, the latter form of dwelling accounted for 68 per cent of the total housing stock with four-storey apartment blocks accounting for most of the higher density housing. In short, East Perth has become a highly desirable and gentrified inner-city neighbourhood as illustrated in Figure 9.1.

Concerns about the social justice objectives of the EPRA expressed by Crawford (2003) have been echoed by other commentators. Gregory (2008), for example, argues that the gentrification of East Perth signifies an obliteration of its rich multicultural and Indigenous past (Gregory 2008). Similarly, Byrne and Houston (2005) contend that despite efforts to include symbolic and aesthetic representations of the past, diversity and difference into the East Perth landscape via public art and Indigenous emblems – what they call 'multicultural consent' – these were manufactured and inauthentic. Ultimately, they served to '"mask" East Perth's alternative Indigenous, immigrant and working-class memories' (Byrne and Houston 2005:322). Indeed, the renewal of East Perth, the lack of affordable housing, and its attendant gentrification has invariably marginalised Indigenous and low-income groups who once defined this suburb.

The redevelopment authority model has proven to be an effective governance vehicle for realising 'policy success' in urban regeneration within Perth. Indeed, as will be shown below, the state government established other redevelopment authorities across Perth between 1994–2000.

The proliferation of redevelopment authorities

In June 1994 the then Liberal government minister for planning, Richard Lewis, announced plans to establish a new redevelopment authority to

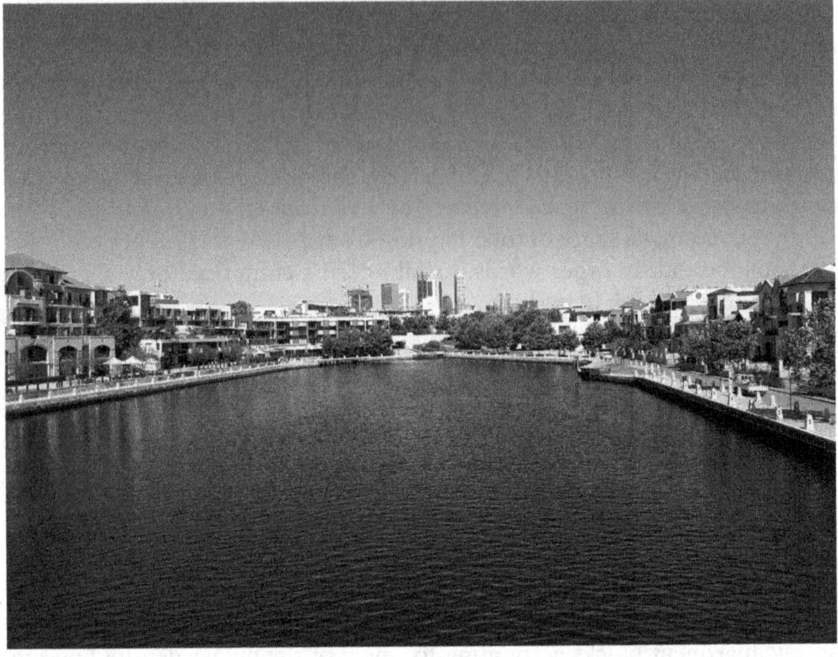

Figure 9.1 Claisebrook Inlet, East Perth (2016) (source: author)

assume responsibility 'for the rejuvenation of more than 70 hectares of under-utilised industrial land [in Subiaco] west of Perth' (Government of Western Australia 1994). This project was supported by funds from the state government as well as the BBCP. However, it is clear from the minister's media statement announcing the establishment of the Subiaco Redevelopment Authority (SRA) that the neo-liberalist turn in WA urban regeneration policy and governance had well and truly taken root:

> The Government is taking the lead in this exciting project and the creation of the Authority is essential to ensuring its timely and successful completion. The involvement of the *private sector* will also be *essential* to the overall realisation of the project and the maximum use will be made of private sector consultants and contractors.
>
> (Lewis, 1994, 1)

Early documentation about plans to revitalise Subiaco indicated that a partnership approach to regeneration was to be adopted. The SRA's (n.d.) initial concept plan for example stated that '[a] unique partnership has been formed between the City of Subiaco and the State Government to develop plans for a vibrant inner-city community in Subiaco'. Similarly, in the SRA's (1995) newsletter, *Outlook*, the then mayor of Subiaco, Tony Costa, stated that 'the Subiaco City Council is delighted with the Premier's recent announcement that the State Government will join with the city of Subiaco in the planning and funding of the Subiaco Redevelopment Project'.

Later, however, mayor Costa proffered a highly critical overview of how the regeneration process had unfolded in Subiaco (Government of Western Australia 2011a). The mayor's comments resonated with the aforementioned critiques of the impacts of renewal and gentrification on the erasure of place attachment and identity associated with East Perth and the imbalance of power within the so-called partnership between the council and the state government:

> Dare I say, it was a saga! There was a lot of heartburn and a lot of tension between Government and the Council. By establishing the [SRA], what we as a Council thought was best and right for Subi was ignored – we were effectively dictated to by the Government of the day. Not inviting us to be involved in the project was, I think, an obstacle that remained a problem throughout the redevelopment process. There were heated debates and public meetings for and against what we were going to do with the area. [...] We were always in favour of the redevelopment of what was a largely industrial wasteland and, in the end, we've got a residential development. But I believe that from a cultural perspective we've lost something. It's changed the social mix of Subiaco.
>
> (Costa cited in Government of Western Australia 2011a: 47)

Frictions between the state government and the local council were also acknowledged by the then minister for planning, Richard Lewis, who noted that there was considerable community opposition to plans to regenerate the area, especially in relation to increasing residential densities and the sinking of the Perth to Fremantle rail line. Furthermore, the local council were vehemently opposed to having lost their planning powers over the site to the SRA and their lack of representation on the SRA Board (Government of Western Australia 2011a). Whilst the council had two representatives on the SRA Board their minority status meant that they did not have the numbers to influence decision-making.

As with the EPRA, the SRA had a board which was responsible for setting strategic directions and monitoring performance. Notably, the SRA did not directly employ any professional (or administrative) staff, other than the CEO (who was also the CEO of EPRA) to implement the project. Rather, a service agreement was established with the EPRA which assumed responsibility for the operational management of the regeneration of Subiaco. The Subiaco regeneration area covered over 80 hectares of former industrial and derelict land and the SRA's overall objectives, as outlined in the initial concept plan, included:

i environmentally friendly urban living combining residential, employment and retail activities;

ii a range of housing options;
iii easy pedestrian access to all areas of the development;
iv high-quality public open spaces; and
v a central focus around the Station Precinct [i.e. 'Subi Centro'].

Subi Centro was the 'jewel in the crown' of the regeneration of Subiaco in that this component of the project was to become a transit oriented development (TOD) premised on new urbanist planning principles. It was necessary to sink part of the Perth to Fremantle rail line at Subiaco in order to generate value capture to enable medium-density residential and commercial development in and around the station square (Figure 9.2). Again, in terms of broad physical and economic performance indicators the regeneration of Subi Centro may be viewed as a 'success story' in that the project has delivered (SRA 2011):

i 1,040 residential units between 1997–2011 – the majority (1,005) of these were completed by 2007 – and is anticipated to deliver 1,975 by the time it is complete;
ii retail floor space increased from 7,900 m^2 in 1998 to 16,202 m^2 by 2011 – an increase of 105 per cent; and
iii commercial floor space increased from 147,400 m^2 in 1997 to 215,461 m^2 in 2011 – an increase of 46 per cent.

As with East Perth, the Subi Centro area, plus the wider Subiaco suburb, has become a desirable and gentrified area. This is reflected in median house prices which peaked in 2013 at $1.275m, some 2.4 times the Perth metropolitan median price of $530,000. Median house prices have fallen as a result of the slowdown in the resources boom and stood at $1.203m at the end of 2016 – this was almost 2.3 times the metropolitan median house price of $527,000. Similarly, apartments/units were also relatively highly valued in Subiaco with a peak median price of $650,000 in 2013 compared to the metropolitan median of $440,000 – a difference of 47.7 per cent (REIWA 2017). Interestingly, however, retailers in the main high street – Rokeby Road – in Subiaco have been struggling in recent years due to high rents and competition from other retail centres.

Outer suburban regeneration

Two further redevelopment authorities were established in outer suburban areas within the Perth metropolitan region. In 1999 the Liberal state government established the Midland Redevelopment Authority to lead the renewal of Midland, a suburb located 18 km east of Perth CBD within the City of Swan. The project area comprised 160 hectares and included the Midland town centre as well as land occupied by former railway workshops and industrial buildings. Whereas the inner-city projects of East Perth and

Figure 9.2 Subi Centro, transit oriented development (2014) (source: author)

Subiaco gave significant emphasis to residential and commercial development, the Midland regeneration project had a more traditional and comprehensive policy remit in the sense that it was concerned with renewing a traditional town centre, developing more residential properties, bulky goods retailing, heritage conservation, public art and urban landscaping.

In 2001, the Armadale Redevelopment Authority (ARA) was established, this time under the auspices of the then recently elected state Labor government – a clear signal that there was cross-political support for this neo-liberalist approach to urban regeneration. Notably, the minister for planning and infrastructure, Alannah MacTiernan, who was instrumental in establishing the ARA also happened to be the elected member for Armadale. Consequently, this gave the impression that the ARA was as much a political decision as a rational policy one. Armadale is located 28 km south-east of the Perth CBD and the ARA regeneration project comprised a total of eight distinct sites totalling some 1,150 hectares and was concerned with enhancing the physical, economic, social and cultural fabric of Armadale via increasing (i) residential development, (ii) commercial development, (iii) retailing, (iv) sporting and recreational facilities, (v) cultural facilities, and (vi) developing transit oriented development in the town centre.

Whilst both of these redevelopment authorities have had some positive impacts on their respective local areas the regeneration process in Midland

and Armadale has moved slowly and incrementally. Relatedly, since these two areas are based in outer-suburban locations they are rendered somewhat marginal and even invisible within the minds of the wider public and political imagination.

The consolidation of redevelopment authorities

By the time the Midland and Armadale redevelopment authorities had been established, the BBCP had long since been terminated by prime minister John Howard following the Liberal Party's federal election victory in 1996. The continuation of the redevelopment authority model in WA, notably by state governments from both sides of the political divide, indicates that it has been seen as an effective mode of urban governance. The four redevelopment authorities continued to operate as distinct organisations until 2011. As part of its policy reform agenda to cut red tape in the planning system the Liberal/National state government consolidated all four redevelopment authorities in a single redevelopment authority, the Metropolitan Redevelopment Authority (MRA) (Maginn and Foley 2014). Hence, the MRA would lead a more co-ordinated and strategic approach to regeneration policy across the Perth metropolitan region. This strategic approach was compounded further by the fact that under the Metropolitan Redevelopment Authority Act 2011 the minister now has the power to establish new redevelopment areas without the need to introduce new legislation into the state parliament as was required in the past.

In September 2013 the premier and planning minister issued a joint statement declaring that the MRA would assume responsibility for the regeneration of a 1.6 km stretch of the coastal suburb of Scarborough Beach located within the City of Stirling, Perth's largest local government area by population (Government of Western Australia 2013). Whilst the MRA is leading this project it is a partnership between the state government and the local council who have contributed \$48m and \$27.4m of funding respectively (MRA n.d).

In a media statement announcing the establishment of the MRA, the then minister of planning, John Day, re-asserted the efficacy of the redevelopment authority model as an agile and entrepreneurial approach to regeneration, and the importance of a partnership approach to regeneration with key stakeholders (Government of Western Australia 2011b). Furthermore, the MRA is also seen as the best vehicle to promote and facilitate the transformation of Perth as a dynamic and growing city via civic boosterism-informed projects:

> In *partnership* with Government, industry and the people of Perth, the MRA will continue to redefine our city through projects such as Perth Waterfront, Perth City Link, Perth Cultural Centre and Riverside. [...] The redevelopment authority model has strengthened the State's planning system since 1991, facilitating the successful redevelopment

> of inner city land and other key strategic centres. [...] The MRA will increase project flexibility, *continue to attract millions of dollars in private sector investment* and create places where people want to live, work and visit.
>
> (Government of Western Australia 2011b)

As outlined in the MRA's (2014) Strategic Plan 2014–2018, the organisation has three fundamental roles: (i) planning regulator, (ii) developer, and (iii) place manager. As such this arguably makes it a powerful planning and regeneration agency. Furthermore, the MRA's overall philosophy and approach to regeneration is underpinned by its *place-making model*, a complex multi-dimensional meta-strategic governance policy framework that informs and guides the MRA's decision-making and actions (Figure 9.3).

Place-making, as noted by Silberberg (2013:2), has:

> Its roots [in] the seminal works of urban thinkers like Jane Jacobs, Kevin Lynch and William Whyte, who ... espoused a new way to understand, design and program public spaces by putting people and communities ahead of efficiency and aesthetics.

This resonates with the MRA's vision to create 'dynamic, authentic and sustainable places' which it seeks to achieve via a range of place-making-related objectives that include: (i) promoting social inclusion, (ii) enhancing urban connectivity, (iii) promoting urban efficiency, (iv) building a sense of place, and (v) improving environmental integrity. That said, it is also clear that the MRA's place-making model is informed by *entrepreneurial managerialism*. This is evident, for example, in discursive utterances such as:

> Our model sees the creation of places through five distinct phases which are influenced and enabled by *external business drivers*, strategic directions and *corporate functions* that support place making.
>
> [Our place management process] ensures that our *assets*, including *properties* and public realm under our management achieve the redevelopment area objectives.
>
> We're building a city that creates *business opportunities* and jobs for the surrounding community. By *attracting private investors* and *promoting business development* tied to market demand – particularly in emerging and niche industries – we are crafting local economies that meet the needs of today and tomorrow. (http://www.mra.wa.gov.au/about-us/our-vision)

In total, the MRA has responsibility for thirteen major regeneration projects across five redevelopment areas: (i) Armadale (2 projects), (ii) Central Perth

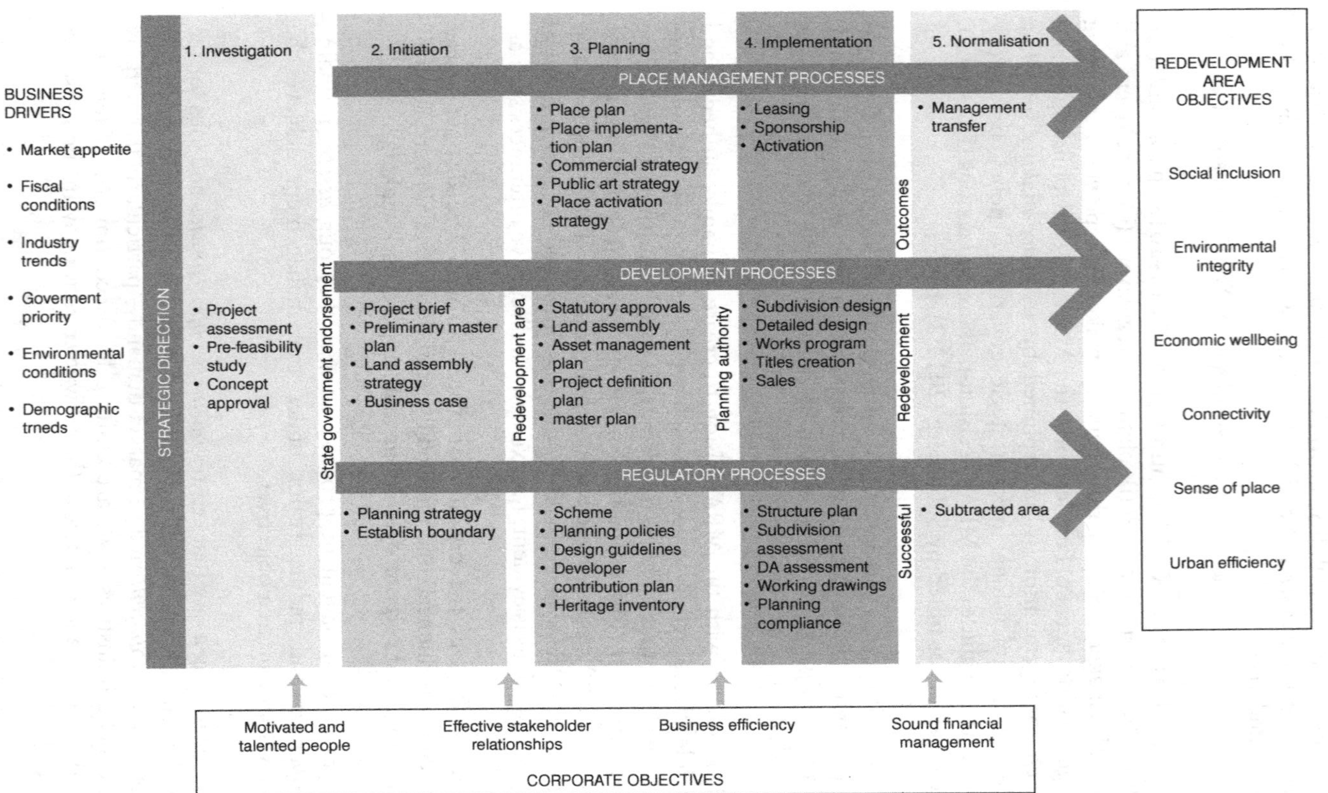

Figure 9.3 Metropolitan Redevelopment Authority's place-making model (source: https://cdn.mra.wa.gov.au/production/documents-media/documents/mra-corporate/file/place-making-model.pdf)

(8 projects), (iii) Midland (1 project), (iv) Scarborough (1 project), and, (v) Subiaco (1 project). Table 9.1 provides an overview of the aspects of each project. It can be seen that the state government will provide over $5.2bn of funding across all projects and this is anticipated to generate almost $16.5bn from the private sector. This equates to a public–private funding ratio of 1:3.1. A number of projects are envisaged to exceed this ratio with Midland, for example, expected to leverage $9.50 in private investment for every $1.00 of public funding. Other highly leveraged projects include: (i) Claisebrook (1:5.39), and (ii) Elizabeth Quay, New Northbridge and Subiaco which are all expected to have a public-private funding ratio of around 1:5.

Despite the broad neo-liberalist policy agenda of the Liberal/National coalition government, the level of public expenditure on urban regeneration projects across Perth demonstrates a highly interventionist policy approach. It can be seen that the level of government funding on a per-hectare basis for each regeneration project is generally high but also quite variable. For example, the Perth City Link project has the highest level of public expenditure at just under $103m per hectare. When projected private-sector funding is taken into consideration then total expenditure on a per hectare basis will be $399m. The high costs of this project are due to a mix of factors that include: (i) complex land ownership arrangements, (ii) land values, (iii) assembly and preparation works, (iv) construction and engineering works relating to the sinking of 1.3 km of the Perth to Fremantle rail line, and a new underground bus terminal.

Whilst the Perth City Link project is at an advanced stage of development, doubts have recently been expressed about its development focus and when the project will be completed. News reports have indicated that there is an oversupply of commercial office space, retail, and high-rise residential developments within the Perth CBD. Furthermore, the softening in the WA economy means that demand for commercial and high-rise residential developments has become somewhat subdued thus exposing developers to higher risks. Consequently, the City Link project witnessed the withdrawal of an international developer, and a partnership between two Australian developers fell through in late 2015. One of these developers continued with its plan to develop commercial offices however they too eventually walked away from the project in mid-2016 (Laschon and Kagi 2016):

> The State Government has rejected claims its Perth City Link project is facing big delays that will hit the Government's already stressed budget bottom line. A Thai property developer has abandoned its plans for a hotel and apartment complex near the Perth Arena and Leighton Properties has backed out of its partnership with Mirvac to develop eight sites in City Link project. The Opposition said it was a clear indication the project was struggling to attract the anticipated level of private investment.
>
> (O'Connor 2015)

Table 9.1 Key facts on redevelopment authority projects, Perth metropolitan region

Redevelopment area/project Site	*Key outputs/outcomes* Expected completion year	*% project normalised*	*Project area (ha)*	*No. new dwellings*	*Additional residential population*	*New floorspace (office/retail/industrial/health)*	*No. new workers*
ARMADALE							
Armadale	2020+	N/A	465	1,150	3,890	2,202,400	20,000
Wungong Urban	2020+	N/A	1,580	18,000	45,000	N/A	N/A
Sub-total A			2,045	19,150	48,890	2,202,400	20,000
CENTRAL PERTH							
Claisebrook Village	N/A	85	138	1,450	2,500	130,000	6,000
Chinatown	N/A	N/A	N/A	N/A	N/A	N/A	N/A
East Perth Power Station	2020+	N/A	9	1,400	2,400	336,000	1,100
Elizabeth Quay	2020+	N/A	10	800	1,400	225,000	N/A
New Northbridge	N/A	70	27	460	1,250	70,000	3,500
Perth City Link	2020+	N/A	14	1,650	3,000	244,000	13,500
Perth Cultural Centre	N/A	N/A	N/A	N/A	N/A	N/A	N/A
Riverside	2020+	N/A	40	4,000	7,000	94,000	6,000
Sub-total B			237	9,760	17,550	1,099,000	30,100
MIDLAND							
Midland	2025+	N/A	161	7,000	14,000	116,000	18,300
SCARBOROUGH							
Scarborough	N/A	N/A	100	2,828	N/A	26,000	N/A
SUBIACO							
Subiaco	N/A	85	85	1,975	3,600	287,000	6,800
TOTAL			2,627	40,713	84,040	3,730,400	75,200
AVERAGE		80	239	3,701	8,404	373,040	9,400

Source: MRA (2015)

Expenditure					
Public funding ($m)	*Private funding ($m)*	*Total funding ($m)*	*Public: private funding ratio*	*Public expenditure per ha ($m)*	*Total expediture per ha ($m)*
165	477	642	2.89	0.35	1.38
265	750	1,015	2.83	0.17	0.64
430	1,227	1,657	2.85	0.21	0.81
127	685	812	5.39	0.92	5.91
N/A	N/A	N/A	N/A	-	-
100	345	445	3.45	11.76	52.35
440	2,200	2,640	5.00	44.00	264.00
60	300	360	5.00	2.22	13.33
1,390	4,000	5,390	2.88	102.96	399.26
508	N/A	508	N/A	-	-
1,428	2,000	3,428	1.40	35.70	85.70
4,053	9,530	13,583	2.35	17.14	57.43
490	4,700	5,190	9.59	3.04	32.24
75	N/A	75	N/A	0.75	0.75
200	1,000	1,200	5.00	2.37	14.20
5,248	16,457	21,705	3.14	2.00	8.26
437	1,646	1,809	3.76	1.83	1.10

> The West Australian Government has signalled a major change in plans to develop the prime inner-city site, foreshadowing a move towards retail and apartment buildings. The move away from office developments at the project comes after the Metropolitan Redevelopment Authority's agreement with developer Mirvac fell through amid tough market conditions. [...] But analyst Gavin Hegney said now was not the right time to bring more retail space and apartments onto the market.
>
> (ABC News, 2016)

Perth foreshore: 'The river. The city. Together again'

The Perth foreshore, or Elizabeth Quay as it is officially known, is arguably the MRA's most significant project within its overall regeneration portfolio. This ambitious project seeks to connect the city and the river to forge a more symbiotic relationship; it also forms part of efforts to link the CBD with Northbridge via the aforementioned City Link project. The Elizabeth Quay site has been a highly contentious project since preparation works on the site began in 2013 with concerted opposition to the project from a small but vocal and articulate group of concerned citizens (Cocks 2016). Notably, there has been a longstanding desire to regenerate the Perth waterfront with over 200 proposals for the site since the early 1990s (Bolleter 2014). However, the failure to take any action until recently is rooted in a deep-seated conservative attitude about the role and meaning of the foreshore area, historically known as the Esplanade and now Elizabeth Quay. As Bolleter (2014: 570) astutely highlights, the site had been seen 'symbolically as the city's "frontyard" [and] [l]ike suburban front yards Perth's foreshore has been typically about ornamental display rather than actual use'.

A determined commitment to regenerate the foreshore area emerged in 2007. As part of an open design competition that sought to 'reimagine the foreshore as "an internationally significant waterfront destination"' the Australian firm Ashton Raggatt McDougall (ARM) tendered a provocative, by Perth standards at least, proposal – the Circle Scheme (Figure 9.4) – designed to bring Perth into the twenty-first century (Ambrose, 2016). There was a mixed reaction to the 'futuristic towers reminiscent of fantasy-like modern architecture' put forward by ARM. There was fairly broad public support for the proposal when it was first announced although there were concerns from two small but vocal and articulate community groups, City Vision and City Gatekeepers, that comprised various built-environment professionals, academics, business people and citizens. These groups were concerned that the proposals for the waterfront were too dramatic, overly commercial in their orientation and neglectful of the local heritage and social and environmental values embedded in the site as a Class A reserve. Ultimately, the Circle Scheme proved to be too much and was pejoratively dubbed 'Dubai on the Swan' within the media (Figure 9.4) (Bolleter 2014; Ambrose 2016).

The criticisms levelled at ARM's circular scheme at the time served to reinforce Perth's 'dullsville' image and signalled that Perth was struggling to come to terms with its emergence as a globalising city as a result of the mining and resources boom (Martinus and Tonts 2013; 2015; Martinus et al. 2015). Ultimately, the Circle Scheme was abandoned and replaced by a comparatively more modest and conservative project – 'the Rectangular Scheme' (Figure 9.5) – with the inlet and buildings assuming an 'orthogonal form and scale of "honest" post-industrial Australian waterfronts' and also 'subservient to the existing form of the city' (Bolleter 2014:583).

Elizabeth Quay opened in January 2016, however none of the high-rise buildings as depicted in Figure 9.5 have yet been built. Construction work has only recently commenced on The Towers, a development comprising 379 luxury apartments, and a Ritz-Carlton hotel that will have 204 luxury suites. The site is currently home to a small number of up-market restaurants/cafes and bars, a new river ferry terminal, a children's playground and a water

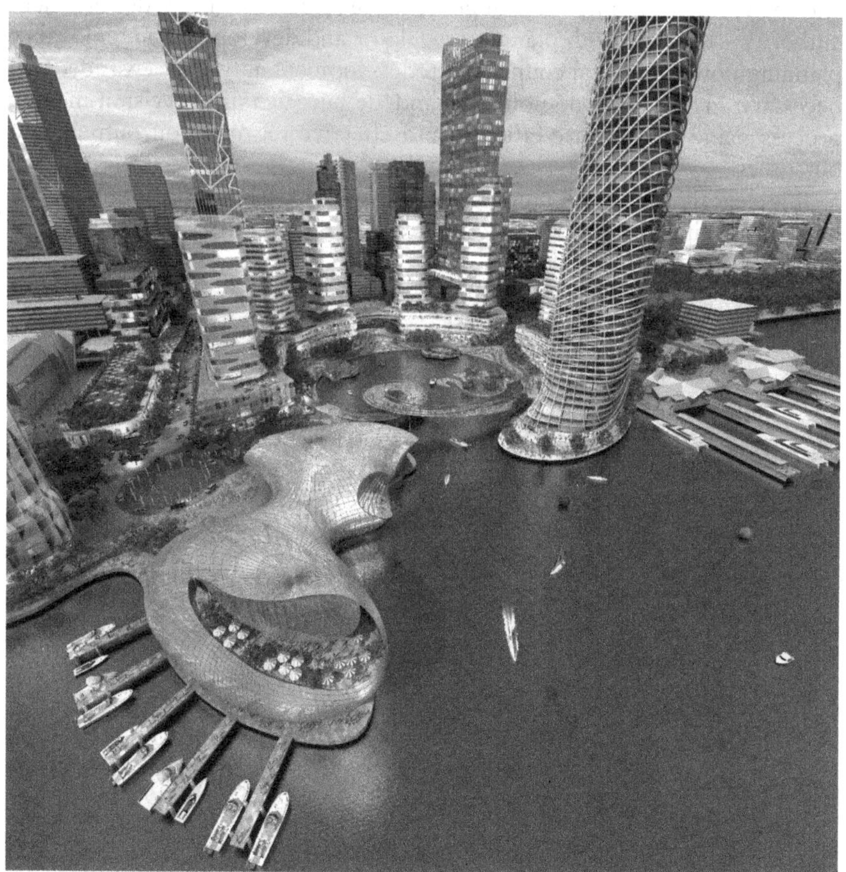

Figure 9.4 'Dubai on Swan' – Perth foreshore redevelopment proposal (c. 2007) (source: ARM Architects)

park – the latter was closed shortly after opening due to environmental health issues – various public open spaces and a number of public art installations. Given the relative lack of commercial, residential and retail developments within the site at present, the MRA has been busy asserting its place management and activation role via hosting and funding a series of arts, entertainment and cultural events on the site. These events have proved popular and played a fundamental role in attracting an estimated two million visitors to Elizabeth Quay during 2016.

Conclusions

Since the late 1980s/early 1990s, urban regeneration in Australia has been shaped by the adoption of neo-liberalist and economic rationalist policies at the federal and state government levels. Over this period, urban regeneration policy in WA has been characterised by a property-led redevelopment approach and the creation of a new urban governance structure – the redevelopment authority – that was both planning regulator and developer. The removal of planning powers of local councils to redevelopment authorities was deemed necessary in order to de-politicise and de-bureaucratise decision-making processes and thus ensure efficient and effective regeneration outputs and outcomes.

It is clear that the MRA (and its predecessors) have had a profound impact in transforming the physical, aesthetic, economic and social character of

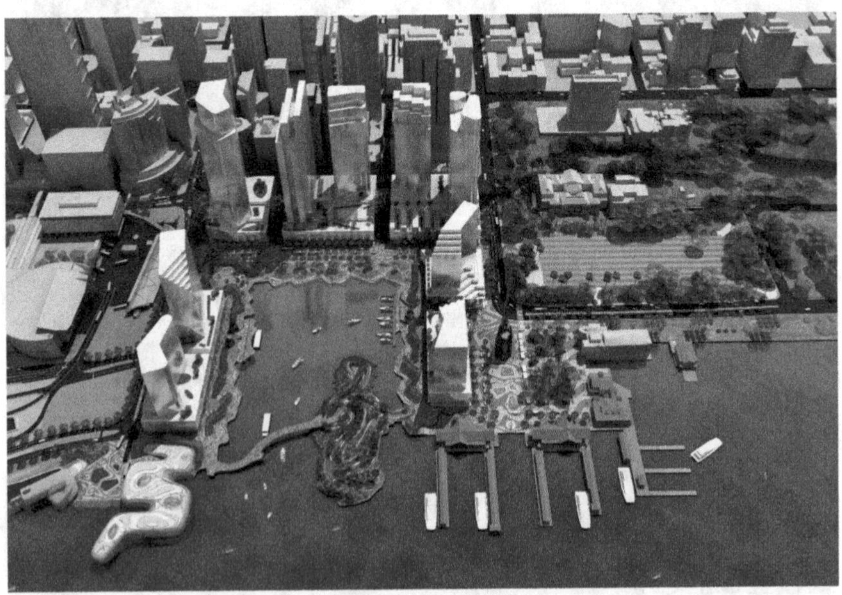

Figure 9.5 Elizabeth Quay – 'The Rectangular Scheme' (source: Metropolitan Redevelopment Authority)

the various redevelopment areas across Perth, especially within the CBD. It would also be fair to say that the redevelopment authority model approach has been instrumental in helping create hundreds, if not thousands, of jobs in construction, retailing, hospitality, and the service sector; enhancing public infrastructure (roads and rail) and the quality of the built environment and streetscapes; and increasing the volume of housing, residential densities and property values. That said, there has been no evaluation to determine just how much of the urban transformation in Perth is due to the redevelopment authorities and/or wider processes and institutions. Moreover, there has been no assessment of the social-equity impacts of the regeneration of East Perth, Subiaco, Midland, Armadale and other regeneration areas (Crawford 2003; Byrne and Houston 2005).

Despite this lack of evaluation the redevelopment authority approach to urban regeneration has endured since the early 1990s. Notably, the neo-liberal underpinnings of this approach to urban regeneration has enjoyed bi-partisan political support in WA. Despite the redevelopment authorities being agents of urban entrepreneurialism the state government has had to inject considerable sums of public expenditure on purchasing, assembling, remediating and servicing land and, in more recent times, organising, managing and funding major cultural events and activities, especially at Elizabeth Quay.

In conclusion, it will be interesting to see what the next evolution in urban regeneration policy will be given that the mining and resources boom in Western Australia is over (at least for the time being) and whether or not the Labor state government elected in 2017 will pursue a business as usual approach or stamp its own imprimatur on urban regeneration policy and governance. The early broad rhetorical policy signs suggest that a more consultative approach to urban regeneration may be on the horizon. In terms of governance, whilst the MRA has thus far escaped being amalgamated as part of the 'machinery of government review' (Government of Western Australia 2017a) there appears to be some policy/political discussions about amalgamating the MRA with LandCorp, the state government development agency. There may also be a spatial refocusing of urban regeneration with a shift in policy emphasis away from the inner-city to the outer suburbs. One clear indicator of this is the new state Labor government's proposed suburban rail network, Metronet, which secured funding from the Commonwealth government in 2017 (Government of Western Australia 2017b; WA Labor n.d.). Furthermore, many of those outer suburbanites affected by the downturn in the mining and resources economy voted for the Labor government. In other words, there seems to be a political imperative to direct policy resources towards the outer suburbs, especially if the economy worsens over the course of this term of government.

References

ABC News (2016) 'Perth City Link "not viable" given market downturn, apartment glut, property analyst says', http://www.abc.net.au/news/2016-06-15/timing-wrong-for-perth-city-link-given-apartment-glut/7512194?WT.ac=statenews_wa (accessed 30 June 2017).

Ambrose, M. (2016) 'Elizabeth Quay: How Perth's biggest project was a dream to change the face of the city', *Perth Now/Sunday Times*, http://www.perthnow.com.au/news/western-australia/elizabeth-quay-how-perths-biggest-project-was-a-dream-to-change-the-face-of-the-city/news-story/534fb67e6a5360dec68f10b04d60d874 (accessed 30 June 2017).

Australian National Audit Office (1996) *Building Better Cities, Department of Transport and Regional Development, Performance Audit*, Canberra: Australian National Audit Office, https://www.anao.gov.au/sites/g/files/net3416/f/ANAO_Report_1996-97_09.pdf (accessed 30 June 2017).

Badcock, B. (1993) 'The urban programme as an instrument of crisis management in Australia', *Urban Policy and Research*, 11(2): 72–80.

Ball, M. and Maginn, P. J. (2005) 'Urban change and conflict: Evaluating the role of partnerships, in urban regeneration in the UK', *Housing Studies*, 20(1): 9–28.

Bolleter, J. (2014) 'Charting a changing waterfront: A review of key schemes for Perth's foreshore', *Journal of Urban Design*, 19(5): 569–592.

Byrne, J. and Houston, D. (2005) 'Ghosts in the city: Redevelopment, race and urban memory in East Perth', in D. Cryle and J. Hillier (eds), *Consent and Consensus: Politics, Media and Governance in Twentieth Century Australia*, Perth: API Network.

Cocks, L (2016) 'Elizabeth Quay: planner says it will struggle to thrive without infrastructure' *WA Today*, 18 March, http://www.watoday.com.au/wa-news/elizabeth-quay-20160317-gnlm2l.html (accessed 15 January 2017).

Cook, I. R. (2008) 'Mobilising urban policies: The policy transfer of business improvement districts to England and Wales', *Urban Studies*, 45(4): 773–95.

Crawford, E. (2003) 'Equity and the city: the case of East Perth redevelopment', *Urban Policy and Research*, 21(1): 81–92.

Dodson, J. (2016) 'Suburbia in Australian urban policy', *Built Environment*, 42(1):23–36.

Dovey, K. (2004) *Fluid City: Transforming Melbourne's Urban Waterfront*, London: Routledge.

EPRA (East Perth Redevelopment Authority) (2003) *EPRA Annual Report 2003*, Perth: EPRA.

EPRA (East Perth Redevelopment Authority) (2011) *EPRA Annual Report 2011*, Perth: EPRA.

Ganis, M., Minnery, J. and Mateo-Babiano, D. (2014) 'The evolution of a masterplan: Brisbane's South Bank, 1991–2012', *Urban Policy and Research*, 32(14): 499–518.

Gleeson, B. and Low, N. (2000) *Australian Urban Planning: New Challenges, New Agendas*, Crows Nest (NSW): Allen & Unwin.

Government of Western Australia (1994) 'Authority to oversee redevelopment of Subiaco land', Ministerial Media Release, https://www.mediastatements.wa.gov.au/Pages/Court/1994/06/Authority-to-oversee-redevelopment-of-Subiaco-land.aspx (accessed 30 June 2017).

Government of Western Australia (2011a) *Recollections and Reflections: Subi Centro* Perth: Government of Western Australia.

Government of Western Australia (2011b) 'New redevelopment authority to transform Perth', Ministerial Media Release https://www.mediastatements.wa.gov.au/Pages/Barnett/2011/12/New-redevelopment-authority-to-transform-Perth.aspx (accessed 30 June 2017).

Government of Western Australia (2013) *Government plans for new-look Scarborough* Ministerial Media Release, https://www.mediastatements.wa.gov.au/Pages/Barnett/2013/09/Government-plans-for-new-look-Scarborough.aspx (accessed 30 June 2017).

Government of Western Australia (2017a) *Public Sector Renewal* https://www.mediastatements.wa.gov.au/MediaDocuments/Public%20sector%20renewal.pdf (accessed 30 June 2017).

Government of Western Australia (2017b) '$2.3 billion jobs and infrastructure boost for Western Australia' Joint media statement https://www.mediastatements.wa.gov.au/Pages/McGowan/2017/05/Joint-media-statement-2-point-3-billion-dollars-jobs-and-infrastructure-boost-for-Western-Australia.aspx (accessed 30 June 2017).

Gregory, J. (2008) 'Obliterating history?, *Australian Historical Studies*, 39(1): 91–106.

Hall, T. and Hubbard, P. (1996) 'The entrepreneurial city: New urban politics, new urban geographies?' *Progress in Human Geography*, 20(2): 153–174.

Harvey, D. (1989) 'From managerialism to entrepreneurialism: The transformation in urban governance in late capitalism', *Geografiska Annaler. Series B, Human Geography*, 71(1): 3–17.

Hastings, A. (1996) 'Unravelling the process of "partnership" in urban regeneration policy'. *Urban Studies*, 33(2): 253–268.

Imrie, R. and Raco, M. (2003) *Urban renaissance? New Labour, Community and Urban Policy*, Bristol: Policy Press.

Laschon, E. and Kagi, J. (2016) 'Perth City Link: Shift to shops, apartments mooted by state government', *ABC News*, http://www.abc.net.au/news/2016-06-14/perth-city-link-shift-signalled-by-wa-government/7511008 (accessed 30 June 2017).

Leary, M.E. and McCarthy, J. (2013) *The Routledge Companion to Urban Regeneration*, London: Routledge.

Lewis, R. (1994) 'Authority to oversee redevelopment of Subiaco land', Media Release: 7 June, https://www.mediastatements.wa.gov.au/Pages/Court/1994/06/Authority-to-oversee-redevelopment-of-Subiaco-land.aspx (accessed 2 October 2017).

Maginn, P. J. (2004) *Urban Regeneration, Community Power and the (In)Significance of 'Race'*, Aldershot: Ashgate.

Maginn, P. and Foley, N. (2014) 'From a centralised to a "diffused centralised" planning system', *Australian Planner*, 51(2): 151–162.

Martinus, K. and Tonts, M. (2013) *Perth as a Globally Connected City*. FACTBase Special Report. Perth: Committee for Perth and The University of Western Australia.

Martinus, K. and Tonts, M. (2015) 'Powering the world city system: energy industry networks and inter-urban connectivity', *Environment and Planning A*, 47(7): 1502–1520.

Martinus, K., Sigler, T., Searle, G. and Tonts, M. (2015) 'Strategic globalizing centres and sub-network geometries: a social network analysis of multi-scalar energy networks', *Geoforum*, 64: 78–89.

McCann, E. and Ward, K. (2013) 'A multi-disciplinary approach to policy transfer research: geographies, assemblages, mobilities and mutations', *Policy Studies*, 34(1): 2–18.

MRA (Metropolitan Redevelopment Authority) (2012) *MRA Annual Report 2011/12*, Perth: MRA.

MRA (Metropolitan Redevelopment Authority) (2014) *Strategic Plan 2014–2018*, Perth: Metropolitan Redevelopment Authority.

MRA (Metropolitan Redevelopment Authority) (2015) MRA Annual Report 2014–15, Perth: MRA.

MRA (Metropolitan Redevelopment Authority) (n.d.) 'Scarborough: An MRA Project', http://cdn.mra.wa.gov.au/production/63844f941eec92126d2cb9b981aab146/scarborough-fact-sheet.pdf (accessed 30 June 2017).

Neilson, L. (2008) ''The "Building Better Cities" program 1991–96: a nation-building initiative of the Commonwealth Government', in J. Butcher (ed), *Australia under Construction : Nation-building – Past, Present and Future*, Canberra: ANU.

Oakley, S. (2009) 'Governing urban waterfront renewal: The politics, opportunities and challenges for the inner harbour of Port Adelaide, Australia', *Australian Geographer*, 40(3): 297–317.

O'Connor, A. (2015) 'WA Government rejects concerns over City Link progress', *ABC News*, http://www.abc.net.au/news/2015-12-29/government-rejects-concerns-over-city-link-progress/7058852 (Accessed 30 June 2017).

Orchard, L. (1999a) 'Shifting visions in national urban and regional policy 1', *Australian Planner*, 36(1): 20–25.

Orchard, L. (1999b) 'Shifting visions in national urban and regional policy 2', *Australian Planner*, 36(4): 200–209.

Peck, J. (2012) 'Austerity urbanism', *City*, 16(6): 626–655.

Pugalis, L. and McGuinness, D. (2013) 'From a framework to a toolkit: Urban regeneration in an age of austerity', *Journal of Urban Regeneration and Renewal*, 6(4): 339–353.

REIWA (2017) 'East Perth suburb profile', http://reiwa.com.au/wa/east-perth/6004/ (accessed 30 June 2017).

Roberts, P. (2000) 'The evolution, definition and purpose of urban regeneration' in P. Roberts and H. Sykes (eds), *Urban Regeneration: A Handbook*, London: Sage.

Sandercock, L. (1977) *Cities for Sale*, Melbourne: Melbourne University Press.

Short, J. R., Benton, L., Luce, D. and Walton, J. (1993) 'The reconstruction of the image of a postindustrial city', *Annals of the Association of American Geographers*, 83(2): 207–24.

Silberberg, S. (2013) *Places in the Making: How Place-making Builds Places and Communities,* Cambridge., MA: MIT. https://dusp.mit.edu/sites/dusp.mit.edu/files/attachments/project/mit-dusp-places-in-the-making.pdf (accessed 30 June 2017).

Steele, W. and Legacy, C. (2017) 'Critical urban infrastructure – Editorial', *Urban Policy and Research*, . 35(1): 1–6.

SRA (Subiaco Redevelopment Authority) (1995) *Outlook Newsletter*, Perth: SRA.

SRA (Subiaco Redevelopment Authority) (2011) *SRA Annual Report*, Perth: SRA.

SRA (Subiaco Redevelopment Authority) (n.d) *Subiaco Concept Plan*, Perth: SRA.
Tallon, A. (2013) *Urban Regeneration in the UK* (2nd edn), London: Routledge.
WA Labor (n.d) 'Metronet: Connecting Perth's suburbs' https://www.markmcgowan.com.au/files/HeavyRailVision-small.pdf (accessed 2 October 2017).
WAPC (Western Australia Planning Commission) (2010) *Directions 2031 and Beyond: Metropolitan Planning Beyond the Horizon*, Perth: WAPC/Dept. of Planning.

10 Pushing the boundaries of sustainable development

The case of Central Park, Sydney

Stuart White, Andrea Turner and Justine Saint Hilaire

Introduction

Central Park, also known as Frasers Broadway, is an iconic precinct development in Sydney, Australia. It forms a new gateway to Sydney's central business district (CBD) and demonstrates how sustainability can be incorporated into the regeneration of our cities, in this case through the revitalisation of a former urban industrial site (Figure 10.1). Its signature green walls provide an immediate statement to residents, workers and visitors that the site aims to embody the principles of sustainability, and go further, towards urban regeneration. This $2 billion staged development was conceived in 2006 and has won a number of awards and accolades (Central Park 2017a). It opened at the end of 2013 and will be completed by the end of 2018.

Now, over a decade after the development began, this chapter has been written by a team at the Institute for Sustainable Futures (ISF), University of Technology Sydney (UTS), led by Stuart White, one of the main authors of the original sustainability strategy for Central Park. With supplementary interviews and research, the chapter provides details of both the vision and reality of this real-world case study and what is possible in urban regeneration. The chapter provides an overview of the development and describes how it progressed from being an abandoned industrial site to becoming Central Park. It highlights some of the difficulties faced, summarises the original visionary sustainability strategy, details two key sustainability components – energy and water management, and discusses some of the key lessons learnt along the way.

Overview

Sydney is growing rapidly. Its population in the mid-2000s was just over four million. Now, a decade later, it has reached over five million (ABS 2017) and is expected to hit eight million by by the middle of the century (Greater Sydney Commission 2017). The NSW government's current strategy to deal with population growth, due to a combination of domestic population growth

Figure 10.1 Artist's impression of Central Park (source: Frasers Property Australia and Sekisui House Australia)

and increased immigration, focuses on urban densification. In the Australian context, this means a greater reliance on high-rise buildings at transport hubs along key transport routes (Greater Sydney Commission 2017, Chapter 2).

Planning for Central Park commenced in the mid-2000s. At that time, the current Greater Sydney Commission, the independent organisation funded by the NSW government to co-ordinate and align planning for development, transport and housing across Greater Sydney, had not yet been created, nor had the associated 20-year plan, 40-year vision and related district plans (Greater Sydney Commission 2017). Yet a decade after the initial concept plans, Central Park provides a demonstration of what new urban densification can look like if sustainability and regeneration principles are taken into consideration. Central Park has helped pave the way for developments in Sydney such as Barangaroo (Barangaroo Delivery Authority 2017), and in Australia more broadly. It is considered an exemplar of various aspects of sustainable urban regeneration, alongside other urban renewal projects such as: Hammarby in Stockholm, Sweden (Gaffney et al. 2007); Sonoma Mountain Village, near San Francisco, CA, USA (BioRegional 2014a); Grow Community, Winslow, WA, USA (BioRegional 2014b); and high-profile cities representing innovation and sustainability including Masdar in the UAE (McGee et al. 2014).

The 5.8 hectare precinct is built on the former Carlton United Brewery (CUB) site on the southern edge of Sydney's CBD and directly adjacent to Ultimo, an area with the highest population density in Australia, some 15,100 people per square km (ABS 2016).

The CUB site offered many opportunities to developers, thanks to its generous size, its close proximity to the CBD and to Sydney's Central Station (Australia's largest transport hub), and its rich historic industrial character. The international property developer Frasers Property bought the site in 2007 and decided to transform it into an innovative development with a mix of residential and commercial buildings, heritage sites, green open spaces and public art. Frasers also decided to make Central Park a flagship development for sustainability.

Frasers engaged international architectural firms, including Foster + Partners and Ateliers Jean Nouvel, which worked with Australian architects Richard Johnson and Alec Tzannes to develop iconic buildings across the site. The best example is the main retail and residential building, One Central Park, which includes an enormous cantilevered structure adorned with a light-reflecting heliostat and green walls consisting of thousands of plants draped down the glass facade.

A team comprising Elton Consulting (Elton) and the Institute for Sustainable Futures (ISF) at the University of Technology Sydney (UTS), whose main campus is adjacent to the development, were engaged to design the sustainability strategy for the redevelopment. A comprehensive strategy (Central Park 2013), aiming to move beyond best practice, was developed based on economic, social and environmental principles. The role of the team was to be visionary and to challenge preconceptions, so that Central Park would set an example for other developments. Even though the whole vision could not be fully realised, it remains an example of the incorporation of a broad range of innovations in the regeneration of a development site, with each individual building achieving a five-star Green Star rating (Chua 2013b).

The whole site will be completed by 2018. The iconic One Central Park opened in late 2013 and was declared the 'best tall building in the world in 2014' by the Council of Tall Buildings and Urban Habitat (CTBUH) (CTBUH 2014a). This award has previously been won by buildings such as the Shanghai World Financial Centre in 2008 (Katz, Robertson and See 2011) and the Burj Khalifa, Dubai, UAE, in 2010 (CTBUH 2014b). Antony Wood, Executive Director of CTBUH and juror of the awards, remarked:

> Seeing this project for the first time stopped me dead. There have been major advances in the incorporation of greenery in high-rise buildings over the past few years – but nothing on the scale of this building has been attempted or achieved. One Central Park strongly points the way forward, not only for an essential naturalisation of our built environment, but for a new aesthetic for our cities – an aesthetic entirely appropriate to the environmental challenges of our age.
>
> (Frasers and Sekisui House 2014)

Table 10.1 provides general facts about the project.

Table 10.1 General facts about Central Park

Value – $2 billion (when finished)
Area – 5.8 hectares
11 new buildings
33 heritage items including the No.1 Gate Portal and Chimney Stack from the former CUB site (Office of Environment and Heritage 2017)
A public art collection valued at $8 million
2,000 parking spaces
38% site coverage
A 6,400 m2 public park
Approximately 2,000 apartments and 869 student dwellings
Approximately 5,300 residents and 1,750 workers
25,000 m2 in commercial space
20,000 m2 in retail space
A minimum 5-star Green Star rating for each building on the site
94% recycling or reuse of materials removed from site during demolition
1,200 m2 of green wall consisting of 35,000 plants from 350 different species.

Source: Authors

Background

Creating Central Park was not easy. It involved long negotiations between stakeholders, and during the initial stages between the developers and the adjacent community. Table 10.2 provides a timeline of some of the key events.

From Carlton United Brewery to Frasers Broadway

In 2005, the owner of the CUB site, Fosters Group, closed its brewery which for over 150 years had dominated the suburb of Chippendale. Some historic buildings, such as the original power station and chimney stack, were identified for preservation and they are now heritage-listed items and thus protected (Office of Environment and Heritage 2017). The City of Sydney, the suburb's local council, was responsible for the planning approvals for the development and had a keen interest in preserving the local heritage, and in the outcomes of the redevelopment, as it had the potential to dramatically change the character of the southern section of the city adjacent to the CBD. However, the site soon became controversial.

The site is part of Chippendale, a small suburb of just over 4,000 people. Chippendale has a long cultural history and strong identity. It consists mainly of small terraced buildings, with limited open space and small warehouses and offices. Due to its historical significance it became a heritage conservation

Table 10.2 Timeline of key events

2005	Closure of the CUB brewery
July 2006	Site listed as 'state-significant'
February 2007	Approval of the first concept plan for the site
April 2007	Land and Environment Court case launched by Drake-Brockman against Fosters and the NSW Minister for Planning
June 2007	Purchase of the CUB site by Frasers for $208 million
August 2007	Elton Consulting and the Institute for Sustainable Futures engaged to develop the sustainability strategy
April 2008	Demolition work starts
May 2008	Presentation of the amended concept plan
February 2009	Approval of the concept plan
October 2013	One Central Park opens
September 2014	Central Thermal and Electrical Plant (CTEP) starts operating
July 2015	CTEP completed
December 2015	CTEP sold to Brookfield Energy
2015	Water recycling plant commences operation
2018	Planned date for recycled water and energy connection to UTS
End 2018	Site development planned to be completed.

Source: Authors

area (HCA) (Office of Environment and Heritage n.d.) through processes instigated by the City of Sydney. Being an HCA provides protection of the character of an area and imposes limitations to the development of buildings within it. Many Chippendale residents were concerned that a large high-rise development on the former CUB site would block their views, increase traffic and change the character of their suburb. Most of the discussion between the local community and the City of Sydney focused on the floor space ratio (FSR) of the proposed development. A compromise between economic value and community wellbeing was needed.

Fosters and the City of Sydney were about to sign a voluntary planning agreement with a mutually acceptable FSR when the state government imposed a levy to raise funds from the development for affordable housing in the area. The only way Fosters could have complied with the levy and made sufficient profit from the development would have been to increase the density (and thus the height) of the proposed buildings. Due in part to the size and mixed-use nature of the site, its central city location and its economic impact, the site was deemed 'state significant' by the NSW Government in June 2006. This meant that the planning approvals process was taken out of the hands of the City of Sydney and passed to the NSW Government, with the minister for planning, Frank Sartor, becoming the approval authority (see Chapter 3). For 'state significant' developments, the height limit and the FSR are far less constrained (Farrelly 2006).

Minister Sartor approved a concept plan developed by Fosters in February 2007. The plan allowed multiple high-rise towers, which would have cast a significant shadow over the site. In addition, the local community considered the proposed green space to be inadequate. The approval therefore led to a case in the NSW Land and Environment Court (*Drake-Brockman* v. *Minister for Planning and Anor* 2007). According to the plaintiff, the approval did not adequately consider the environmental sustainability of the proposed development. The judge ruled in favour of the minister in August 2007. It was in June 2007, in the middle of these proceedings, that Frasers Property Australia, a subsidiary of an international development company, purchased the site from Fosters (Draper 2007).

From Frasers Broadway to Central Park: a new vision: design and sustainability excellence

Frasers recognised community concerns and acknowledged the need to change the redevelopment. Frasers started to plan its own concept for the site, which it renamed 'Frasers Broadway'. Frasers' vision was to develop an urban precinct that was innovative in its architecture, landscape and design (Watpac 2015).

Frasers acknowledged that this vision could not become a reality unless engagement with the local community and other stakeholders improved. Frasers appointed Elton Consulting, experts in community engagement and conflict resolution, to lead the consultation process. ISF, part of the UTS, whose main campus is adjacent to the CUB site, was contracted to design the sustainability strategy. In November 2007, Frasers invited internationally recognised architectural firms Foster + Partners, Ateliers Jean Nouvel, Richard Johnson and Tzannes to collaborate in the planning and design of the site. Together, the teams created a substantially modified concept plan that was approved by the NSW Department of Planning in April 2008 (Frasers and Sekisui House 2009).

The demolition and site preparation started in late 2008, and were completed a year later. Construction began in 2009 but was halted due to the global financial crisis (GFC). The GFC made it difficult to attract tenants for the commercial office space, and Frasers decided to prioritise retail, residential and student housing instead. The viability of student housing benefited from the growing student population, with over 100,000 students in the local area.

While the GFC impacted on the timing and priorities of the development, the commitment to sustainability was retained and augmented through a joint venture. In July 2011, Frasers announced a joint venture with the Japanese construction company Sekisui House to develop most of the high-profile mixed-use development site. Frasers' parent company in Singapore had worked with Sekisui House on several other joint ventures in the past incorporating key sustainability principles (Frasers and Sekisui House 2011).

A precinct is born

When it is completed, Central Park will be an $2 billion mixed-use development of 11 buildings, with nearly 50,000 m^2 of commercial, hospitality and retail space, and 6,400 m^2 of green space – the 'Chippendale Green' public park.

The most architecturally significant feature of the development, One Central Park, opened in October 2013. It comprises two residential towers linked by a shopping centre. The east tower is the tallest point of the site at 117 metres. The enormous cantilevered structure extending from this tower is the French architect Jean Nouvel's signature. The cantilever features a light-reflecting heliostat on its lower surface. Its 320 mirror panels redirect the sunlight from the top of the lower tower onto the retail atrium below during the day. At night, hundreds of colourful LED lights installed on the reflectors light up to display a representation of the nearby Darling Harbour through the seasons. This artwork was created by the French artist Yann Kersale (Chua 2013a). The green walls that surround the building's facade were designed by Patrick Blanc and provide an immediate visual clue to the sustainable character of the precinct. These features contributed to One Central Park being declared the 'best tall building in the world 2014' (Frasers and Sekisui House 2014).

The sustainability strategy

The original sustainability strategy for Central Park was developed by ISF in 2007. The aim was to develop an innovative project that would demonstrate what can be achieved in sustainable infrastructure, and where possible, the strategy takes account of economic, social and environmental dimensions. The design process included significant engagement with the local community, organised by Elton Consulting, through workshops and open days. Communication from the developers to the community was also maintained through information posted on the development's website during the planning and construction period. Through the process implemented by Elton and ISF, Frasers aimed to develop a sustainability framework that it could replicate in subsequent projects. The sustainability strategy went beyond environmental parameters to include social and cultural dimensions (Wang et al. 2015).

The goals of the sustainability strategy included creating a flagship for commercial success and sustainability in urban redevelopment at the precinct scale. The development needed to incorporate design quality and a healthy social and environmental landscape for residents, visitors and the local community. It also aimed to differentiate itself from competing projects through its forward-looking approach to sustainability and community engagement, and its use of cutting-edge technology.

In the strategy, key initiatives were initially identified in order to meet these goals and go beyond baseline compliance requirements in an effort to meet more ambitious targets (Critchley and White 2008). These initiatives included:

- zero net carbon output and zero net potable water consumption;
- the equivalent of a six-star Green Star rating for the whole precinct;
- a gas tri-generation system for power, heating and cooling;
- an integrated water-recycling scheme to capture and use roof water and reuse treated black water for non-potable uses;
- the removal of a significant component of surface traffic from the precinct, prioritising bicycle and pedestrian movements and services in street and building design;
- developing a series of social and economic sustainability initiatives;
- partnering with the UTS to pioneer a long-term longitudinal sustainability monitoring and evaluation program.

However, as is often the case with sites that aim to introduce sustainable infrastructure, some of the original aims were not met. A six-star Green Star rating for the entire site was not achieved, due in part to the challenges associated with the institutional arrangements for selling electricity from the tri-generation plant back into the grid, a barrier that is not unique to this development (Dunstan and Langham 2010). As an alternative Frasers decided that each building would aim for a five-star Green Star rating in both design and built form (Chua 2013b). A number of initiatives were considered and not actioned, principally due to the novelty of some of the proposals, including the idea for combined collection of organic waste and biosolids from sewage with anaerobic digestion of the combined waste onsite. All of these technologies have been applied in various contexts, but rarely, if at all, together, and so at the time it was not possible to implement them.

Greenhouse gas emissions

The following measures were implemented to meet the aim of a zero net carbon emissions.

Building efficiency

It was estimated that energy efficiency measures alone could achieve up to fifty per cent savings in energy demand. Access to natural sunlight in buildings, and natural ventilation, were maximised. For thermal efficiency, special attention was paid to insulation and glazing, and many systems were installed to ensure electrical efficiency. Many apartments have air-conditioning systems which prevent residents from turning on the air-conditioning simultaneously in the living room and bedroom. The lighting systems are highly efficient, and all lighting in common areas and the car park is controlled by timers or sensors.

Efficient appliances

Installing efficient appliances in residential buildings minimises energy consumption. All apartments have high star rated energy- and water-using appliances including fridges, washing machines, cooktops, ovens and dryers (Energy rating n.d.; WELS n.d.).

Tri-generation plant

The development's tri-generation produces electricity as well as hot and chilled water. Waste heat from the cooling system for the natural gas-powered generator and from flue gases is used to heat water, some of which is then converted to chilled water through an absorption chiller. Greenhouse gas emissions are less than fifty per cent of the emissions that would be produced to create the same energy using coal-fired electricity. The City of Sydney has investigated the feasibility of producing renewable gas to supply a network of such tri-generation plants as part of its Sustainable 2030 Strategy (Leow 2013).

Car-share facilities, active transport and low-emission transport

Car-share facilities have been created for Central Park and neighbouring residents, and bicycle storage and amenities are provided for commercial workers. Central Park is also fortunate to be located within 500 metres of Australia's largest public transport hub, Sydney's Central Station. In addition, the car parking areas have been separated from the ownership structure of the premises, which enables their subsequent adaptive use for other purposes and reduces their 'institutional lock-in' as car parking spaces (Chancellor 2014).

Water-cycle management

The following measures were implemented to assist in achieving the goal of zero net mains water use.

Water efficiency

Efficiency improvements alone can provide a reduction of over fifty per cent in mains potable water demand (Turner and White 2017). Hence, attention was focused on installing best-practice water-efficient cooling towers, plumbing fixtures, appliances, irrigation systems, infra-red tap controllers and shower heads (Flow Systems 2013).

Using different water sources

The planning process considered all four sources of water available to the site: mains water from the local public utility (Sydney Water); rainwater that falls on the roofs of the buildings; onsite stormwater; and wastewater generated onsite. It was decided that all the sources could be utilised and treated using the onsite water-recycling facility. The site is therefore supplied with high-quality recycled water, further reducing the amount of potable water used. The recycled water is used for cooling towers, toilets, irrigation (including irrigation of the green walls) and laundry cold-water use. The rainwater is used for hot-water production, with heating provided by the central thermal and electrical plant.

Exporting treated effluent

The use of recycled water more than offsets any mains water used for the development, and will enable the site to become water positive, a significant and unusual feature in such a development. The onsite water recycling plant has the capacity to export treated water to neighbouring sites and, beginning in 2018, recycled water will be provided to the UTS across the road from the precinct (Sustainability Matters 2014).

General sustainability initiatives

Green roofs and walls

The green walls of One Central Park were not part of the initial sustainability strategy. However, they are now the most visible sustainability feature of the precinct. The green walls and roofs have many benefits, including good thermal insulation (Downton 2013). They reduce heating and cooling loads, as they also reduce the impact of the urban heat island effect and improve air quality in the local area (Irga et al. 2015).

The green walls were designed by the French botanist Patrick Blanc. The aim was to increase the open park space by creating gardens where space was available, on the building's facade. The plants are grown hydroponically, using recycled water. The vertical gardens cover a total area of 1,200 m^2 using 35,000 plants, and 350 different species (Blanc 2013).

Waste

Over 94 per cent of the demolition waste, including industrial materials from the former brewery, was recycled or reused. These rates are high, even by European standards (European Commission 2011). At the time, this resulted in Central Park being one of the largest recycling projects in Australia (Central Park 2013).

Building materials

Frasers used environmentally preferred materials and products for the building. These included low volatile organic compound (VOC) paints, green star-compliant concrete which uses fly ash, green star-compliant plasterboard and green star-compliant structural steel and reinforcement. In addition the project minimised the use of PVC and formaldehyde (Watpac 2015).

Community sustainability

Sustainability is not just about the latest green technologies; it is also about community wellbeing, health and liveability. The sustainability strategy also focused on public space, parkland, architecture, public art and the cultural dimensions of Chippendale. The vision of Central Park was of a vibrant place day and night, with active streetscapes, urban laneways and outdoor cafes. The best illustration of this is the redevelopment of Kensington Street on the eastern side of the precinct. Its heritage terrace houses and warehouses were transformed into bars, cafes, restaurants and galleries.

Central Park also supports the creative industries. Frasers implemented this creative aspect from the commencement of the construction work by establishing Fraser Studios in 2008. For nearly four years during the construction, three vacant warehouses on Kensington Street were made available on application, to local artists and performers. In addition, an $8 million public art collection is displayed throughout the site. Among the most famous are the 'Halo' wind-activated kinetic sculpture in the middle of the park, and the colourful and changing LED display on One Central Park's enormous cantilevered structure.

Energy: central thermal and electrical plant[1]

The onsite tri-generation power plant, which provides low-carbon thermal energy and electricity, comprises two 1.1 MW reciprocating engines. The plant uses waste heat from the engines and from exhaust gases to heat water, with some of that hot water used to produce chilled water in an absorption chiller, making the whole plant highly efficient. Even though it uses a fossil fuel, the plant emits significantly less greenhouse gas than it would if it used coal-fired grid electricity. In future, biogas derived from municipal solid waste, could replace the natural gas which now powers the plant (CoS 2014).

The tri-generation central thermal and electrical plant (CTEP) is located in a basement beneath the old brewery yard building. It provides all the thermal energy needs of the site, including the refrigeration units of the supermarket in One Central Park. It also supplies electricity to the central thermal plant and to selected buildings. Such plants are relatively common, but tri-generation has rarely been implemented at this scale for a commercial development. Central Park is the first development with a precinct-wide central thermal plant with a tri-generation facility for multiple buildings with multiple owners. This $80 million facility has been operating since September 2014, and was awarded the 'best co-generation or district energy project 2014' by the Australian Energy Efficiency Council (Kelly 2013).

Institutional arrangements

Frasers decided to produce its own power onsite, and, partly due to the requirements for obtaining a six-star Green Star rating for the whole precinct at that time, decided to use tri-generation. ISF proposed a tri-generation plant with a 10 MW capacity, enough to supply electricity to all the buildings on the site and enable the development to be independent of the grid, thereby avoiding expensive connection fees.

However, regulatory requirements made it compulsory for Frasers to be connected to the grid so that the retailer could offer a 'choice' of electricity supplier to tenants. Frasers therefore had to rethink its plans. Meanwhile the city electricity network had been augmented and this undermined the commercial case for exporting power. Hence the capacity of the tri-generation plant was reduced to 2.2 MW, with only the commercial buildings closest to the CTEP supplied with power from the plant. At first, Frasers just wanted to excavate a basement and find a private owner to build and run the plant. However, it did not find a company willing to take the risk on such a project, mainly because at that time it was difficult to assess the future loads of the buildings and to size the plant accurately. Frasers therefore decided to build the plant itself.

The CTEP started operating in September 2014, and was run by Frasers who had not yet found a utility operator. Around that time, Brookfield Energy which operates as a registered electricity retailer, showed an interest in the CTEP. The negotiation process took a long time, but in December 2015 responsibility for operating the plant was passed to electricity retailer Brookfield Energy under a 50-year lease. The third-party provision and billing of electricity and hot water by embedded operators represents a model which could prove disruptive to traditional infrastructure provision. However, there remain significant regulatory barriers in Australia to the operation of local power generators, including tri-generation plants and distributed rooftop photovoltaic systems (Dunstan and Langham 2010).

Plant operation

The CTEP is highly efficient in terms of energy, costs and space. With an efficiency of over 80 per cent, the tri-generation plant is twice as energy efficient as a coal-fired power plant, because of the use of waste heat and the reduction in transmission and distribution losses. Another important efficiency feature is the balance between residential, retail and commercial peaks created by the diversity of uses in the precinct. This has the potential to spread the peak demand and obtain greater optimal use of the equipment.

The centralised facility saves on both capital cost and space, and reduces the total power consumption through reduced losses compared to those that would be incurred if separate systems were used. The grouping of all the equipment in one basement eliminates the need for duty and standby equipment at each location. While 11 buildings would usually require up to 36 chillers, there are only eight in the whole precinct. Similarly, the number of boilers needed was reduced from 24 to 5 (Central Park 2013). This means the roofs of the individual buildings are not needed for cooling towers and other equipment, enabling them to be used for other purposes such as green roofs. Reducing the amount of equipment has also significantly reduced maintenance costs.

In case of a major grid outage or a gas supply interruption, a diesel backup generator will turn on. The main purpose of this generator is to eliminate the risk of a failure to the refrigeration units in the supermarket in the main retail building. It is not needed to support the thermal production of the plant, since the pumps and equipment used to convey the hot or chilled water rely on power from the grid.

Environmental impact

Heat that cannot be recaptured is eventually vented as steam from six cooling towers on top of the old Brewery Yard. However, the tri-generation plant also emits pollutants, the most significant being nitrogen oxide (NO_x). To comply with air pollution regulations, Frasers decided to reuse the heritage chimney stack to vent gases. At 55 metres it provides a good solution to limit the ground concentrations of pollutants. Frasers also installed an advanced catalytic converter that brings the NO_x level in the exhaust down to one-tenth of the permitted level.

The CTEP provides significant CO_2 savings relative to a business-as-usual supply system, with an estimated reduction of 136,000 tonnes of CO_2 (equivalent) in greenhouse gas emissions over its expected 25-year life. This is equivalent to taking more than 62,000 cars off the road. Local generation helps to reduce the need for costly augmentation of the electricity supply to the city.

The CTEP is a major innovation in terms of precinct-scale sustainability, and the facility is working very well. However, now that the regulatory arrangements for private electricity supply are starting to evolve, it has become clear to Frasers that it could have increased the size of the plant and supplied electricity to more buildings, as originally intended. The thermal plant is slightly oversized, because the plant was built ahead of the rest of the site and it was hard to predict the exact loads of each future building. Also, since Frasers wanted to sell the CTEP, it had to invest in and install the plant at an early stage of the construction. However, Frasers managed to make good use of the oversizing in its thermal energy linkage with UTS. By around 2018, pipes that are already under the road separating Central Park and UTS will be connected to a new building at the university, enlarging the network of the CTEP, providing chilled water to a new building on the UTS campus.

Water management

Central Park uses high-quality water from various sources, distributed via a network which provides multiple grades of water quality. The water-recycling facility in the basement of the main retail building is operated by the private utility, Flow Systems, now part-owned by Brookfield. Onsite wastewater, rainwater and stormwater are recycled to meet most of Central Park's non-potable water needs. In future, wastewater from a neighbouring site will also be used. The combination of high water efficiency and the use of recycled water for approximately 95 per cent of non-potable uses allows for very high potable-water savings and contributes to Central Park's five-star Green Star rating.

The planning process

The precinct's original concept design, developed by ISF, took into account a full range of options for water efficiency and recycling. After modelling the water needs of the precinct, ISF investigated different options and considered their viability in technical, economic and policy terms. Discussions with Sydney Water, the public utility which provides water and sewerage services in the area, revealed that there was no likelihood of the utility being able to provide a centralised supply of recycled wastewater within the required time frame.

Hence in 2010, a private water supplier, Flow Systems, was brought in to implement the chosen onsite recycled water design in the basement that Frasers had started excavating under the main retail building. Flow Systems had already worked on similar schemes, having developed a plant and recycling network for 940 homes in Pitt Town in North West Sydney (Flow Systems 2017a). Flow Systems created a new water company, Central Park Water, as a fully owned subsidiary (Central Park Water 2017). The biggest challenge was how to fit the treatment facilities into the already excavated basement. Flow's solution was to build a vertical system over four basement levels to minimise the footprint of the equipment used. When fully operational the plant will be the world's largest water recycling facility in the basement of a residential building (Flow Systems 2017b).

The plant started supplying recycled water in 2015, and as at mid-2016 it provided enough water for irrigation, toilet flushing, washing-machine cold water and cooling towers – just over 50 per cent of all water use. The cost of the facility is estimated at around $13 million, with $8–9 million for all the fit-outs, and $5 million for administrative and management costs. The latter is a significant sum due to the novelty of the plant. The plant is being constructed in two stages to provide a gradual increase in capacity. Stage 1 was completed in 2015, with a capacity of 0.5 Ml/day. Stage 2 will take the final capacity up to 1 Ml/day, enough to deal with seasonal fluctuations in demand and export water to neighbouring developments. By 2018 the facility will be connected to the new buildings opposite Central Park at the UTS. The plant will treat a portion of the wastewater from the university and supply the university with recycled water.

Water-recycling plant operation

The facility recycles onsite wastewater. During the initial stages of the development when occupancy rates were low, sewage was obtained from an adjacent Sydney Water sewer via sewer mining. The recycled water network also harvests rainwater, irrigation runoff and car park drainage from across the precinct. Each roof collects rainwater in a 200 kl water tank. There are 10 tanks with a combined capacity of 2 Ml. At the present stage of development, not all this rainwater is required and overflows are passed to the stormwater system.

The facility produces two grades of recycled water: water for irrigation, toilet flushing and washing machines; and higher-quality water for the cooling towers. The main difference is the salt level, which needs to be lower for the cooling towers to increase the number of times the water can be used before being discharged. Water for the cooling towers undergoes an additional purification step (reverse osmosis). All the recycled water has an equivalent quality to potable water, but is only used for non-potable purposes.

Sydney Water provides potable water across the site and this source is also available as a backup if the recycled water and rainwater storage on site, equivalent to one day of demand, are insufficient. Trade waste generated by the recycled water plant is discharged into an adjacent Sydney Water sewer under a trade waste agreement. The wastewater undergoes eight purification processes, ranging from biological and mechanical to chemical. This multi-barrier approach, with multiple critical control points, ensures that the quality of the recycled water meets the Australian guidelines for water recycling (DSEWPAC 2017).

The recycling process is highly reliable, both in terms of quality and continuity of supply. Measurements are taken all along the purification process to ensure that the water reaches the quality required. Most of the important components of the plant are duplicated, so that the facility can continue to function if there are minor breakdowns. If a major breakdown occurs, the whole precinct can be supplied by mains potable water from Sydney Water if the one-day buffer of supply is insufficient. Electricity for the system is drawn from two points on the grid for reliability purposes.

Licence and contractual arrangements

As a private utility, Flow Systems has two licences administered by IPART (the NSW Independent Pricing and Regulatory Tribunal) under the Water Industry Competition Act (2006):

- a retail supplier's licence for the supply of sewage, recycled water and drinking water services;
- a network operator licence for the operation of the underlying network.

To ensure the quality of its water and the reliability of its processes, Central Park Water is audited by IPART. Central Park Water directly bills its customers, both for its recycled water and the potable mains water it buys from Sydney Water. While Central Park Water matches Sydney Water's regulated retail price for drinking water, the recycled water produced on site is 10 cents cheaper per kilolitre than drinking water.

The edges of innovation

The design and planning process for Central Park was constrained by time, finances and regulations, and it involved a large number of professionals and stakeholders. Under these circumstances, it was inevitable that some ideas for improving the sustainability of the development were not incorporated. Some of these are described below, principally to provide guidance as to the possible features of future precinct developments of this type.

Geothermal heat rejection

In Central Park, cooling towers are used to vent waste heat from air-conditioning systems and from the central thermal plant. Cooling towers are more energy efficient than air cooling, so towers are usually preferred, even though they are large consumers of water. In some developments it is possible to avoid having to choose between high water and high energy consumption through the use of ground source heat rejection, which relies on the heat dissipation capacity of the earth. For two reasons, a ground source, or geothermal heat exchange system, was not chosen for the site. First, the layout of the site made the installation physically difficult and so the cost would have been significantly higher than a conventional cooling tower system. Second, the water usage is mitigated by the use of recycled water for cooling towers.

Smart metering systems

Smart metering systems measure energy and/or water consumption in short time intervals (e.g. every 15 minutes or more frequently depending on specifications) unlike traditional meters that only measure accumulated consumption, and which are only read typically once a quarter or once a month depending on the property type. Smart meters can provide consumption and price information that can be made available to retailers and customers via in-home displays.

The original intent for Central Park was to install smart meters in every apartment. Such meters could display detailed data on electricity, hot water, potable drinking water and recycled water use. The data provided enabling residents to monitor their energy and water consumption, and learn how to make better use of resources.

Individual apartment smart meters and data displays did not come to fruition at Central Park. However, such technology is becoming cheaper and more reliable and will likely become a standard feature of commercial and multi-residential dwelling apartments in the near future.

Organic waste collection and treatment

A precinct like Central Park generates a significant amount of organic material. Such materials include, for example, food waste from the residential units and commercial and retail businesses and fats, oils and grease discharged to certified grease traps for collection and offsite treatment. In addition, there are clippings and prunings from the parks and gardens and green walls, as well as the organic content of sewage generated onsite. In most places, these wastes end up in landfill (in the case of food waste) unless separate collection for offsite processing is arranged as in the case of sewage biosolids, unless they are partially separated at centralised sewage treatment plants and used for methane generation. In many cases they are dried and sent to landfill. In the case of central Sydney's sewage treatment, a large proportion of the biosolids are discharged through an ocean outfall.

Localised collection of organic waste and sewage biosolids, and anaerobic digestion of these wastes were considered for the project. This would have reduced the volume of material, and the system could have generated biogas including methane which, with suitable scrubbing, could supplement the local natural gas supply for cooking or to power the tri-generation system. If a vacuum system were installed, food waste could be easily and reliably collected from residential units, commercial and retail outlets and the adjacent Kensington Street, now a thriving restaurant venue. In addition, vacuum toilets could be installed to reduce water use dramatically, with the space requirements less than they are for gravity sewer systems.

While elements of this option had been implemented in various sites around the world when Central Park was being designed, the combined use of these collection, transport and treatment systems was rare at the time, and to this day, there are still only limited examples in dense urban settings (Turner and White 2017; Turner et al. 2017). Hence, it was agreed that such a system would be a step too far. Since then, however, there have been more examples and reference cases developed, such as the anaerobic digester at Federation Square, a sustainability precinct in the heart of Melbourne that treats up to 800 kg of waste food a day (Victorian Government 2016). With the current focus on reducing food waste to landfill, and in some locations banning it, there is now significant incentive to include innovative organic waste management in both commercial and multi-residential buildings. This includes retrofitting such systems and practices in buildings like Central Park post construction. Although, incorporation in new buildings is preferable and more cost effective (Turner et al. 2017).

Neighbourhood emissions reduction scheme: 'the green wave'

In high-density urban developments, it becomes physically impossible to achieve a zero net carbon development through onsite renewables. This leaves developers with a few options. They can reduce energy use through foundational design principles including energy-efficient design and efficient equipment (including tri-generation), and they can maximise rooftop photovoltaic installation and purchase sufficient renewable energy offsite.

Purchasing offsite renewables has the disadvantage of requiring electricity to be supplied through the transmission and distribution system, often over a great distance. This contributes to peak demand which increases the cost of the distribution network. Half the cost of electricity in Australia comes from distribution costs. Another approach would be to invest in energy efficiency, peak demand management and distributed generation projects in the immediate surroundings of the precinct. These projects could involve installing: efficient lighting and appliances; energy and water-efficient showerheads; insulation and improved air conditioning; and solar photovoltaic systems. Such an approach would mitigate electricity transmission and distribution constraints and promote energy and carbon neutrality, while at the same time building community strength and local economic and social benefits – especially if it prioritised low-income households. The low cost of energy efficiency measures compared to renewable electricity and certified carbon offsets suggests that this could be a cost-effective alternative. This option was not taken up due to complexity and timing issues, but remains a likely cost-effective option for any development wishing to implement an alternative and innovative approach to offsetting residual energy use and greenhouse gas emissions.

Conclusion

Many lessons have been learnt from the Central Park development. Reflections include:

- Significant opportunities become possible in a precinct development, compared to the development of a single building, including shared infrastructure such as a central thermal energy plant, a common basement and, in design terms, an improved mix of green space and other design opportunities.
- The fact that the site is a mixed-use development provides opportunities and benefits, including the variation in hourly, weekly and seasonal energy and water demands which can balance loads and increase the cohesion of the development.

- There are also opportunities for synergies between the different sustainability aspects of the development. For example, while the green walls and park areas use a significant amount of water, this water can be provided by the recycling system. Similarly, the tri-generation system waste heat can be used to heat the collected rainwater to supply hot water to residents and businesses. A systems view of the development and its needs is crucial for these benefits and solutions to be achieved.
- The additional capital cost of some of the onsite infrastructure such as tri-generation or wastewater treatment plant can be offset by the reduced or avoided capital costs of upgrades to supply the development through conventional centralised means. The level of reduced or avoided costs varies with the specifics of the development and the institutional arrangements of the energy, water, waste and transport providers. In this case this was not a major factor, but in some cases can make a significant difference to the business case for a development and for more sustainable options to be considered such as water recycling (Turner et al. 2016).
- In this development, the institutional arrangements in place in NSW provided some opportunities. For example, NSW laws made it possible to have privately operated utilities for energy and water. Such laws are less well-developed elsewhere in Australia. Similarly, the development was able to take advantage of a specific local government rates-based loans scheme for the capital cost, called an environmental upgrade agreement (Kelly 2013).
- Sustainability is about more than the environmental performance, and should include social, cultural and economic considerations – in particular, the design impacts on the level of activation and liveliness of the precinct. Similarly, planning for sustainability should not just be about 'doing less harm' but should also be about regeneration or restoration, making the site better than it was before and generating net benefits (Mitchell 2008).
- During such regeneration projects it is essential to engage effectively and respectfully with the surrounding community and enable them to voice their concerns and their ideas to minimise conflict and maximise community acceptance and engagement.
- In the planning and design phase of a major project like this, the path is never linear, rational or straightforward. It is a complex process involving many stakeholders with sometimes competing goals. There are multiple critical paths and inevitably events that cannot be anticipated. In this case, the global financial crisis set the timing back considerably and moved the focus of the initial build onto the residential and student housing component. It is worthwhile accepting the uncertain nature of planning and building in as much flexibility and adaptability as possible.
- The complex and often rushed nature of the planning, design and development process can make it difficult to innovate. This requires good processes of multi-disciplinary engagement of the design team, and

also an agile and rapid response to scope ideas to either reject them or take them forward.

- Instead of rejecting the risky or novel ideas with limited precedents in other locations, there is an opportunity to trial or pilot them in smaller contexts. This could be done in parts of a building, to assist in driving innovation forward and potentially incorporate new designs in later parts of developments, especially if they are implemented over long periods of time such as Central Park.

While Central Park did not encompass the full range of innovative options that were available, it is a unique and substantial exemplar, both in its planning and design processes and in its outcomes and provides a demonstration of what new urban densification can look like if sustainability and regeneration principles are taken into consideration.

Acknowledgements

The authors would like to thank the many staff of Frasers Property Australia Pty Ltd and Flow Systems Pty Ltd who assisted with data and information, including Andrew Horton and Frier Bentley of Flow Systems, Scott Clohessy and Michael Goldrick, Frasers Property Australia Pty Ltd and Dr Chris Riedy, Institute for Sustainable Futures, at the University of Technology Sydney.

Note

1 All the information in this section comes from an interview with Scott Clohessy of Frasers Property Australia except when indicated otherwise by a citation.

References

ABS (2016) '3218.0 – Regional population growth, Australia, 2014–15', http://www.abs.gov.au/ausstats/abs@.nsf/lookup/3218.0Media%20Release12014-15 (accessed 5 May 2017).

ABS (2017) '3218.0 – Regional population growth, Australia, 2015–16', http://www.abs.gov.au/ausstats/abs@.nsf/mf/3218.0 (Accessed 5 May 2017).

Barangaroo Delivery Authority (2017) 'Overview',, http://www.barangaroo.com/discover-barangaroo/overview.aspx (accessed 5 May 2017).

BioRegional (2014a) 'Sonoma Mountain Village 2020 sustainability action plan report', http://www.bioregional.com/wp-content/uploads/2014/10/sonoma-mountain-village-opap-website.pdf (accessed 5 May 2017).

BioRegional (2014b) 'One planet action plan for Grow community', http://bioregional.com.au/wp-content/uploads/2014/10/GROW-OPAP-Final-Summary.pdf (accessed 5 May 2017).

Blanc, P. (2013) 'One Central Park vertical gardens', Curating Cities, http://eco-publicart.org/one-central-park-vertical-gardens/ (accessed 5 May 2017).

Central Park (2013) 'Green transformation', http://www.centralparksydney.com/explore/a-sustainable-habitat (accessed 5 May 2017).

Central Park (2017a) 'Awards won', http://www.centralparksydney.com/explore/awards (accessed 5 May 2017).

Central Park Water (2017) 'Who we are', http://centralparkwater.com.au/about-us/who-we-are/ (accessed 5 May 2017).

Chancellor, J. (2014) 'GoGet car share for Central Park inner city residents', *Property Observer*, 28 May http://www.propertyobserver.com.au/forward-planning/investment-strategy/property-news-and-insights/31785-go-get-car-share-for-central-park-inner-city-residents-2.html (accessed 7 May 2017).

Chua, G. (2013a) 'French lighting artist Yann Kersale unveils first Australian artwork at Central Park Sydney', *Architecture & Design*, 6 December, http://www.architectureanddesign.com.au/news/french-lighting-artist-yann-kersale-unveils-first (accessed 7 May 2017).

Chua, G. (2013b) '5 green star development one Central Park welcomes first residents', *Architecture & Design*, 21 June http://www.architectureanddesign.com.au/news/5-green-star-central-park-welcomes-first-residents (accessed 7 May 2017).

CoS (City of Sydney) (2014) 'Advanced waste treatment master plan 2013–2030', http://www.cityofsydney.nsw.gov.au/__data/assets/pdf_file/0014/215204/2014-429946-Advanced-Waste-Treatment-master-plan-FINAL-amended-as-per-Council-resolution.pdf (accessed 7 May 2017).

Critchley, V. and White, S. (2008) *Frasers Broadway: Key Sustainability Initiatives.* Report prepared by Elton Consultants and the Institute for Sustainable Futures for Frasers Property, Sydney: University of Technology.

CTBUH (Council of Tall Buildings and Urban Habitat) (2014a) 'Awards events herald vertical greenery, retrofits and skyscrapers that twist, shift and lie down', http://ctbuh.org/Awards/SymposiumsCeremonies/Awards2014Symposium/Awards2014Symposium/tabid/6673/language/en-US/Default.aspx (accessed 5 May 2017).

CTBUH (Council of Tall Buildings and Urban Habitat) (2014b) 'Burj Khalifa honored as first recipient of CTBUH's new Tall Building 'Global Icon' Award', http://www.ctbuh.org/Events/Awards/2010Awards/2010AwardsDinner/tabid/1710/language/en-US/Default.aspx (accessed 5 May 2017).

DSEWPAC (Department of Sustainability, Environment, Water, Population and Communities) (2017) 'National Water Quality Management Strategy: Australian guidelines for water recycling: managing health and environmental risks (Phase 1)', http://webarchive.nla.gov.au/gov/20130904202017/http://www.environment.gov.au/water/publications/quality/water-recycling-guidelines-health-environmental-21.html (accessed 5 May 2017).

Downton, P. (2013) 'Green roofs and walls', Australian Government, http://www.yourhome.gov.au/materials/green-roofs-and-walls (accessed 5 May 2017).

Drake-Brockman v *Minister for Planning & Anor* (2007) NSWLEC 490, 13 August 2007, http://www.austlii.edu.au/au/cases/nsw/NSWLEC/2007/490.html (accessed 7 May 2017).

Draper, M. (2007) 'Residential development brewing on Sydney's old CUB site', *The Age*, 13 June, http://www.theage.com.au/news/business/residential-development-brewing-on-sydneys-old-cub-site/2007/06/12/1181414299787.html (accessed 07 May 2017).

Dunstan, C. and Langham, E. (2010) *Close to Home: Potential Benefits of Decentralised Energy for NSW Electricity Consumers.* Report prepared by the Institute for Sustainable Futures, University of Technology Sydney Sydney: City of Sydney, http://cfsites1.uts.edu.au/find/isf/publications/dunstanlangham2010closetohome.pdf (accessed 7 May 2017).

Energy rating (n.d.) 'Save energy, save money, reduce emissions', http://www.energyrating.gov.au/ (accessed 5 May 2017).

European Commission (2011) 'CDW: Material recovery and backfilling', http://ec.europa.eu/environment/waste/studies/pdf/CDW%20Statistics%202011.pdf (accessed 5 May 2017).

Farrelly, E. (2006) 'Once-bullied Sartor grows up and becomes a bully', *Sydney Morning Herald,* 14 June http://www.smh.com.au/news/opinion/oncebullied-sartor-becomes-a-bully/2006/06/13/1149964529995.html?page=fullpage#contentSwap1 (accessed 7 May 2017).

Flow Systems (2013) 'Local water utility recycling water in the basement of a residential building', *Sustainability Matters,* 5 November http://www.sustainabilitymatters.net.au/content/water/article/local-water-utility-recycling-water-in-the-basement-of-a-residential-building-768331193 (accessed 7 May 2017).

Flow Systems (2017a) 'Pitt Town water', https://flowsystems.com.au/communities/pitt-town-water/ (accessed 5 May 2017).

Flow Systems (2017b) 'Central Park Plus', https://flowsystems.com.au/communities/central-park-plus/ (accessed 5 May 2017).

Frasers and Sekisui House (2009) 'Premier announces approval of concept plan for Frasers Broadway site', http://www.centralparksydney.com/what-s-on/news-events/2009-feb-premier-announces-approval-of-concept-pla (accessed 7 May 2017).

Frasers and Sekisui House (2011) 'Frasers Property Australia announces joint venture at Central Park'. https://www.dropbox.com/s/ba3cpbuudeocqfr/Frasers%20Property%20and%20Sekisui%20House%20announce%20Central%20Park%20JV_F_7July2011.doc?dl=0 (accessed 7 May 2017).

Frasers and Sekisui House (2014) 'The best tall building in the world', http://www.centralparksydney.com/what-s-on/news-events/ctbuh-best-tall-building-award (accessed 5 May 2017).

Gaffney, A., Huang, V., Maravilla, K. and Soubotin N. (2007) *Sjostad Stockholm, Sweden: A Case Study,* http://www.aeg7.com/assets/publications/hammarby%20sjostad.pdf (accessed 5 May 2017).

Greater Sydney Commission (2016a) 'Our Greater Sydney 2056: A metropolis of three cities – connecting people', https://gsc-public-1.s3.amazonaws.com/s3fs-public/draft_greater_sydney_region_plan_web.pdf (accessed 4 December 2017)

Greater Sydney Commission (2016b) 'Strategic planning', https://www.greater.sydney/strategic-planning (accessed 5 May 2017).

Greater Sydney Commission (2017) 'Strategic planning', https://www.greater.sydney/strategic-planning (accessed 4 December 2017)

Irga, P., Burchett, M. and Torpy, F. (2015) 'Does urban forestry have a quantitative effect on ambient air quality in an urban environment?', *Atmospheric Environment*, 120: 173–181

Katz, P., Robertson, L. and See, S. (2011) 'Shanghai World Financial Center, China', http://www.ctbuh.org/TallBuildings/FeaturedTallBuildings/Archive2008/ShanghaiWorldFinancialCenter/tabid/2810/language/en-US/Default.aspx (accessed 5 May 2017).

Kelly, D. (2013) 'City of Sydney's first EUA at Central Park announced', *The Fifth Estate*, http://www.thefifthestate.com.au/articles/city-of-sydneys-first-eua-at-central-park-announced/45885 (accessed 7 May 2017).

Leow, D. (2013) 'Sydney's Central Park residents to get energy efficient heating and cooling system', *Property Observer*, 3 April, http://www.propertyobserver.com.au/finding/residential-investment/new-developments/21456-friday-march-22-feature-sydneys-central-park-residents-to-get-greenest-heating-and-cooling-as-construction-begins-on-new-carbon-efficient-energy-plant.html (accessed 7 May 2017).

McGee, C., Wynne, L., Milne, G. and Dovey, C. (2014) *Guiding World-Class Urban Renewal: Framework for UrbanGrowth NSW*. Report prepared by the Institute for Sustainable Futures, University of Technology Sydney, Sydney: UrbanGrowth NSW.

Mitchell, C. (2008) 'Restorative water: beyond sustainable', *Waste Management and Environment Magazine* July: 20–21

Office of Environment and Heritage (NSW) (2017) 'Search for NSW heritage', http://www.environment.nsw.gov.au/heritageapp/heritagesearch.aspx (accessed 5 May 2017).

Office of Environment and Heritage (NSW) (n.d.) 'Chippendale Heritage Conservation Area', http://www.environment.nsw.gov.au/heritageapp/ViewHeritageItemDetails.aspx?ID=2421466 (accessed 5 May 2017).

Sustainability Matters (2014) 'Self-sufficient village in Sydney', 28 January, http://www.sustainabilitymatters.net.au/content/sustainability/article/self-sufficient-village-in-sydney-338430307 (accessed 7 May 2017).

Turner, A., and White, S. (2017) *Urban Water Futures: Trends and Potential Disruptions*. Report prepared by the Institute for Sustainable Futures, University of Technology Sydney, Sydney: Water Services Association of Australia (WSAA) https://opus.lib.uts.edu.au/handle/10453/77441 (accessed 7 May 2017).

Turner, A.J., Mukheibir, P., Mitchell, C., Chong, J., Retamal, M., Murta, J., Carrard, N. and Delaney, C. (2016) 'Recycled water: Lessons from Australia on dealing with risk and uncertainty', *Water Practice & Technology* 11(1): 127–138. http://wpt.iwaponline.com/content/11/1/127 (accessed 7 May 2017).

Turner, A., Fam, D., Liu, A., and Madden, B. (2017) *Pyrmont Ultimo Precinct (PUP) Scale Organics Management Scoping Study*. Report prepared by the Institute for Sustainable Futures, University of Technology Sydney, Sydney: Sydney Water Corporation and the NSW Environment Protection Authority.

Victorian Government (2016) *Victorian Food Organics Recycling. A Guide for Small-Medium Food Services Organisations*, Melbourne: Department of Health and Human Services.

Wang, S., Williams, P., Shi, J. and Yang, H. (2015) 'From green to sustainability: Trends in the assessment methods of green buildings', *Frontiers of Engineering Management*, 2(2): 114–121

Water Efficiency Labelling and Standards (WELS) Scheme (n.d.) 'Water rating', http://www.waterrating.gov.au/ (accessed 5 May 2017).

Watpac (2015) *Central Park, One Central Park, Park Lane and The Mark*, Technical paper, Australian Construction Achievement Award http://www.acaa.net.au/wp-content/uploads/2015/03/One-Central-park-in-Sydney.pdf (accessed 7 May 2017).

11 Rethinking culture-led urban regeneration

The creative (re)assembling of inner-city Newcastle

Nathalie Gentle and Pauline McGuirk

Introduction

The purpose of this chapter is to amplify and extend knowledge of urban creativity to suggest the diverse and contextually contingent ways in which creativity can contribute to regenerating urban spaces. The chapter presents a generative reading of the varied and multiple relationships between creativity and urban regeneration (Dovey 2010; Farias and Bender 2010; Brownill 2013; McGuirk et al. 2016), underpinned by assemblage thinking, a framing increasingly deployed to capture the multiple dimensions and forms of agency at work in the urban (Färber 2014: 121). The chapter proceeds by, first, making the case for why such a generative reading is productive, followed by outlining how we use assemblage thinking to reposition creative-led regeneration. Then, through this lens, we analyse a case study of the Hunter Street Mall in inner-city Newcastle, NSW and Renew Newcastle – a community-based initiative finding temporary creative uses for vacant inner-city properties – and the ways creativity is contributing to Newcastle's regeneration.[1] Finally, we conclude with reflections on assemblage thinking's potential for contributing to conceiving and enacting alternative forms of urban regeneration.

From critical to critical/generative understandings of culture-led urban regeneration

Culture-led urban regeneration, the deliberate harnessing of culture and creativity for urban revitalisation, has become a widespread tool for urban policymakers, reflecting recognition of the cultural economy as a potential driver of urban economies (Gibson and Stevenson 2004; Waitt and Gibson 2009; Comunian 2010). The cultural and/or creative industries[2] (CCIs) and CCI clusters have been understood as integral to the success of these strategies and widely incorporated in attempts to revitalise urban spaces experiencing economic downturn, often in relation to the 'creative city' thesis, popularised by Landry (2000) and Florida (2002). Strategies range from large-scale redevelopments that involve public–private partnerships and new cultural infrastructure, for example Docklands Melbourne, Darling Harbour Sydney and Perth's Cultural Centre, through to smaller-scale *ad hoc* temporary urban uses which engage CCI clusters to activate disused spaces, such as Renew Newcastle and its variant Renew Australia (see Waitt and Gibson 2009; Atkinson and Easthope 2009; Colomb 2012; Lashua 2013; Shaw 2014). And indeed culture is mobilised in a variety of ways through these strategies: as a flashpoint for reanimating under-utilised urban spaces (e.g. inner Newcastle); as a driver for cultural infrastructure investments aimed to attract international events and visitors (e.g. Perth's Cultural Centre, Sydney's Darling Harbour); as the rationale for renewing brownfield sites as new urban districts imagined through the lens of culture and innovation (Melbourne's Docklands; Sydney's Bay Precinct). In a variety of forms, then, support for culture and cultural producers is fused with development strategies hinged on the notion that the creative, culturally-infused city produces a self-reinforcing cycle of regeneration by attracting creative economy workers, firms and further rounds of cultural amenities and economic investment.

Culture-led urban regeneration, and more critically urban creativity as expressed in the 'creative city' thesis, has been critically associated with neoliberalised urban planning and governance as a culture-infused form of urban entrepreneurialism in which creativity is harnessed for commercial purposes designed to drive economic competitiveness (Peck 2005; 2011; Atkinson and Easthope 2009; Colomb 2012; Borén and Young 2013). These critiques resonate with recent analyses of temporary or DIY urbanism in which vacant or transitional urban spaces are temporarily deployed by creative/cultural initiatives,[3] often encouraged by urban authorities and property owners as a 'second-best' or interim option to activate space while waiting for more profitable uses to materialise (Munzner and Shaw 2014; Andres 2013). Critiques foreground temporary urbanism's potential to be a seed-bed for gentrification, as they re-image abandoned spaces as attractive spaces for cultural consumption, that are easily appropriated by city branding strategies (Colomb 2012; Shaw 2014). Thus local community, grassroots or small-to-medium enterprise alternatives to large-scale, top-down regeneration are reframed as strategies to market the city to capital and 'the creative class' as vibrant and liveable (Lashua 2013).

Whilst these critical accounts provide valuable insights into the pitfalls of mobilising culture and creativity in urban regeneration, they can risk producing a reductive understanding of *how* creativity may be contributing to urban regeneration, by assuming urban creativities are inevitably embroiled in neoliberal urban projects (Waitt 2008; Waitt and Gibson 2009; Borén and Young 2013; Breitbart 2013; Van Schipstal and Nicholls 2014). First, a methodological focus on business and political elites can narrow the lens on creativity and so miss out on generating knowledge of alternative forms of creativity in urban regeneration (McGuirk et al.. 2016; Borén and Young 2013; Waitt 2008). Second, an analytical privileging of the neoliberal urban project can risk prioritising a top-down, linear casual formation in which commodified forms of culture and creativity are inevitably dominant in (re)forming the urban, such that other ways that the urban may be (re)forming through diverse forms of creativity are obscured. Finally, urban materiality can be viewed as 'dull, immutable and passive' (Swanton 2013: 286) in resultant formations of capitalist socio-economic activities. Thus the agency of a multiplicity of creative urban social and material actors and their potential to legitimise new forms of urban formations can be hidden from view (Larner 2011).

This chapter seeks to bring the multiple possibilities of urban creativity more forcefully into the analytical frame. Certainly, approaches that point to these multiple possibilities are not novel[4] and there is a rich history of creative practices being mobilised in post-industrial cities as people and communities seek to improve their lives (Breitbart 2013). Creativity is multidimensional, being both a product and a producer of the social, cultural, political and economic (Waitt and Gibson 2013). Far from being a homogenous subject, it is a practice negotiated through 'different scales of government/governance, locally contingent conditions and power struggles, artistic scenes and networks, and complex material[alities]' (Borén and Young 2013: 1805). While creativity may indeed be amenable to co-option by neoliberalised urban regeneration, it may also generate multiple forms of experimentation and opportunity (Van Schipstal and Nicholls 2014; Waitt 2008; Iveson 2013) and allow for power to be transferred to smaller-scale, community-based enterprises and individuals such that new urban practices and opportunities, once deemed unimaginable, may emerge (Andres 2013; Lashua 2013; Iveson 2009; Waitt 2008). The analytical lens provided by assemblage theory is well-geared both to conceptualising and revealing urban creativity's multiplicity and the possibilities it can entail for urban regeneration.

An assemblage approach to the city and urban regeneration

Assemblage theory conceives of the socio-material world – and the city – as produced through the co-functioning of multiple, heterogeneous, socio-material elements that gather and hold together through relationships, without forming a new organic whole or system. Each component retains its heterogeneity and autonomy to act in the formation (Dovey 2010; McFarlane 2011a; Anderson et al. 2012).[5] As an assemblage, the city is shaped through multiple heterogeneous assemblages and comprised of multiple processes of composition (Farias 2011). So heterogeneity, multiplicity and socio-material relationality are at the forefront of analysis (Brownill 2013). The city is revealed as continually being made and remade through the re/assembling of these formations. Urban spaces are understood as always emergent, contingent and multiplex: continually being reworked through the (re) alignment of socio-material relations (McFarlane 2011a; Dovey 2012).

Assemblage thinking (AT) opens up space for new questions to be posed of creativity and how it may be regenerating the urban (McGuirk et al. 2016). Three particular aspects are especially productive for our analysis. First, AT's commitment to heterogeneity extends our attention beyond the elements and actors normally considered in human-centred, structural analyses of urban process. AT brings into view the wide array of social and material elements involved in the context-specific formation of creative assemblages to include formal and informal actors and institutions; documents, models and visualisations, contracts, concepts and imaginaries, infrastructure, buildings, streetscapes and affects (see Jacobs et al. 2007, 626). The enlarged cast of elements allows for emergent possibilities in how creativity might be contributing to regeneration to be recognised. Second, AT does not privilege human agency but positions it as always distributed between and emergent from heterogeneous social and material elements in relation (Bennett 2005). For creative urban regeneration, this allows us to inquire into how urban materiality and its agentic capacities enable or constrain particular forms of creativity.

Third, AT understands the city, and its regeneration, as an always-emergent 'work in progress' (McCann 2011: 145) as its multiple heterogeneous assemblages continually shift as their socio-material components align, disperse and realign, in specific time-space contexts (Farias 2011). Furthermore, AT suggests that any component of an assemblage – social or material – can be 'detached and plugged into' different assemblages, where the interactions change to produce other capacities by virtue of what the component is drawn into relation with (Anderson et al. 2012). Thus the city, and its regeneration, is always open to being reworked in unpredicted ways (McGuirk et al. 2016). As an orientation to urban regeneration – as heterogeneous, agentic and always emergent – AT attunes us to recognising where new trajectories of regeneration might be taking shape and, in the case of inner Newcastle, to teasing out the distinct and contingent ways in which creativity might be contributing to the re-assembling (and regeneration) of urban spaces.

Undertaking assemblage-informed research requires ontological commitments to heterogeneity, socio-materiality, distributed agency and contingent emergence. Research methods must be practised in ways that are attuned to these commitments and to capturing and storying the emergent and relational composition of the social and material. As it is not possible to have advance knowledge of all the socio-material actors, relationships and 'labours of composition' involved in assembling urban regeneration through creativity prior to conducting research (Farias 2011: 367), an open, explorative form of inquiry is required. Such inquiry is well-suited to case study as methodology. Thus our inquiry involved empirical investigation, focused on producing 'thick empirical descriptions' of how relational creative socio-material assemblages are forming and holding together in the regeneration of inner Newcastle (see Anderson et al. 2012; McFarlane

2011b). Reflecting this, our research employed a qualitative multi-method approach involving analysis of policy/planning and media documents, participant observation with a creative fund-raising event and 20 semi-structured interviews with policy/planning actors in government and non-government sectors, small creative/cultural business owners/operators and small not-for-profit creative/cultural organisations. The rich empirical accounts these assemblage-informed methods enable, can bypass top-down or bottom-up interpretations of urban formations and reveal how varied forms of creativity can contribute in diverse ways to Newcastle's culture-led urban regeneration.

The Hunter Street Mall, located in the East End of inner Newcastle, emerged through the data as a key site of assemblage. In the following, we trace the mall's history, leading to a detailed consideration of the socio-material assembling of its creative regeneration and the centrality of the Renew Newcastle initiative to this.[6] As no single account can include every actor implicated in the history and context of a place, our description represents a purposefully gathered account of the social and material actors assembled in the mall's creative regeneration (see McCann 2011).

Newcastle and the Hunter Street Mall

Newcastle, NSW, is Australia's largest regional city, approximately 160 km north of Sydney. Its coastal and riparian location and significant coal deposits saw it emerge as a major port. The establishment of the BHP steelworks in 1915 cemented its role as a major industrial centre for much of the twentieth century, specialising in steel production and manufacturing (DoPI 2012). The inner city, particularly Hunter Street (which runs parallel to the harbour), became a strategic commercial and retail core strongly aligned with this industrial role (Figure 11.1).

Inner Newcastle has struggled in the city's post-industrial transition. Suburbanisation and the 1980s global economic recession resulted in manufacturing closure, and declining employment and population (McGuirk et al. 1996). A major earthquake in 1989 and the eventual closure of the BHP steelworks in 1999 led to subsequent degradation of building and urban infrastructure. Abandoned industrial sites along the harbour fell into neglect and disrepair. In the early 1990s, the formal redevelopment of the former industrial harbourside began. The federally funded, state-owned Honeysuckle Development Corporation (now Hunter Development Corporation – HDC) was formed to oversee and manage the redevelopment aimed to catalyse revitalisation for the wider inner city (McGuirk et al. 1996). The ongoing Honeysuckle project now mirrors many attributes of culture-led urban regeneration strategies with the harbourside area now comprising mixed residential, commercial, cultural and entertainment facilities (HDC 2012). Yet despite contributing to an increase in both population and employment (HDC 2012), Honeysuckle did not kick-start a wider urban renewal for the

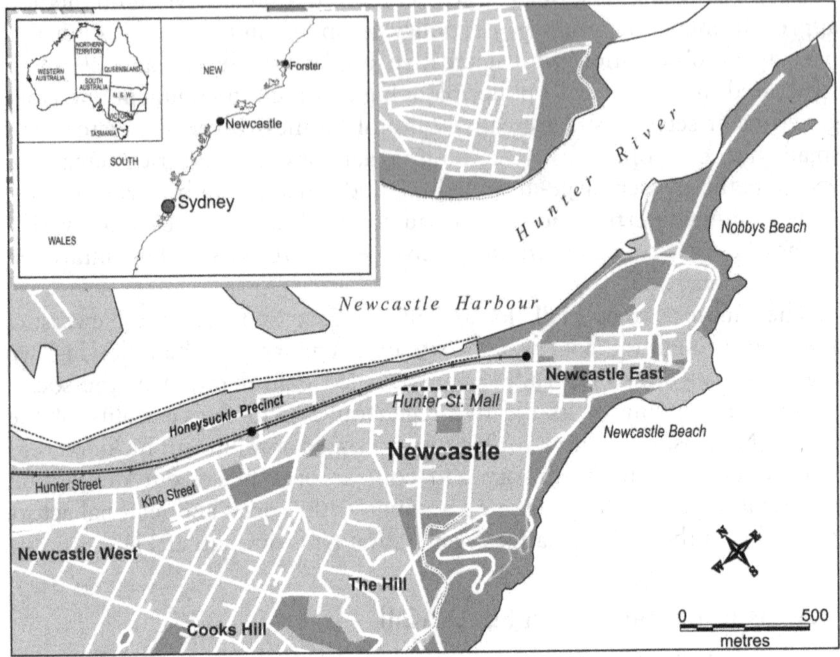

Figure 11.1 Inner Newcastle (source: Olivier Rey-Lescure)

CBD. By 2006, Hunter Street remained a stubborn artefact of Newcastle's industrial past. Through two decades of deindustrialisation, suburbanisation and subsequent disinvestment by property owners, developers and retail traders, the Hunter Street Mall, once a bustling retail core for the wider Hunter region, declined and became embroiled in narratives of abandonment and decay. By the late 2000s, it was known as a problem space increasingly characterised by vacant, derelict buildings, night-time alcohol-fuelled violence, and poor safety perceptions (HDC 2009) (Figure 11.2). One local creative business owner recalled:

> walking down the mall and ... there was not one person in any shop apart from the shopkeeper ... I wanted to put little speakers all up and down the mall and get buckets of sand and have eerie music going out, and blow tumbleweeds down the mall ... it was deserted.

Since then local and state governments have made concerted efforts to revitalise the mall, especially by assembling a suite of actors and imaginaries thought capable of leveraging culture/creativity to enhance regeneration. In the past decade approximately 50 reports, plans and strategies have been produced that include measures to address mall revitalisation, regularly invoking the regenerative powers of culture and creativity (Table 11.1).

Figure 11.2 Hunter Street Mall, 2008 (source: Marcus Westbury)

These have acted as linchpins, drawing together into new relations actors with formal authority to implement large-scale regeneration strategies to rework the materiality of the inner city, often through invocations of culture and the creative: Newcastle City Council (NCC); the NSW Department of Planning and Infrastructure (DoPI);[7] the Hunter Development Corporation (HDC); inner-city business improvement association, Newcastle Now (NN); and more recently, UrbanGrowth NSW.[8]

Yet the mall's regeneration is worked upon not only by actors in formally authoritative positions, nor indeed exclusively by social actors. Through the prism of Renew Newcastle (RN), we provide an account that reveals the heterogeneity of social and material actors drawn into relation in emergent new assemblages around the mall's distinctive form of creative regeneration. The cast of actors involved is opened out considerably by positioning the 'social and material on the same explanatory plane' (Jacobs et al. 2007), such that the mall's materiality is understood to have agentic capacities, opening up and closing down opportunities, revealing the activation of new capacities it is (re)configured, and shaping diverse and emergent possibilities for varied forms of creative regeneration.

Assembling Renew Newcastle

Struck by the 150 vacant properties on Hunter Street during a visit from Melbourne, Newcastle-raised Marcus Westbury created Renew Newcastle in

Table 11.1 Key regeneration strategies Newcastle's inner city – 2006 to 2016, by date of publication

Document	*Imaginaries/practices of culture and creativity*
The Lower Hunter Regional Strategy 2006–2031 (DoPI 2006)	• Newcastle as a 'lifestyle' and 'cultural' city.
Revitalising Newcastle City Centre Plan (NCCP) Vision – (Regional Cities Taskforce, 2006)	• Cultural economy as growing economic sector and source of employment. • The Mall to become heart of the 'old city', revitalised through day and night economies. • Revitalised Mall as space for major events and markets.
Newcastle City Centre Renewal Report (HDC 2009)	• Newcastle cultural offering as visual and performing arts, night culture and economy. • Revitalised public domain to create sense of place; safer more active spaces; spaces for outdoor expression (visual and performing arts) to attract investment.
Hunter Street Revitalisation Master Plan Final Strategic Framework (SCAPE Strategy 2010, for NCC)	• Support and expand Renew Newcastle-style initiatives; enable outdoor dining; develop night-time economy. • Positive branding strategy; annual program of events along Hunter Street.
Newcastle Urban Renewal Strategy (DoPI 2012)	• Revitalise Hunter Street Mall as a catalyst for renewal. • Potential to build as a unique shopping destination. • Improve Mall public domain to improve its performance as a public space. • Development of cultural industries to stimulate economic growth and build 'uniqueness' of city centre. • Cultural industries to be fostered through local initiatives such as Renew Newcastle to build city centre 'uniqueness'.
Newcastle 2030 – the Newcastle Community Strategic Plan (NCC 2013)	• Arts and culture contribute to a liveable city. • Cultural activity supports the city's renewal through creating safe, activated spaces. • Expand Newcastle Art Gallery as a key destination in the city's revitalisation.
Revitalising Newcastle (UrbanGrowth NSW 2015)	• Seek temporary activation activities for areas in transition. • Hunter Street and Mall as a cultural hub. • Develop Newcastle's cultural economy, including art studios and art retail spaces. • Lists Renew Newcastle as a business/industry stakeholder in revitalisation.
Economic Development Strategy 2016–2019 (NCC 2015a)	• Support creative industries which stimulate innovation and attract businesses.
Newcastle City Council Cultural Strategy 2016–2019 (NCC 2015b)	• Links culture and creativity with economic and tourism values. • Acknowledges Renew Newcastle's role supporting the arts/creative community • Advocates the creation of new art spaces to promote the city's revitalisation.

2008 with the primary purpose of activating empty spaces through a form of creative temporary urbanism (see Westbury 2015 for a full account). RN was created as a not-for-profit[9] which 'borrows' empty vacant commercial properties from property owners and 'lends' them to creative and cultural individuals, projects and community groups to undertake their activities while the properties remain available for commercial leasing[10] (RN n.d.). In founding RN, Westbury brought to the mall his own creative background as festival director, cultural project manager, writer and television presenter (RN n.d.).

Westbury's promotion of RN as a short-term solution to inner-city vacancy and an opportunity for creatives brought him to the attention of GPT, a publicly-listed property group primarily concerned with commercial and residential development. GPT became associated with the Mall in 2007 when it acquired multiple properties on its western end. This new actant[11] mobilised a new imaginary of the mall through a major planned redevelopment of its landholdings. However, the mall's existing socio-materiality resisted this new imaginary; for instance, Newcastle's heavy rail line, originally built to facilitate industry, separated Hunter Street and the revitalised Honeysuckle precinct. GPT, alongside many business leaders and some government agencies, viewed it as a barrier to redevelopment (Ruming et al. 2016). By 2008, GPT's proposed redevelopment was being hampered by the combined agency of the *in situ* rail line[12] and the wider assemblage of the global financial crisis. In response GPT was seeking transitory or temporary uses to reactivate its vacant landholdings,[13] and maintain their property values by gathering new socio-material components to the mall, when it came in contact with RN (Empty Spaces n.d.).

GPT began to engage with RN to determine a contractual basis to allow creative and community groups who lacked capital to temporarily occupy vacant landholdings whilst the properties remained available for commercial leasing[14] (Figure 11.3). The organisation was driven by the motivation to provide 'opportunities for creatives that wouldn't ordinarily be there' (Interview, RN). The contracts acted as a legal mechanism for cohering a new socio-material assemblage, drawing together property owners, the materiality of the mall's vacant space, and various kinds of creative and community ventures that have substantively regenerated the mall. A distinctive form of grass-roots, arts-led creativity was being assembled in the mall, shifting its socio-material composition by cohering new actants, effecting new material spaces and realising new capacities for mall regeneration.

Figure 11.3 GPT-owned property contracted to Renew Newcastle, Market Square, Hunter Street Mall, 2013 (source: authors)

In contrast to the materiality of vacant, boarded-up retail spaces, by late 2015, the mall had emerged as an eclectic mix of small-scale specialty retail and dining spaces, business and financial services, RN projects and community-run spaces. For example, the ground floor of the former iconic David Jones building was transformed into multiple boutique-style spaces, collectively known as The Emporium (Figure 11.4). Established in 2012, it became a 'treasure trove of locally made art, fashion, furniture and design' (RN n.d.). In 2014, the administrative and management function of the not-for-profit Newcastle Youth Orchestra was drawn into relation with the mall through RN and to occupy first-level space of the former David Jones building. In 2015, under-utilised garden-bed space and a community grant through NCC brought the social enterprise Victory Gardens to the mall. The Hunter Street Mall Victory Garden was situated within a 'creative hub' in the mall positioned close to former RN projects such as Studio Melt and Roost Creative[15] alongside the newly established cafes and bars, Three

Figure 11.4 Hunter Street Mall, 2016 (source: authors)

Bears Café and Basement on Market Street. The garden – a community-run space – was to be enhanced with outdoor furniture and murals, and through reciprocal relationships with the surrounding businesses (Victory Gardens n.d.). These new actants in the mall once again shifted its socio-material composition, and contributed to its contingent and creative regeneration. When RN began in late 2008, more than half the mall's retail space was vacant. By late 2015, few vacant retail spaces remained, and new RN projects were being established in first floor vacant office space (Westbury 2015). To date, RN has supported in excess of 200 creative and/or community projects in approximately 74 properties across the CBD, with more than 20 of the projects having 'graduated' to full commercial leases (Westbury 2015).

Beyond Renew Newcastle: emergent creative regenerations

The impact of a small-scale, community-based, creative and community actor, contingently aligned with the lively materiality of vacant space, in creatively (re)assembling and regenerating the mall, could not easily have been envisaged. The local media cast the changes as 'the miracle on Hunter Street ... Newcastle's main drag has a pulse ... once declared dead ... there is a flicker of life and with that a sense of hope and renewal' (Jameson 2009). Local and regional planners and developers never imagined the capacities that were realised when the lively properties of vacant space came into relation with the assemblage of RN:

> [RN] have done a fantastic job ... and such damn hard work to actually get in and have those spaces activated again.
>
> (Interview, State government agency)

Even its founders were surprised at how the project evolved:

> By the end of the first year we were pretty surprised ... We weren't planning on it being the next big thing ... There were lots of things that I think we underestimated about it.
>
> (Interview, RN)

RN has enabled Newcastle's creative and cultural community to be 'plugged in' to a new assemblage, centred on the mall, wherein new socio-material relationships have generated new capacities which have, in turn, opened up emergent opportunities for that community and for the creative regeneration of the mall. For the most part, this creative regeneration has emerged through flows of generative agency rather than through the explicit exercise of authority, power or the outlay of large-scale capital investment. As such this emerging socio-material assemblage has been able to enact its own form of grassroots 'creative' city (Gibson 2015).

RN's leader is conscious of the vibrant, emergent nature of the mall as it (re)assembles: it 'becomes something else – it's deepened or it's changing all the time, because the landscape of the city changes and other players come into the city' (Interview, RN). And through this (re)assembling, the mall itself has been invested with new capacities and mobilised new actants and materialities, in turn enabling new rounds of assembling that continue to shape Newcastle's regeneration. Referring to the burgeoning of speciality retail and food stores, public events, cafés and the wider animation of the mall's streetscape, RN's leader reflected:

> There's so much going on at the moment, you can't even keep track of things ... and I think they're a really good indication ... of how healthy, dynamic the place is.
>
> (Interview, RN)

Other creative actors in the inner city also recognise the socio-material changes as a catalyst for further activity. A not-for-profit music association interviewee commented:

> [RN] has changed the mall ... It's activated spaces that just would have just been laying dormant. It's bringing in a different mix of artisans or visitors which is great ... So it's been a real success story.

The distinctive character of community-based, grass-roots creativity in the mall has proliferated in creative initiatives and enterprises across the inner city that are contributing to regeneration (Table 11.2). These initiatives, involving a diversity of actors, have been enabled by the materiality of vacant or under-utilised commercial and/or public space and inspired by the distinctive atmosphere, affect and streetscape arising from the mall's creative regeneration. They draw on culture and creativity in a multiplicity of ways to enact diverse visions and objectives, further transforming the inner city's socio-materiality and enabling new trajectories for creative regeneration.

New capacities and further emergent regeneration pathways also emerge from the way RN is being plugged into formal planning frameworks for inner city regeneration. The Newcastle Urban Renewal Strategy positions RN as nurturing creative industries that contribute to the 'uniqueness' of inner-city Newcastle (DoPI 2012: 113). HDC acknowledges that the ongoing 'low-key yet effective' work of RN complements the Newcastle Urban Renewal Strategy (HDC 2015). NCC's Economic Development Strategy 2016–2019 aims to 'support creative business development through Renew Newcastle-style initiatives to encourage start-up businesses' (2015: 9). And, UrbanGrowth, recognising RN's role in the creative revitalisation of the city, became a funding partner of RN in 2015 (UrbanGrowth NSW 2015). Becoming 'plugged in' to other inner-city assemblages, constituted by business and political elites pursuing development agendas may well warrant

Table 11.2 Grass-roots creative ventures in inner Newcastle

Initiative	*Description*
The Royal Exchange Hybrid Theatre	A small creative arts establishment (predating Renew Newcastle) modelled on the basement-like art community houses of New York and Europe. The owner sought to utilise the vacant space as a for-profit enterprise which has evolved as a performance space for music, poetry, cabaret, experimental performances, foreign and cult films and drawing classes.
The Underground @ The Grand Hotel	Established in March 2011 by the not-for-profit Newcastle Improvised Music Association, the Grand Hotel provides weekly jazz and improvised music performance opportunities in under-utilised basement space.
Tower Cinema Film Festivals	With declining patronage at the Tower Cinema's inner-city complex, a cluster of community-run, not-for-profit film festivals have emerged since 2011 including: French, Italian, and German language film festivals, the Shout Gay and Lesbian film festival, Silent Film, Extreme Sports film festivals, Travelling Film festival and Real film festival.

Source: Authors

caution as a route to co-option (Colomb 2012). Yet such co-option of RN and its wider assemblage of community-based creativities and regeneration trajectories, cannot be predetermined nor assumed. These multiple actors are conditioned though never fully determined through the socio-material compositions in which they are embedded, retaining their heterogeneity and autonomy (Swanton 2013; Anderson et al. 2012). RN has never seen itself aimed at (re)configuring the mall, or the wider inner city, and indeed, is clear on the limits of plugging in to formal networks of authority:

> We're a very, very small organisation and we've had to kind of strip it back to 'no, our thing is getting access to empty space and putting arts and creative projects into it', and if we can do anything else above that, then it's just got to radiate from that.
>
> (Interview, RN)

Nonetheless, RN can be interpreted, in some senses, as a purposively created assemblage aimed to bring people, things and knowledge together to energise and resource the purposive creation of a particular kind of creative urban regeneration. It has, unpredictably, been able to claim authority in the fields of decision-making that formally seek to shape urban regeneration possibilities (Müller 2015, Iveson 2013), both in Newcastle and further afield, becoming a model enacted now across many Australian cities through the RN-based organisation, Renew Australia. The particular forms of creative regeneration it assembles is allowing grass-roots community-based groups continued visibility in Australian cities.[16] Moreover, the multiple, heterogeneous social and material actors gathered into assemblages behind the distinctive regeneration of the mall have rematerialised the inner city and its affective resonances (McGuirk et al. 2016). The mall has drawn together impassioned community support for its newly regenerated materialities and its small-scale, piece-meal variety of 'creative regeneration' and a commitment to maintaining it. Perhaps ironically given GPT's role in kick-starting the RN project, this support ignited strong community reaction against the proposed high-rise, large-scale redevelopment of the mall by GPT and UrbanGrowth NSW. In late 2015, GPT/UrbanGrowth released a revised, scaled-down masterplan where proposed building heights and commercial floor space have been significantly reduced and public space maintained. The trajectories of Newcastle's creativity-led regeneration are, clearly, emergent, influenced by an expansive cast of heterogeneous actants, and far from pre-determined.

Conclusion

The heterogeneous creative assemblages that have driven the regeneration of Hunter Street Mall are not the causal effects of policy visions and formal strategies for culture-led regeneration. Nor are they limited to the commodified renderings of creativity justifiably the focus of critique in critical analyses of culture-led urban regeneration. These creativities, continually fuelled by the mall's lively materiality, are maniform and multidimensional, as are the ways in which they contribute to urban regeneration. How creativity is envisioned and enacted, shifts as new actors and elements are gathered into relation in specific time–space contexts, and power is transferred to smaller-scale, community-based enterprises and individuals beyond traditional business and political elites. New urban performances have emerged that were once deemed unimaginable, creating a culture-led regeneration that had eluded the best efforts of urban planners and developers (see Waitt 2008; Andres 2013).

Westbury (2015: 144–145) argues that RN has been embraced by a diversity of urban actors across the political spectrum. Its malleability allows it to be co-opted by a diversity of urban stakeholders for multiple purposes. As such, it remains a paradox – is it another form of neoliberal urban entrepreneurialism temporarily maintaining property values while developers wait in the wings, or a grassroots movement that mobilises culture and creativity to enact an alternative politics and an alternative creative regeneration? Emergent research suggests such temporary creative movements have the potential to be either, and that the extent to which alternative pathways are enacted depends, in part, on political and material contexts and how far these can be worked to allow normalised discourses and practices around 'rights to the city' to be challenged (Iveson 2013; 2009; Novy and Colomb 2013; Van Schipstal and Nicholls 2014). To date, RN remains a resourceful organisation, seeking to create openings and space for Newcastle's grass-roots creative and cultural communities. Its strategies and tactics are suggestive of the progressive socio-material assemblages that can be drawn together to enable more inclusionary forms of creative-led regeneration than those that have rightfully attracted critique as the creative face of neoliberal urbanism.

This chapter has sought to expand knowledges of urban creativity and of how creativity regenerates urban spaces in contingent ways, using assemblage thinking to reach beyond a focus on elite actors and the assumption that neoliberal modes of regeneration will prevail, and attending to the agency of social and material actors in realising regeneration. Assemblage thinking may also generate insight into how we might begin to assemble particular socio-material compositions to enact wider and more inclusive forms of regeneration. Such a stance might encourage us not to dismiss creative temporary urbanisms as 'second-best' options for regeneration but to consider how they might be more effectively mobilised as opportunities for communities to co-opt space and enact different forms of socio-economic urban regeneration that exceed those of the developmental imaginaries of property, business and political elites.

Acknowledgement

Thanks are due to Jill Sweeney for research assistance associated with this chapter, and to the reviewers and editor for helpful suggestions on chapter revisions. Particular thanks are also due to a variety of participants in Newcastle's urban regeneration who kindly gave their time to take part in the research on which this chapter in based.

Notes

1 Following Lagendijk et al. (2014) we approach urban regeneration as an open conception, without assuming predefined prepositions about actors, roles, practices, materials or mechanism. Rather, we regard it as assemblage of processes centred on producing ongoing improvements in the social, economic and physical conditions of places and communities experiencing aspects of decline (adapted from Leary and McCarthy, 2013: 9).
2 Industries that produce goods and services that have symbolic and economic value (see Scott 1997).
3 These can be of grassroots, not-for-profit or commercial in nature.
4 For instance Jane Jacobs' *The Death and Life of Great American Cities* (1961) argued that creativity and innovation were key tenets of a successful social *and* economic urban space.
5 Assemblage theory and actor network theory share common conceptualisations in their relational, networked approach to urban phenomena. For detailed discussion on the convergences and differences between assemblage theory and ANT see Anderson et al. 2012 and Müller 2015.
6 This is not to imply that Renew Newcastle has been the only creative enterprise contributing to Hunter Street's creative regeneration and re-assembling. Of course, a multiplicity of creative actors are involved but there is not the space here to elaborate on these.

7 Now NSW Department of Planning and Environment.
8 The NSW state government's agency for managing major development projects (formerly Landcom).
9 RN has since attracted funding from Arts NSW, the NSW Department of Trade and Investment, HDC, NCC and NN.
10 Under a 30-day rolling contract, RN takes on insurance and basic maintenance responsibilities and loans the properties to 'custodians' –'artists, cultural projects and community groups' (RN, n.d.). These pay a minimal weekly fee to RN and are responsible for maintaining building appearance; however the building is occupied rent free. If a property under a contract attracts a commercial viable lease, the project is given 30-days' notice.
11 Following Latour (2004), the term actant refers to anything (human or non-human) that modifies what other actors do.
12 In 2012 the NSW state government announced truncation of the rail line to the west of the inner city, to be replaced with buses and light rail system. The line was finally truncated on 25 December 2015.
13 At this point, GPT owned around 25 buildings in the mall (approximately 90 tenancies) with many vacant.
14 For RN, this initial creation of a legal contractual arrangement has become its standard practice for the wider organisation founded by Westbury after the RN success, Renew Australia.
15 Studio Melt is an artisan jewellery store that graduated to a full lease in 2012. The Roost Creative is a collaborative space (where individuals engaged in the creative industries can share ideas, equipment and space) and graduated to a full lease post-2013.
16 Despite this expansion, Westbury (2015: 161–162) acknowledges that the RN model is not a panacea for all cities experiencing decline and that some cities have barriers that cannot be addressed through its model.

References

Anderson, B., Kearnes, M., McFarlane, C. and Swanton, D. (2012) 'On assemblages and geography', *Dialogues in Human Geography*, 2(2): 171–189.

Andres, L. (2013) 'Differential spaces, power hierarchy and collaborative planning: A critique of the role of temporary uses in shaping and making places', *Urban Studies*, 50(4): 759–775.

Atkinson, R. and Easthope, H. (2009) 'The consequences of the creative class: The pursuit of creativity strategies in Australia's cities', *International Journal of Urban and Regional Research*, 33(1): 64–79.

Bennett, J. (2005) 'The Agency of assemblages and the North American blackout', *Public Culture*, 17(3): 445–465.

Borén, T. and Young, C. (2013) 'Getting creative with the "Creative City"? Towards new perspectives on creativity in urban policy', *International Journal of Urban and Regional Research*, 37(5): 1799–1815.

Breitbart, M. M. (2013) 'Introduction: Examining the creative economy in post-industrial cities: alternatives to blueprinting Soho', in M. M. Breitbart (ed), *Creative Economies in Post-Industrial Cities. Manufacturing a (Different) Scene*, Aldershot: Ashgate Publishing.

Brownill, S. (2013) 'Just add water: Waterfront regeneration as a global phenomenon'. In: M. Leary and J. McCarty (eds), *The Routledge Companion to Urban Regeneration*, London: Routledge.

Colomb, C. (2012) 'Pushing the urban frontier: Temporary uses of space, city marketing, and the creative city discourse in 2000s Berlin', *Journal of Urban Affairs*, 34(2): 131–152.

Comunian, R. (2010) 'Rethinking the creative city: The role of complexity, networks and interactions in the urban creative economy', *Urban Studies*, 48(6): 1157–1179.

Department of Planning (2006) *The Lower Hunter Regional Strategy 2006–2031*, Sydney: Department of Planning.

Department of Planning and Infrastructure (DoPI). (2012) *Newcastle Urban Renewal Strategy*, Sydney: Government, Department of Planning and Infrastructure.

Dovey, K. (2010) *Becoming Places: Urbanism, Architecture, Identity, Power*, New York: Routledge.

Dovey, K. (2012) 'Informal urbanism and complex adaptive assemblage', *International Development Planning Review*, 34(4): 350–367.

Empty Spaces. (n.d.) 'A landlord's story: GPT', http://emptyspaces.culturemap.org.au/page/gpt, (accessed 28 August 2013).

Färber, A. (2014) 'Low-budget Berlin: Towards an understanding of low-budget urbanity as assemblage', *Cambridge Journal of Regions, Economy and Society*, 7: 119–136.

Farias, I. (2011) 'The politics of urban assemblages', *City*, 15(3–4): 365–374.

Farias, I. and Bender, T. (2010) *Urban Assemblages: How Actor Network Theory Changes Urban Studies*, New York: Routledge.

Florida, R. (2002) *The Rise of the Creative Class: And How it's Transforming Work, Leisure, Community, and Everyday Life*, New York: Basic Books.

Gibson, C. (2015) 'Negotiating regional creative economies: Academics as expert intermediaries advocating progressive alternatives', *Regional Studies*, 49(3): 476–479.

Gibson, L. and Stevenson, D. (2004) 'Urban space and the uses of culture', *International Journal of Cultural Policy*, 10(1): 1–4.

HDC (Hunter Development Corporation). (2009) *Newcastle City Centre Renewal Report to NSW Government*, Newcastle: Hunter Development Corporation.

HDC (Hunter Development Corporation). (2012) *Honeysuckle Celebrating 20 Years*, Newcastle: Hunter Development Corporation.

HDC (Hunter Development Corporation.) (2015) *Annual Report 2014–2015*, Newcastle: Hunter Development Corporation.

Hunter Development Corporation (2009) *Newcastle City Centre Renewal Report*, Newcastle: Hunter Development Corporation.

Iveson, K. (2009) 'Responding to the financial crisis: From competitive to cooperative urbanism, *Journal of Australian Political Economy*, 64: 211–221.

Iveson, K. (2013) 'Cities within the city: Do-it-yourself urbanism and the right to the city', *International Journal of Urban and Regional Research*, 37(3): 941–956.

Jacobs, J. (1961) *The Death and Life of Great American Cities*, New York: Random House.

Jacobs, J. M., Cairns, S. and Strebel, I. (2007) '"A tall storey ... but, a fact just the same": the red road high-rise as a black box', *Urban Studies*, 44(3): 609–629.

Jameson, N. (2009) 'The can-do guy', *Newcastle Herald*, 12 ~December http://renewnewcastle.org/media/the-can-do-guy/ (accessed 2 October 2017).

Lagendijk, A, Van Melik, R., De Hann, F., Ernste, H, Ploegmakers, H. and Kayasu, K. (2014) 'Comparative approaches to gentrification: A research framework', *Tidjschrift voor Economische en Social Geografie*, 10(3): 358–365.

Landry, C. (2000) *The Creative City: A Toolkit for Urban Innovators*, London: Comedia.

Larner, W. (2011) 'C-change? Geographies of crisis', *Dialogues in Human Geography*, 1(3): 319–335.

Lashua, B. (2013) 'Pop-up cinema and place-shaping: urban cultural heritage at Marshall's Mill', *Journal of Policy Research in Tourism, Leisure and Events*, 5(2): 123–138.

Latour, B. (2004) *The Politics of Nature*, Cambridge, MA: Harvard University Press.

Leary, M, and McCarthy, J. (2013) *The Routledge Companion to Urban Regeneration*, London: Routledge.

McCann, E. J. (2011) 'Veritable inventions: cities, policies and assemblage', *Area*, 43(2): 143–147.

McFarlane, C. (2011a) 'The city as assemblage: dwelling and urban space', *Environment and Planning D: Society and Space*, 29(4): 649–671.

McFarlane, C. (2011b) 'Encountering, describing and transforming urbanism', *City*, 15(6): 731–739.

McGuirk, P., Winchester, H. P. M. and Dunn, K. M. (1996) 'Entrepreneurial approaches to urban decline: The Honeysuckle redevelopment in inner Newcastle, New South Wales', *Environment and Planning A*, 28(10): 1815–1841.

McGuirk, P., Mee, K. and Ruming, K. (2016) 'Assembling urban regeneration? Resourcing critical generative accounts of urban regeneration through assemblage', *Geography Compass*, 10(3): 128–141.

Müller, M. (2015) 'Assemblages and actor-networks: Rethinking socio-material power, politics and space', *Geography Compass*, 9(1): 27–41.

Munzner, K. and Shaw, K. (2014) 'Renew who? Benefits and beneficiaries of Renew Newcastle', *Urban Policy and Research*, 33(1): 17–36.

Newcastle City Council (2013) *Newcastle 2030 – the Newcastle Community Strategic Plan*, Newcastle: Newcastle City Council.

Newcastle City Council (2015a) *Economic Development Strategy 2016–2019*, Newcastle: Newcastle City Council.

Newcastle City Council (2015b) *Newcastle City Council Cultural Strategy 2016–2019*, Newcastle: Newcastle City Council.

Novy, J. and Colomb, C. (2013) 'Struggling for the right to the (creative) city in Berlin and Hamburg: New urban social movements, new "spaces of hope"?', *International Journal of Urban and Regional Research*, 37(5): 1816–1838.

Peck, J. (2005) 'Struggling with the creative class', *International Journal of Urban and Regional Research*, 29(4): 740–770.

Peck, J. (2011) 'Creative moments: working culture, through municipal socialism and neoliberal urbanism', in E. McCann and K. Ward (eds), *Mobile Urbanism: Cities and Policymaking in the Global Age*, Minneapolis, MN: University of Minnesota Press.

Regional Cities Taskforce (2006) *Revitalising Newcastle City Centre Plan (NCCP) Vision*, Newcastle: Department of Planning.

RN (Renew Newcastle). (n.d) 'About' http://renewnewcastle.org/about, (accessed 31 January 2013).

Ruming, K., Mee, K. and McGuirk, P. (2016) 'Planned derailment for new urban futures? An actant network analysis of the "great [light] rail debate" in Newcastle, Australia', in Y Rydin and L Tate (eds) *Actor Networks of Planning: Exploring the influence of Actors Network Theory*, London: Routledge.

SCAPE Strategy (Scape). (2010) *Hunter Street Revitalisation Final Strategic Framework*' Newcastle: Newcastle City Council.
SCAPE Strategy (2010) *Hunter Street Revitalisation Master Plan Final Strategic Framework*, North Sydney: SCAPE.
Scott, A. J. (1997) 'The cultural economy of cities', *International Journal of Urban and Regional Research*, 21(2): 323–339.
Shaw, K. (2014) 'Melbourne's Creative Spaces program: Reclaiming the 'creative city' (if not quite the rest of it)', *City, Culture and Society*, 5: 139–147.
Swanton, D. (2013) 'The steel plant as assemblage', *Geoforum*, 44: 282–291.
UrbanGrowth NSW. (2015) *Revitalising Newcastle*, Sydney: NSW State Government.
Van Schipstal, I. and Nicholls, W. (2014) 'Rights to the neoliberal city: The case of urban land squatting in "creative" Berlin', *Territory, Politics, Governance*, 2(2): 173–193.
Victory Gardens. (n.d.) 'Hunter Street Mall Victory Garden', http://202020vision.com.au/project/?id=306, (accessed 10 December 2015).
Waitt, G. (2008) 'Urban festivals: Geographies of hype, helplessness and hope', *Geography Compass*, 2(2): 513–547.
Waitt, G. and Gibson, C. (2009) 'Creative small cities: Rethinking the creative economy in place', *Urban Studies*, 46(5-6): 1223–1246.
Waitt, G. and Gibson, C. (2013) 'The Spiral Gallery: Non-market creativity and belonging in an Australian country town', *Journal of Rural Studies*, 30: 75–85.
Westbury, M. (2015) *Creating Cities*, Melbourne: Niche Press.

Part III
Middle-ring and suburban regeneration

12 Greyfield regeneration

A precinct approach for urban renewal in the established suburbs of Australia's cities

Peter Newton and Stephen Glackin

Introduction

With bipartisan political support for a big Australia and historically high levels of international immigration, Australia's major cities are experiencing rates of population growth of the order of 1.6 per cent pa, unmatched in comparable OECD countries (IA 2015). Approximately 80 per cent of the nation's forecast population growth to 30.5 million by 2031 is expected to be absorbed by its biggest cities; and Australian Bureau of Statistics (ABS) projections foreshadow a further concentration of Australia's increasing population within the capital cities: from 66 per cent in 2013 to 72 per cent in 2051. By mid-century, Sydney and Melbourne are projected to have populations of almost 8 million each, while Perth is projected to reach the current size of Sydney (4.8 million) and Brisbane to match Melbourne's current population of 4.3 million. City-planning strategists have the challenge of developing metropolitan plans capable of accommodating this additional population (Chapter 2).

Three principal pathways exist. The first involves a continuation of the well-established process of *greenfield* development on the urban fringe of Australian cities that perpetuates, through low-density master planned residential estates, a continuation of the garden cities models of development (Chapter 17). The associated sprawl of Australian cities has proven to be unproductive economically, socially and environmentally (Newton et al. 2012). It has been unbalanced development. The failure here has been to develop (mostly detached) housing that relies almost exclusively on private transport in locations that require commutes to jobs and services at increasing distance from the place of residence. The second has involved attempts to disperse population growth to regional cities. Here the failed decentralisation policies of an earlier era need to be replaced by those that recognise the agglomeration economy benefits that accrue to large cities and to ensure regional cities can become *part* of a functional mega-metropolitan region – a 'system of cities' – via *fast* rail. Here 350 km/hr high-speed rail services linking major capital cities and their provincial cities would transform these regions into the equivalent of present day middle-ring metropolitan suburbs where 30-minute commutes to work are typically the norm (Newton et al.

2017b). The third pathway involves redirecting population and investment inwards to the established middle (and inner) suburbs.

Here it is critical to differentiate urban infill redevelopment taking place in *brownfield* arenas as distinct from *greyfields* (Newton 2010; Newton and Glackin 2014). Brownfields represent abandoned or under-utilised industrial or commercial sites associated with earlier eras of economic activity. In Australian cities the docklands districts developed in the eighteenth and nineteenth centuries have proved to be an attractive focus for renewal in recent decades (e.g., Sydney's Barangaroo, Melbourne Docklands – Chapters 6 and 7). The relative absence of significant heavy industry in Australia's capital cities has meant that there are not the vast tracts of brownfields characteristic of North American and European cities. In contrast, greyfields constitute areas located between the CBD (and inner city housing markets) and the more recently developed greenfield suburbs. They are well-located with respect to public transport access, jobs and specialist health services and tertiary education, but their residential stock is ageing physically and technologically and is environmentally poor-performing. The value of property in the greyfields lies predominantly in the land, not the built asset. As such, the real estate is under-capitalised. In an era where Australian cities need to transition from suburban to more urban forms, and to regenerate to more sustainable, productive and resilient built environments, new models need to be developed to assist with the renewal and intensification of the greyfields in Australia's large cities.

Attempts at intensification in the greyfields are playing out in contested space (Ruming 2014), and it is clear more needs to be done in relation to engagement between property owners and state and local-government planners for greyfield infill to occur in a more effective, strategic and regenerative fashion. Outlining the new processes, instruments and rationale required for precinct-scale redevelopment in the established suburbs of Australian cities is the focus of this chapter. On the basis of research outlined in the sections that follow, greyfield renewal precincts (GRPs) have been recommended as a new instrument for urban redevelopment in *PlanMelbourne Refresh* (DELWP, 2015). We provide a brief overview of greyfield redevelopment to date (focusing principally on Melbourne), identifying where previous metropolitan planning processes are failing to deliver the necessary volume, mix and location of new housing in the face of increasing demographic, environmental, social and economic pressures on Australia's major cities. The chapter then proceeds to outline the case for precinct scale regenerative redevelopment, and the arenas where urban innovation is required to enable governments to instantiate GRPs within their metropolitan planning processes.

Planning strategies for urban intensification in the greyfields

In the most recent strategic metropolitan plans for the major capital cities in Australia, a number of different spatial frameworks have been introduced to encourage more compact forms of city development (Chapter 2). These include activity centres and transport corridors. Meanwhile, piecemeal infill has become the major vehicle for residential redevelopment in the greyfields. *Activity centres*, ranging in scale from the CBD to principal, major and neighbourhood activity centres, provide a basis for realising the multiple-nuclei city, first articulated by Harris and Ullmann (1945) and subsequently reconceptualised as the poly-centred city – an urban form capable of enabling more integrated transport and land-use development (Newton 2000). The more recent and related 'city of cities' and '20-minute city' concepts feature in the metropolitan planning strategies for both Sydney and Melbourne (MAC, 2015; NSW Department of Planning 2005). Major outer suburban activity nodes are becoming targets for transit oriented development (TOD) projects as a vehicle for enhancing mobility in car-dominated regions of Australian cities (eg. stations on Sydney's North West Rail Link: see Newton et al. 2017b). *Transport corridors* act as a focus for higher density redevelopment along major road arterials, and in cities like Melbourne, road and tram corridors. Intensification of residential and commercial development along these routes (see Adams et al. 2009) represents an attempt to inject greater mixed use activity within the urban fabric (as with activity centres).

For both activity centres and transport corridors, special growth zones have been established to attract higher density development 'as of right'. To date, however, neither has proven to be the strong magnets for new residential development anticipated in the strategic metropolitan plans of Melbourne (Newton and Glackin 2014) or Sydney (Troy 2016), with the exception of high-rise apartments in the CBD/inner city. Activity centres and transport corridors are both necessary but not sufficient foci for achieving infill housing targets of the order 70 per cent of total new housing development, generally recognised as the quantum needed to put a brake on greenfield development (MAC 2015). Study of urban infill development in Melbourne between 2004 and 2010 revealed that neither activity centres nor transport corridors were performing as strong attractors for more intensive housing development as envisaged in recent metropolitan strategic plans. Rather, housing redevelopment was occurring most strongly in the greyfields as piecemeal 'knock-down-rebuild' (KDR) (Wiesel et al. 2013), a process that is common across all Australian cities due to the fact that it is readily accommodated within existing planning and building regulations. It is a sub-optimal form of redevelopment with respect to housing yield (almost 80 per cent of greyfield housing redevelopment projects were 1:1 or 1:2–4 single lot redevelopments; Newton and Glackin 2014). Environmental innovation (water, energy and waste) is poor; and loss of private green space is not being compensated (Hurley et al. 2016). The fragmented nature of KDR is also fast becoming a future inhibitor to the prospect of more economically, environmentally and socially attractive precinct-scale regeneration in the greyfields.

Precinct-scale residential regeneration

Precinct-scale residential regeneration has been advanced as a new model for urban renewal in Australia's greyfields (Newton 2010; Newton et al. 2011). It represents the *scale* capable of contributing most significantly to transformative (regenerative) sustainable urban development goals (Newton et al. 2013). Its premise rests on the multiple benefits that can be gained from creation of precincts through the amalgamation of individual contiguous greyfield properties, enabling more creative urban re-designs that can positively activate ageing neighbourhoods. This provides for: more housing supply and mixed-use development capable of meeting the needs of population growth that is now being directed inwards rather than outwards; enhanced built environment performance associated with local renewable energy generation (Newton and Newman, 2013); water-sensitive design and waste management (Waller et al. 2016); and more eco-efficient retrofitting of urban infrastructure services using distributed technologies applicable to precinct scale (Swinbourne et al, 2016). In order for it to be realised, however, *innovation* is required in several different arenas related to: spatial analysis, urban planning, urban design, stakeholder engagement (especially local communities of landowners), legal and financial models for property redevelopment, modular construction and governance (Figure 12.1). Drawing on several Australian Housing and Urban Research Institute (AHURI) funded workshops with over 70 thought leaders and specialists in city planning and development we have developed a model of greyfield regeneration (full details in Newton et al. 2011). This framework has become the basis for an extended program of applied research (Greening the Greyfields[1]), directed towards developing the new tools and processes required to underpin a major transition in urban practice in the established suburbs of Australian cities. A new development model for precinct-scale residential regeneration in suburban greyfields is required – akin to the model for brownfield precinct development that was established during the Building Better Cities Program of the early 1990s. The innovations required in relation to greyfields precinct regeneration are outlined in the following sections.

Where to focus?

Geospatial tools (such as ENVISION; Glackin 2013) have been created to characterise all property parcels in a city in relation to their redevelopment potential, either in basic market value terms (e.g., value of land as percentage of total property value; above 70 per cent land value typically triggers KDR following sale) or in respect of broader regenerative planning considerations using multi-criteria analysis (e.g., focusing more intensive development on land with high redevelopment potential that is close to public transport, schools, public green space, etc.). The objective is to increase active transport; introduce more mixed use; reconfigure precinct land previously associated with roads and car parking. This provides a sound basis for municipal-scale

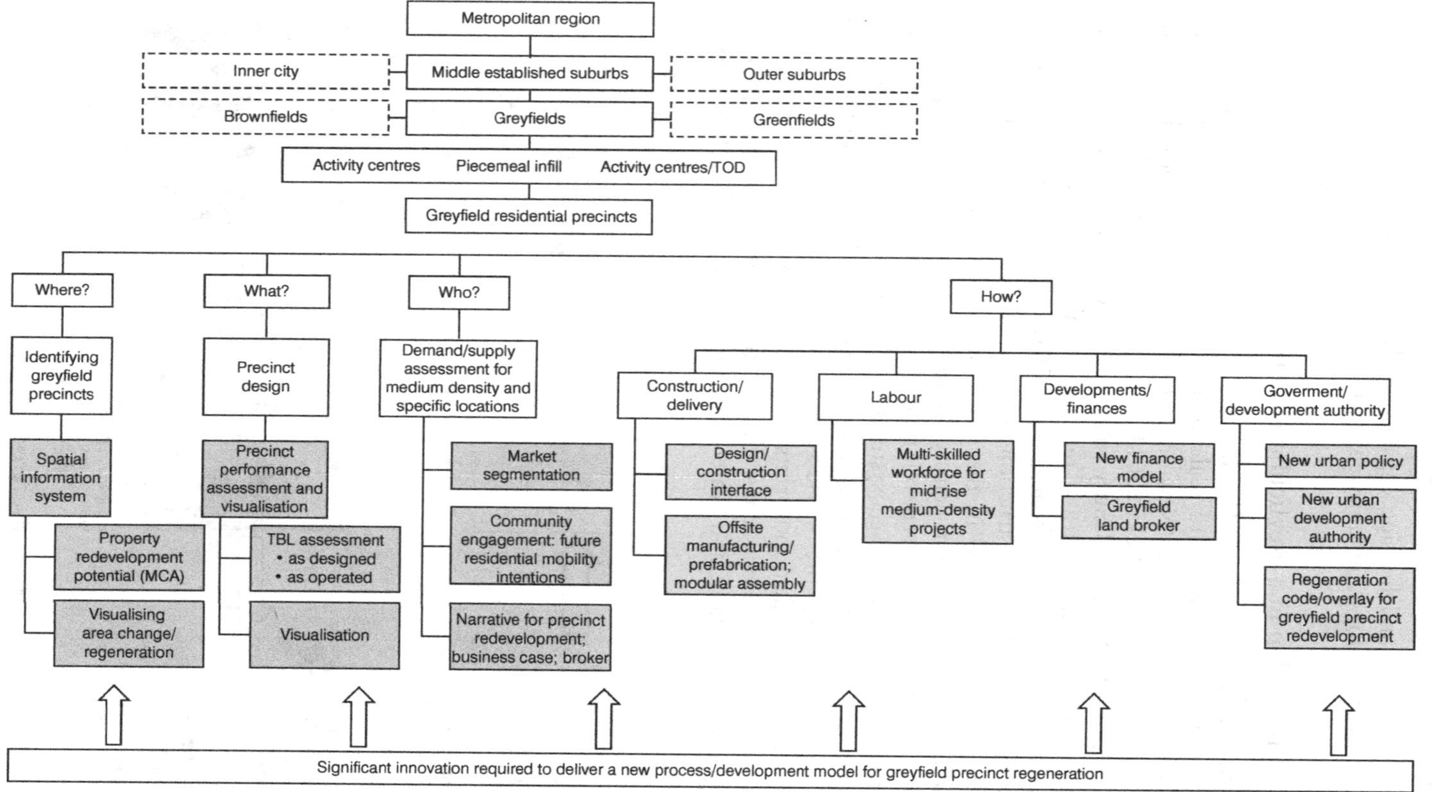

Figure 12.1 Innovation arenas for greyfield precinct regeneration (source: adapted from Newton et al. 2011)

strategic planning and housing capacity analysis (see items 1–3 in Figure 12.2) – a platform for community engagement.

What to redevelop in a greyfield precinct?

Medium density is the focus for this model. There has been a significant undersupply of this type and scale of residential redevelopment in the established suburbs to date (Newton and Glackin 2014), notwithstanding the latent demand which Kelly (2011), Newton et al. (2011) and Newton and Glackin (2015) clearly identify. In terms of new residential development in Melbourne, detached housing remains a major segment, but high-rise apartment construction has soared since 2010 – controversially in some respects (e.g., quality of construction and size and liveability of apartments). Most has been constructed in the CBD, surrounding suburbs and well-located brownfields. Medium-density housing construction is lagging.

It is critical that greyfield redevelopment is regenerative: that is, renewing and restoring the wider ecosystems underpinning urban settlement (Girardet 2015). Newton (2012) has revealed that while Australia's cities are classed as among the world's most liveable, they also exhibit the world's largest ecological footprints – due to the resources that their built environments and residents consume and the emissions to air (especially greenhouse gases), land and water that they generate. The low-density

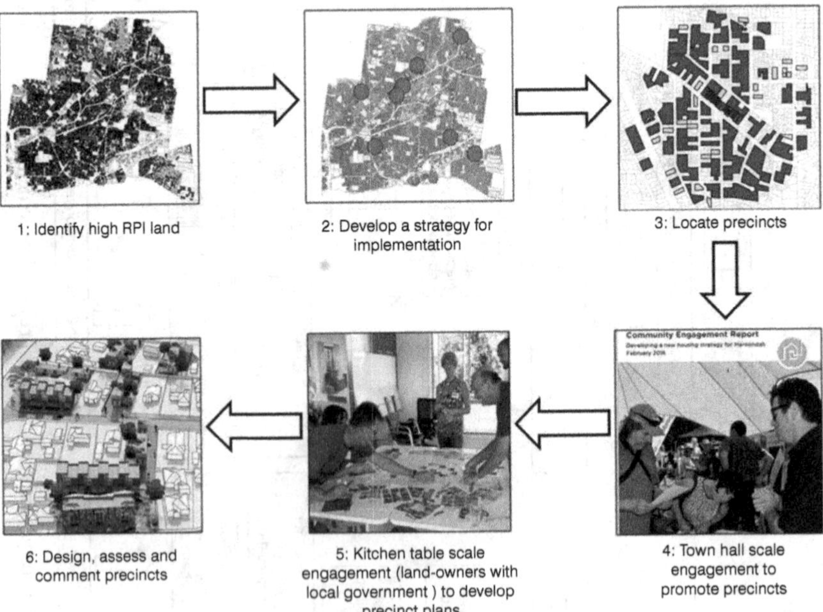
1: Identify high RPI land
2: Develop a strategy for implementation
3: Locate precincts
Community Engagement Report
4: Town hall scale engagement to promote precincts
5: Kitchen table scale engagement (land-owners with local government) to develop precinct plans
6: Design, assess and comment precincts

Figure 12.2 Locating prospective greyfield precincts for regeneration (source: authors and Monash Architecture (box 6))

(suburban) form of Australian cities is a major contributing factor (Newton 2000). The piecemeal KDR characteristic of middle-ring redevelopment to date is not regenerative (Chapter 14). It does not provide the optimal pathway for carbon neutrality (despite recent uptake in rooftop solar; Newton and Newman 2013), water sensitive urban design (CRC for Water Sensitive Cities 2015), management of food waste (Waller et al. 2016), offsetting loss of private green space, or reconfiguring road spaces for walkability and community interaction. Precinct-scale redevelopment can positively address these urban design challenges and advance the transition of established residential areas from suburban to urban in a manner that *enhances* the character of the neighbourhood (Newton et al. 2013). It should be a requirement of the development approval process (see later section on planning) that a greyfield precinct redevelopment *adds benefit* to the locality, or in Birkeland's (2014) term is an *eco-positive* development. A new class of precinct information modelling platform is emerging to drive this type of urban design innovation by enabling more comprehensive (in terms of range of performance attributes considered) and faster forms of design assessment (Newton et al. 2017a; Trubka et al. 2016).

Who wants to live in medium-density housing in the established suburbs?

The established suburbs in Australia's major cities represent an increasingly attractive location to live, given their high level of residential amenity and accessibility to jobs, higher education, specialist health services and the CBD (Rawnsley and Spiller 2012). Recent housing surveys have revealed significant levels of preference for medium-density accommodation as a place to live: 40 per cent in Sydney and 38 per cent in Melbourne (Kelly 2011); and in Melbourne 60 per cent opted for 'compact city living', characterised by medium-density housing in a locality well served by public transport, compared to a 'garden city' environment – detached house and yard in car-dependent outer-suburban location (Newton and Glackin 2015).

Currently, however, new housing supply is not matching demand. Studies of dwelling occupancy by age continue to reveal that there is massive under-occupancy of (largely detached) housing in Australian cities (Newton et al. 2011; Newton and Glackin 2015): with 60 per cent or more of households aged between 60 and 80 years of age living in accommodation with two or more spare bedrooms. They are ageing in place; seemingly a preferred government outcome for an ageing Australia (Productivity Commission 2015: 6): 'small households, big houses – not necessarily a problem'. This also appears aligned with a Property Council of Australia (2015) report that advocates property owners unlocking their home equity in retirement by entering into a reverse mortgage or the shared-sale agreement with a bank or financial institution. The prospect of downsizing and remaining in the same district (a persistent preference for households relocating within cities) and having a reasonable cash surplus from the sale is thwarted by the current similarity in price points for older detached houses and new medium-density located in the same area (Newton et al. 2011). Under these circumstances for the average household, why move?

This is the challenge that 'Greening the Greyfields' (Newton et al. 2012) is attempting to address: creating a new model process for precinct-scale redevelopment in the established suburbs which will become a strong inducement to property owners of detached greyfield housing to become pro-active in engaging with neighbours to consolidate their properties into a *greyfields renewal precinct* (GRP). The economics of this process are proving attractive to growing sets of neighbours in cities such as Sydney and Melbourne who are typically getting 50–100 per cent more for their property if they can sell collectively rather than individually. From this point, a number of development options are available to the group of property owners: sell to a property developer and move elsewhere (but with enhanced financial capacity to downsize and update within same area), or participate in a public–private co-housing type of development partnership and occupy one of the new dwellings in the medium–density precinct.

Currently, however, there is no GRP process that has been clearly articulated by government planning agencies at state and local level that can encourage the innovative medium–density designs that would deliver a higher dwelling yield as well as the additional economic, social and environmental benefits that can potentially accrue to precinct scale redevelopment. This is a topic we now turn to.

How can greyfield precinct regeneration be delivered in established suburbs? The role of the planning process

In revisiting *Plan Melbourne 2014* and publishing *Plan Melbourne Refresh* (DELWP 2015), the current state government in Victoria has flagged its intention to consider some new policy objectives related to the challenge of urban infill. These are:

- establishing a 70/30 target for new infill/greenfield housing development;
- creating residential zones that identify areas where more intensified redevelopment is encouraged in contrast to those areas requiring appropriate protection from 'inappropriate development';
- injecting more diversity in new housing;
- introducing a policy statement in *Plan Melbourne 2016* capable of supporting greyfield renewal precincts (GRPs); seen as 'a form of innovation which will become more feasible and attractive as the city's housing stock ages' (DELWP 2015: 57). This is a direct response to the research undertaken in Melbourne for 'Greening the Greyfields' (MAC 2015: 34–35).

The landscapes of cities are typically framed by two levels of planning: one to do with urban plans, the other to do with codes – that is, strategic planning and statutory planning (Marshall 2011). In Australia, the long-term strategic planning of cities has largely been the preserve of state governments, and the statutory planning implementation and development approval of individual urban (property) projects almost entirely in the hands of local government. The alignment of these processes and tiers of government can often be problematic – especially during periods where attempts are being made to steer a transition in urban planning to more compact cities. The vision of desired changes in relation to future urban housing strategies – as outlined in *Plan Melbourne Refresh* – needs to be supported by an aligned set of land-use codes (zones).[2] In the sections that follow, we analyse the latest data available on new housing development across Melbourne in relation to patterns of urban infill and the influence that the recent (re)zoning may play in achieving residential intensification in the established suburbs.

Infill patterns and zoning

Over the period 2005–2014, urban infill accounted for 53 per cent of all (net) new dwellings built in Melbourne (Table 12.1). This represents a significant shift in the pattern of metropolitan residential development over a relatively short period in the city's history, but still falls short of the 70 per cent urban infill target in *Plan Melbourne Refresh*. Infill is largely undifferentiated in most Australian metropolitan planning documents. However, it contains both brownfields and greyfields redevelopment arenas: greyfields containing both activity centres and transport corridors which are zoned in a manner designed to attract higher-density residential

Table 12.1 New housing development in Melbourne, 2005–2014: infill versus greenfield arenas

Development Arena	*Net new residential construction, 2005–2014*	
	Number of new dwellings (net)	*Percentage of new development: 2005–2014*
Greenfield	142,819	47
Brownfield	78,714	26
Total Greyfield	82,312	27
• Activity centres	10,120	3
• Transport corridors	15,123	5
• Ad hoc infill (KDR)	57,069	19
Total Infill (brownfield + greyfield)	161,026	53
Total metropolitan area	303,845	100

Source: Derived by the authors from Housing Development Data (HDD) supplied by DELWP

growth; with lower-density ad hoc knock-down-rebuild everywhere else. New Geographic Information System (GIS) modelling (see Newton and Glackin 2014 for details) using latest state government housing data and planning overlays reveals that the three principal foci in strategic metropolitan plans designed to concentrate residential redevelopment (and associated populations) in Melbourne – brownfield precincts, activity centres and transport corridors – are attracting 34 per cent of all (net) new housing. In contrast, fragmented knock-down-rebuild contributes one fifth of the total new dwelling construction – and 35 per cent of all infill. Greenfield, brownfield and greyfield housing developments have different geographies as well as underlying property development models. For Melbourne, greenfield development is concentrated in the outer suburban growth corridors; brownfields are most strongly concentrated in the inner municipalities of Melbourne, Yarra, Port Phillip and Stonnington. New housing development in the greyfields is more widely spread across the city, although it can be seen to be most heavily concentrated in the middle-ring suburbs.

Zoning of residential areas is meant to be an instrument of government that enables councils and their communities to better direct the location and scale of residential change within the context of a broader metro-wide strategy. With the release of new residential zones by the state government of Victoria in July 2013 (DELWP 2016), municipal councils were given 12 months to amend their planning schemes to implement the new residential zones. The result has been a mixed process with mixed outcomes. For those municipalities which completed the process it is clear that for a number there has been a clear attempt to 'lock up' a significant percentage of residential properties in the neighbourhood residential zone (NRZ), the zone that permits only minimal levels of redevelopment (one-for-one replacement of individual dwellings and 'dual occupancies under some circumstances'). These are the municipalities that also generate the largest number of applications to the Victorian Civil and Administrative Appeals Tribunal by property owners challenging planning development applications in their neighbourhoods (Newton and Glackin 2013). Most councils have assigned zero or limited parcels of properties to residential growth zones (RGZs). Others have assigned most property to the general residential zone (GRZ), with zero or limited property to NRZ. Attempts to 'game' a residential zoning system by local governments can undermine a metropolitan strategy attempting to re-shape the city by encouraging higher levels of urban infill.

This has not gone unnoticed – from the press (Donegan 2015) or from the state government. The latter subsequently announced: 'Differences and inconsistencies in the application of the zones are now subject to review by the Victorian government which has appointed the Managing Residential Development Advisory Committee for this purpose' (DELWP 2016: 1).

Greyfield housing redevelopment potential

Earlier research (Newton 2010) has revealed that residential properties with high residential redevelopment potential (i.e. when the percentage of land value to total property value is above 70 per cent) tend to be redeveloped within a short span of time after coming onto the market: typical KDR. The residential redevelopment potential analyses for each municipality in Melbourne reveal a clear housing life-cycle across cities: youthful and maturing in the greenfields, ageing and redeveloping in the greyfields (Newton and Glackin 2014). The analyses indicate:

- Which municipalities have greatest housing redevelopment potential – fundamental information for metropolitan planning agencies in gauging future infill housing supply potential (Glackin and Newton 2016). Evidence has shown that some municipalities are already facing very significant change brought about by fragmented lot-by-lot redevelopment. In most cases, however, this scale of redevelopment is not providing sufficient (net) new housing supply nor is it creating enhanced neighbourhood amenity. It is therefore critical that these municipalities begin the process of strategically engaging with the issue of larger-scale medium-density urban infill projects. The new Victorian State Government requirement that all Melbourne municipalities prepare housing strategies capable of responding to the future metropolitan plans for the city represents an attempt to better align state/local as well as strategic/statutory planning processes (DELWP 2015).
- Which areas within a municipality have greatest potential for residential redevelopment. This can be identified through using tools such as ENVISION, mentioned earlier. By using spatial tools to identify areas containing significant concentrations of highly redevelopable lots, municipalities can begin to identify areas that could potentially be open to precinct scale re-zoning. Here it is clear that there is scope for progressing beyond KDR *if* precinct-scale regeneration based on lot amalgamation can become a model for retrofitting greyfields (Figure 12.3).

The manner in which residential redevelopment potential is being constrained by current zoning practices by a number of municipalities in Melbourne is further illustrated in Table 12.2. Several local government areas have over 50 per cent of their properties with high redevelopment potential (where over 70 per cent of total value is in the land) zoned as NRZ.

Conclusion: towards a planning pathway for greyfield precinct regeneration

Progress is being made towards the articulation of a statutory planning framework capable of accommodating precinct scale greyfield residential

Table 12.2 Residential redevelopment potential by residential zone 2014

Municipality	*Residential growth zones*			*General residential zones*			*Neighbourhood residential zones*		
	Total Dwellings	*High RPI*	*Percentage*	*Total Dwellings*	*High RPI*	*Percentage*	*Total Dwellings*	*High RPI*	*Percentage*
Banyule	196	31	16	32281	20006	62	13848	7111	51
Bayside	46	33	72	5229	2854	55	32352	21939	68
Boroondara	2089	208	10	16312	6559	40	45360	32412	71
Brimbank	2288	1195	52	27208	12649	46	37392	11683	31
Cardinia	0	0	0	14213	1994	0	2552	512	0
Casey	4	3	75	83947	14676	17	0	0	0
Darebin	1244	662	53	49779	29285	59	6747	4351	64
Frankston	0	0	0	50770	14727	0	0	0	0
Glen Eira	1941	587	30	10576	3260	31	43578	23794	55
Dandenong	7596	2654	35	28884	15340	53	13412	8557	64
Hobsons Bay	0	0	0	35305	20287	0	0	0	0
Hume	0	0	0	48024	17653	0	0	0	0
Kingston	2	2	100	58543	30805	53	0	0	0
Knox	0	0	0	4308	1133	0	50332	14906	0
Manningham	2195	732	33	33690	21152	63	1669	1147	69
Maribyrnong	806	21	3	21921	10419	48	4811	2624	55
Maroondah	863	257	30	26670	10435	39	15008	7874	52
Melbourne	1833	189	10	17081	3435	20	299	134	45

Melton	3325	458	14	35459	3830	11	1395	206	15
Monash	206	0	0	65749	42328	64	905	713	79
Moonee Valley	0	0	0	44807	24626	0	0	0	0
Moreland	2177	812	37	18676	11226	60	39504	26257	66
Mornington	0	0	0	81266	31315	0	0	0	0
Nillumbik	200	76	38	9205	3113	34	4967	2036	41
Port Phillip	0	0	0	44999	12013	0	0	0	0
Stonnington	5867	643	11	26292	7220	27	13247	6728	51
Whitehorse	3873	1076	28	29183	17858	61	28994	18069	62
Whittlesea	1005	608	60	57538	30180	52	863	711	82
Wyndham	0	0	0	66684	6940	0	0	0	0
Yarra	0	0	0	12174	2487	0	19654	7829	0
Yarra Ranges	1366	374	27	2815	421	15	23257	3748	16

Source: Derived by the authors from HDD data provided by DELWP

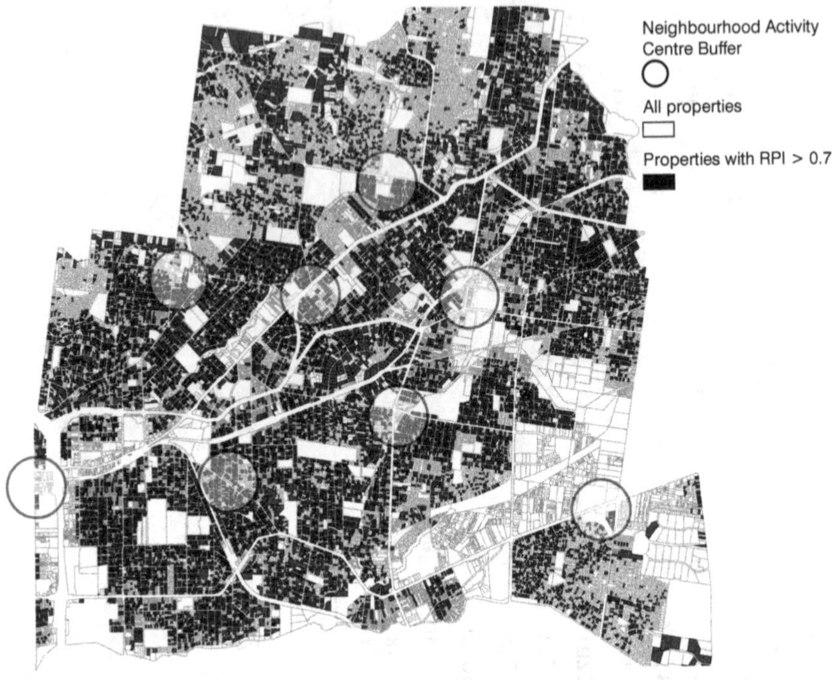

Figure 12.3 Residential properties with high redevelopment potential in the context of neighbourhood activity centres (source: results of ENVISION analyses based on HDD data provided by DELWP

regeneration. Statutory codes are currently being used, however, to prohibit the necessary types and scales of redevelopment required in the established suburbs. They are too standardised and restrictive, stifling creativity and innovation in urban design (there is an equal challenge to the urban designers here as to the code makers to create more eco-efficient medium-density housing precincts). Notwithstanding, at a strategic policy level, there is a recognition that *greyfield precinct renewal* (GPR) is a necessary new policy direction in the next metropolitan strategic plan for Melbourne. It features in the precursor document *Plan Melbourne Refresh* (DELWP 2015: 57).

In the current review of Melbourne's residential zones it will be important that new codes (zones) are developed that support an urban vision involving greyfield precinct regeneration: alignment of plan and code is powerful. They need to be part of a proactive vision for reshaping our urban landscapes: from suburban to urban in character and from ecosystem consumers to more sustainable eco-positive contributors. In short, they need to ensure that urban redevelopment (or retrofitting as it is now being termed: Eames et al. 2017; Hodson and Marvin 2016) is regenerative – *adding benefits* that were lacking in the built environment that is being replaced. Coding embraces a variety of elements, principles and relationships that must be capable of guiding ongoing development at urban (eg street, precinct) and building (eg. typology, massing etc.) scale. Statutory approaches that can be used to deliver greyfield precinct renewal (GPR) on large or amalgamated lots could involve developing a *new GPR zone* to be applied to this purpose or develop a special purpose *GPR overlay* capable of being applied with existing zones and zone schedules.

An overlay, by its nature, applies an extra layer of control in relation to the land against which it applies. It overlays the zoning of any land, and would have an effect of modifying (influencing) a development application. Areas most suited to a GPR overlay would be in current general residential zones or neighbourhood residential zones. Since the regulatory regime over any land parcel is the combination of the zone, overlays, general provisions and policies, the effect of the total package is important. A GPR overlay would need to incorporate elements that speak to the objectives of greyfield precinct renewal:

- define areas where policies to encourage GPR are to be applied (Figure 12.2, box 3);
- require consistency with a (municipal) plan for the preferred form of development intended for the area;
- set out neighbourhood character objectives: neighbourhood character provisions typically set out requirements intended to preserve an existing or former neighbourhood character; however, for an area where GPR is to be encouraged, they could be used very differently, to set out a *desired future character* based on redevelopment, encouraging integrated development and discouraging single lot projects (that would be capable of impeding achievement of the overall scheme);
- overcome aspects of the existing planning scheme which are important to manage the interface between new and existing development in the context of lot-by-lot redevelopment, but which may be unnecessarily restrictive within a GPR project (i.e. permitting gradation of heights within the precinct);
- provide for GPR to take place over time, in stages, not necessarily in a defined order (see Figure 12.2, box 6); this has elements in line with the *hybrid* greyfield precinct renewal model outlined in Newton et al. (2011).

A development contributions plan could set project development contributions required to fund infrastructure retrofits and upgrades required with redevelopment of a precinct. Municipal councils may also provide additional support for GPR projects. For developments consistent with an incorporated plan, or which meets criteria specified in the overlay, a simplified and expedited planning process could apply – in particular, applications could be made exempt from notice, objection, and third-party review processes *if* they adhere to the municipal design guidelines for such projects. The EcoDistricts Protocol recently released in North America provides a useful framework for achieving more sustainable neighbourhood-scale redevelopments (EcoDistricts, 2016).

At this stage, there is no need to promote one approach (special zone, combination of overlays, special overlay) above the other. The important issue is to define the *outcomes* to be achieved through controls and demonstrate the effect of required control systems on development project outcomes. Here, the capacity for rapid assessment and visualisation of precinct-development plans using state-of-the-art digital design platforms (Glackin et al. 2016; Newton et al. 2017a) means that different stakeholder ideas and interests can be examined by more effective engagement processes than is currently the case (Glackin and Dionisio 2016). 'Greening the Greyfields' is currently directing additional attention to:

- developing legal, financial and management structures to facilitate citizen-led lot consolidations and sales needed to achieve precinct-scale regeneration and to market the product;
- addressing other potential barriers to GPRs, such as, but not limited to, the effect of restrictions from covenants;
- defining and promoting the potential benefits to property owners of participating in GRP schemes rather than undertaking lot-by-lot redevelopments;
- investigating the possibility of tax incentives for developers who generate social good, or landowners who, through obtaining additional income, are not prohibitively taxed and financially less well-off than if they didn't become part of the precinct;
- ensuring that there is ample provision to promote 'good design', with benefits of redevelopment extending beyond that of additional housing yield to wider neighbourhood amenity.

These touch on a number of the innovation arenas outlined in Figure 12.1. To be successful, urban regeneration requires co-ordinated innovation on multiple fronts.

Postscript

Since this chapter was written, 'greyfield redevelopment' is now Policy 2.2.4 in the metropolitan planning strategy – *PlanMelbourne 2017–2050* – released in March 2017: a seven-year transition from concept to policy.

Acknowledgements

The authors would like to acknowledge the CRC for Spatial Information and AHURI for the funding they have provided over the past six years while we have been researching this topic, including colleagues at Curtin University and more recently University of Canterbury, Monash University and RMIT University. For his insights into the urban planning process in Melbourne we are grateful to John Manton (Victorian Planning Scheme Services).

Notes

1 http://www.crcsi.com.au/research/4-5-built-environment/4-55-greening-the-greyfields/
2 For metropolitan Melbourne, the residential zones established in 2013/4 to guide residential redevelopment can be found at http://www.dtpli.vic.gov.au/__data/assets/pdf_file/0003/229854/Reformed-Residential-Zones-fact-sheet-1_07_2014.pdf

References

Adams, R., Eagleson, S., Goddard, S., Przibella, S., Sidebottom, T., Webster, F. and Whitworth, F. (2009) *Transforming Australian Cities: For a More Financially Viable and Sustainable Future: Transportation and Urban Design*, Melbourne: City of Melbourne and Department of Transport.

Birkeland, J. (2014) 'Resilient and sustainable buildings'. In L. Pearson, P Newton and P Roberts (eds), *Resilient Sustainable Cities*, London: Routledge.

CRC for Water Sensitive Cities (2015) 'Ideas for Fishermans Bend'. http://watersensitivecities.org.au/new-publication-ideas-for-fishermans-bend/ (accessed 23 April 2016).

DELWP (2015) *Plan Melbourne Refresh*, discussion paper, Melbourne: Victoria State Government.

DELWP (2016) *Residential Zones State of Play*, Melbourne: Victoria State Government.

Donegan, P. (2015) 'The selfishness that's tearing Melbourne apart', *The Age*, 29 April, p. 45.

Eames, M., Hunt, M., Lannon, S. and Dixon, T. (eds) (2017) *Retrofitting Cities for Tomorrow's World*, Chichester: Wiley.

EcoDistricts (2016) 'EcoDistricts protocol' https://ecodistricts.org/get-started/the-ecodistricts-protocol/ (accessed 6 June 2017).

Girardet, H. (2015) *Creating Regenerative Cities*, Routledge: London.

Glackin, S. (2013) 'Redeveloping the greyfields with ENVISION: using participatory support systems to reduce sprawl in Australia', *European Journal of Geography* 3(3): 6–22

Glackin, S. and Dionisio, M. (2016) 'Deep engagement and urban regeneration: tea, trust, and the quest for co-design at precinct scale'. *Land Use Policy* 52: 363–373.

Glackin, S. and Newton, P. (2016) 'Assessing the capacity for urban infill in Australian cities', paper presented at Future Housing: Global Cities and Regional Problems Conference, 9–10 June, Melbourne.

Glackin, S., Trubka, R. and Dionisio M. (2016) 'A software-aided workflow for precinct scale residential redevelopment'. *Environmental Impact Assessment Review* 60: 1–15.

Harris, C. and Ullman, E. (1945) 'The nature of cities', *The Annals of the American Academy of Political and Social Science* 242: 7–17.

Hodson, M. and Marvin, S. (eds) (2016) *Retrofitting Cities*, London: Routledge.

Hurley, J., Parmeter, E., Phelan, K., Amati, M., and Livesly, S. (2016) 'Does higher density city development leave urban forests out on a limb?' *The Conversation*, 14 April, http://theconversation.com/does-higher-density-city-development-leave-urban-forests-out-on-a-limb-57106 (accessed 6 June 2017).

IA (2015) *Population Estimates and Projections*, Sydney: Infrastructure Australia.

Kelly, J.-F. (2011) *Getting the Housing We Want*, Melbourne: Grattan Institute.

MAC (2015) *Plan Melbourne 2015 Review*, Report of the Ministerial Advisory Committee, Melbourne, http://refresh.planmelbourne.vic.gov.au/ (accessed 6 June 2017).

Marshall, S. (ed) (2011) *Urban Coding and Planning*, New York: Routledge.

Newton, P. (2000) 'Urban form and environmental performance', in K. Williams, E. Burton and M Jenks (eds), *Achieving Sustainable Urban Form*, London: E&FN Spon.

Newton, P. (2010) 'Beyond greenfields and greyfields: The challenge of regenerating Australia's greyfield suburbs', *Built Environment*, 36(1): 81–104.

Newton P. (2012) 'Liveable and sustainable? socio-technical challenges for 21st century cities', *Journal of Urban Technology*, 19(1): 81–102.

Newton, P. and Glackin, S. (2013) 'Using geo-spatial technologies as stakeholder engagement tools in urban planning and development', *Built Environment*, 39(4): 480–508.

Newton, P. and Glackin, G. (2014) 'Understanding infill: Towards new policy and practice for urban regeneration in the established suburbs of Australia'slities', *Urban Policy and Research* 32(2): 121–143.

Newton, P. and Glackin, S. (2015) 'Regenerating cities: Creating the opportunity for greyfield precinct infill development', in R. Leshinshy and C. Legacy (eds), *Instruments of Planning: Tensions and Challenges for more Equitable and Sustainable Cities*. New York: Routledge.

Newton, P. and Newman, P. (2013) 'The geography of solar photovoltaics (PV) and a new low carbon urban transition theory', *Sustainability*, 5: 2537–2556.

Newton, P. Murray, S. Wakefield, R. Murphy, C. Khor, L-A and Morgan, T. (2011) *Towards a New Development Model for Housing Regeneration in Greyfield Residential Precincts*, Final Report no. 171, Melbourne: Australian Housing and Urban Research Institute.

Newton, P. Newman, P. Glackin, S and Trubka, R. (2012) 'Greening the greyfields: Unlocking the redevelopment potential of the middle suburbs in Australian cities', *World Academy of Science, Engineering and Technology*, 71: 658–677

Newton, P., Marchant, D., Mitchell, J., Plume, J., Seo S. and Roggema R. (2013) *Performance Assessment of Urban Precinct Design*, Sydney: CRC for Low Carbon Living.

Newton, P., Plume, J., Marchant, D., Mitchell, J. and Ngo, T. (2017a) 'Precinct information modelling: a new digital platform for integrated design, assessment and management of the built environment', in A. Sanchez, K. Hampson and G. London (eds), *Integrating Information across the Built Environment Industry*, London: Routledge.

Newton, P., Taylor, M. A. P., Newman, P., Stanley, J., Rissel, C., Giles-Corti, B. and Zito, R. (2017b) 'Decarbonising suburban mobility', in H. Dai (ed), *Low Carbon Mobility for Future Cities: Principles and Applications*, Stevenage: IET.

NSW Department of Planning (2005) *City of Cities: A Plan for Sydney's Future*. NSW Government's Metropolitan Strategy, Sydney: NSW Department of Planning.

Productivity Commission (2015) 'Housing decisions of older Australians', research paper, Canberra: Productivity Commission.

Property Council of Australia (2015) 'Unlocking home equity for senior Australians to free up extra retirement income' http://bettertax.gov.au/files/2015/09/Property-Council-of-Australia-Submission-3.pdf (accessed 6 June 2017).

Rawnsley, T. and Spiller, M. (2012) 'Housing and urban form: a new productivity agenda' in R. Tomlinson (ed), *Australia's Unintended Cities*, Melbourne: CSIRO Publishing.

Ruming, K.J. (2014) 'Urban consolidation, strategic planning and community opposition in Sydney, Australia: Unpacking policy knowledge and developing a typology of support/resistance', *Land Use Policy*, 39: 254–265.

Swinbourne, R., Hilson, D and Yeomans, W, (2016) *Empowering Broadway, Co-operative Research Centre for Low Carbon Living Phase 1 Report*, Sydney: Cooperative Research Centre for Low Carbon Living.

Troy, L. (2016) *Projections, targets, forecasts and models: how do Sydney's metro strategies try to anticipate our future housing needs?*, Sydney: City Futures Research Centre.

Trubka, R., Glackin, S., Lade, O. and Pettit, C. (2016) 'A web based 3D visualisation and assessment system for urban precinct scenario modelling', *ISPRS Journal of Photogrammetry and Remote Sensing*, 117: 175–186.

Waller, V., Blackall. L. and Newton P.(2016) 'Composting as everyday alchemy – producing compost from food waste in 21st century urban environments' in R. Crocker and K. Chiveralls (eds), *Reuse in an Accelerated World: Mining the Past to Reshape the Future*, London: Routledge.

Wiesel, I., Pinnegar, S., Freestone, R. (2013) 'Supersized Australian dream: Investment, lifestyle and neighbourhood perceptions among "knockdown-rebuild" owners in Sydney', *Housing, Theory and Society*, 30: 312–329.

13 Shopping centre-led regeneration

Middle-ring town centres and suburban regeneration

Kristian Ruming, Kathy Mee, Pauline McGuirk and Jill Sweeney

Introduction

Regeneration is conventionally associated with inner-city environments. However the ageing of middle-ring suburbs has encouraged a new round of activities aimed at suburban regeneration, including mixed-use retail-led regeneration focused on a town centre (Randolph and Freestone 2008; Ruming et al. 2010; Newton 2010). Such strategies involve strengthening the town centre through master planning retail redevelopment, improvements to public transport and the public domain, and increasing the density of housing around the shopping centre and transport hub. In the Australian context, this overlaps with a thrust for polycentric cities (more recently Malcolm Turnbull's '30-minute city') driven by the use retail development as a lever for the formation or revitalisation of a town centre, creating financing vehicles for public infrastructure investment and public domain improvements, providing employment opportunities, increasing housing supply (including affordable housing), and transit oriented development (Chapter 2).

Mixed-use retail-led regeneration can involve main street-style retail revitalisation but also can be triggered by shopping centre redevelopment (Southworth 2005). Such regeneration can emerge as an opportunity presented by private centre developers seeking to expand and enhance their retail investments. The wider (and modest) literature and policy discussions around mixed-used retail-led regeneration have largely been conducted in general terms, outlining the policy logics and aspirations (Southworth 2005; Goodman and Coiacetto 2012). Generally lacking from this literature has been a careful appraisal of the importance of contextual specificities in which middle-ring regeneration is being mobilised (Ruming et al. 2010). In this chapter, drawing on an analysis of attempts to induce the retail-led regeneration of Charlestown, a middle-ring suburb of Newcastle in NSW's Hunter region, we demonstrate the importance of contextual specificity to configuring the opportunities and limits of retail-led regeneration, particularly that focused on shopping centres. We show that, despite literature which tends to treat shopping centres as homogenous reflections of suburban development, they have specific connections to locally based patterns of retail development, population change dynamics, and transport developments. The types of regeneration that result may vary considerably across a city's middle-ring suburbs. Our analysis of Charlestown thus reveals the importance of more carefully considering the complexities of Australian cities' middle-ring suburbs and their regeneration.

Background

The suburban shopping centre is a ubiquitous feature of the Australian urban landscape (Goodman and Coiacetto 2012). Privately owned, enclosed, and increasingly large-scale developments, these shopping centres began to spawn in the 1950s and 1960s as Australian cities rapidly expanded via outward suburban development (Guimaraes 2014). They soon became one of the defining features of urban sprawl. Shopping centres have also been at the core of 'retail suburbanisation', the exodus of retail functions from central urban regions, following outward population movement (Espinosa and Hernandez 2016; McGreal and Kupke 2014). They have been critiqued as the antithesis of good planning and development, being embedded in large car parks, detached from local communities with limited (if any) pedestrian access, enclosed and inward-oriented, and of poor urban design and aesthetics (Southworth 2005). Moreover, they have been criticised for their tendency to impact negatively on traditional high-street shopping strips, characterised by independent retailers, and smaller centres (Griffiths et al. 2008; Instone and Roberts 2006; Thomson, Benson and McDonagh 2015). Further, out-of-centre commercial and retail developments have been criticised for competing with designated centres and inducing traffic congestion in locations lacking adequate road infrastructure (Guy 2008; Goodman and Coiacetto 2012).

Despite this dominant discourse of urban design deficit, these centres play an important role in the lives of many urban citizens. Research has outlined the important roles shopping centres play in the social and community lives of young people, older populations and parents with children (Fobker and Grotz 2006; Matthews et al. 2000; Tyndall 2010; Vanderbeck and Johnson 2000). Community facilities (such as libraries, daycare centres, health centres) have increasingly been integrated into shopping centres as state and local government budgets were stretched, leaving partnership with commercial retail centres as a cost-effective form of provision. Importantly, these partnership arrangements often involve a trade-off between the provision of community infrastructure and negotiated planning conditions aligned with the commercial interests of centre owners. In the 1980s, shopping centres diversified to incorporate entertainment and leisure functions including restaurants, cinemas, bowling alleys and ice rinks. Relatedly, they have become employment centres in their own right.

Centre developers, owners and major retailers are important players in the politics of urban change and exert a significant influence over urban planning and built form and thus, on the social and economic geographies of their locales (Emery 2006). Economically, shopping centres present a profitable business model for large development and retail companies alike (such as Stockland and Westfield) (Goodman and Coiacetto 2012; McGreal and Kupke 2014). Profit levels, accompanied by ageing building stock and a new strategic position in urban policy, have prompted many companies to embark on ambitious, large-scale renovation, retrofitting and expansion projects across their portfolio. Increasingly, there has been a shift away from the large block development model towards one which mobilises the logic of the "town centre". More problematically, in the UK, some have suggested that large retailers, such as Tesco, have a history of actively targeting regeneration projects in an attempt to expand market share (Imrie and Dolton 2014; Wrigley et al. 2002). Indeed one of the critiques of UK high street urban regeneration has been the dominance of a small number of large retail chains and their capacity to direct development options (McNeill 2011). A number of authors have been critical of the level of influence large developers and retailers have in shaping urban environments, in particular, the extent to which commercial development promotes a landscape of consumption, which reinforces social and economic inequalities and poor urban design (McNeill 2011; Guimaraes 2014; Raco 2003).

Within the context of the regeneration of Australian cities, the process of renovation and retrofitting suburban shopping centres represents an important and under-researched element of urban change. The continued expansion of Australian cities has altered the relative positioning of older suburban shopping centres, with many now located in middle-ring suburbs identified in metropolitan strategic plans as sites of dwelling densification and population growth as new mixed-used 'town centres' (Chapter 2), a trend also observed internationally (Lowe 2000; Tallon 2013). These are clearly activities critical to regenerating Australian cities. For example, the South East Queensland Regional Plan (QLDDIP 2009: 20) states that:

> activity centres at Southport and Robina [large shopping centre precincts] are expected to expand their roles as commercial, retail, and administrative and specialist centres, and evolve into mixed-use centres.

Middle-ring suburban shopping centres are repositioned within contemporary strategic planning policy, then, as catalysts for mixed-used regeneration. Importantly, this regeneration is thought to be both advanced by the shopping centre itself, as owners redevelop and retrofit existing assets, *and* by the promotion of a wider urban regeneration in planning policy, by catalysing the development of housing, employment, transport, improved aesthetics and built form, for which the re-imagined shopping centre becomes a localised town centre. Growing demand for commercial and retail services, as well as the option to diversify into housing provision, means that large retail property interests (property developers and shopping centre owners) are well placed to leverage their assets to secure investment funds to contribute to this wider regeneration. Yet to meet the regeneration aspirations they are freighted with, these centres are typically assumed to adopt a transit oriented development (TOD) model, characterised by quality public transport, mixed land use and high relative population/dwelling densities (Newman 2009). The traditional inward looking, self-contained retail centre model is not well aligned with these strategic objectives.

Furthermore, the faith placed in the capacity of such centres to catalyse broader regeneration as town centres, is largely untested (Guy 2008). The logic underpinning the economic and financial claims on shopping centre-led town centre regeneration is that shopping centre redevelopment, as a cornerstone development, can attract other development representing a level of commercial confidence to the market as well as drawing customers who will, in turn, create demand for other retail and, ultimately, commercial and residential development. Thomas and Bromley's (2003) study of town-centre regeneration in Wales identified the extent of functional integration between a new shopping centre complex and the existing town centre as the central component of successful renewal. Yet, there is an inherent commercial tension between the large shopping centre and retailers located outside and not paying (typically high) rents to the shopping centre owner. They effectively represent commercial competition and can induce a resistance to improving economic performance, public amenity, access and services that might draw customers away from the shopping centre (Voyce 2006; Tyndall 2010). Retail-led regeneration, particularly in the form of large shopping centre-led regeneration, represents a complex balancing act between the commercial objectives of the shopping centre owner and the broader regeneration objectives of local governments and communities, particularly in terms of the scope for mixed-use development, increased housing density and provision of public transport.

In this chapter we explore the redevelopment of one middle-ring shopping centre precinct – Charlestown, in Newcastle, NSW – to illustrate the broader processes, issues and implications of suburban shopping centre redevelopment and the impact on middle-ring regeneration. In order to unpack the processes and outcomes of regeneration at Charlestown, we draw upon analysis of planning documents spanning the last decade, and interviews with the local council, planning consultants and major landowners, and residents.[1]

Charlestown

Charlestown is a middle-ring suburb in 'Greater Newcastle', the regional centre of NSW's Hunter region (Figure 13.1). Newcastle's urban area spreads across a number of local government areas, particularly the local government areas of Newcastle (population 161,225 in 2015[2]) and Lake Macquarie (population 204,166 in 2015[3]). Charlestown is in Lake

Macquarie. Lake Macquarie has a significantly older population than the state average with 25 per cent of residents aged 60 or over (compared to 20 per cent for NSW).[4] This is partly due to the region's attraction as a retirement destination being in close proximity to Sydney and Newcastle. The population of Lake Macquarie is projected to grow by '60,000 people between 2006–2031 creating demand for 36,000 new dwellings and 12,200 new jobs',[5] leading to increased demand for medium-density and senior housing developments.[6]

Charlestown Square, opened in 1979, was the first major 'big box' shopping centre to be developed in Lake Macquarie and represented a departure from smaller-scale retail centres which had dominated the region (Hassell 2007a). It is located in the midst of a wider town centre, a cluster of mixed-use business, retail, employment and social services alongside medium- and high-density housing. Moreover it is the hub of sub-regional road and transport networks (Figure 13.2).

Charlestown plays a significant role in regional strategic planning, identified in the Lower Hunter Regional Strategy as Lake Macquarie's only 'regional centre' and part of a renewal corridor for higher-density residential and mixed-use redevelopment (NSWDoP 2006). Projected

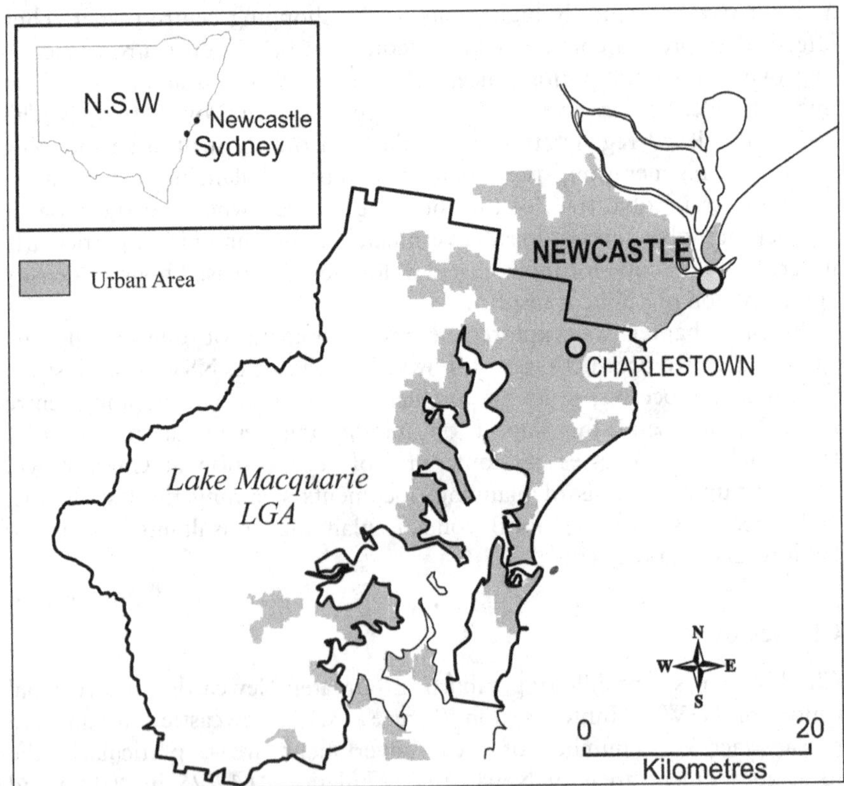

Figure 13.1 Charlestown, NSW (source: Olivier Rey-Lescure)

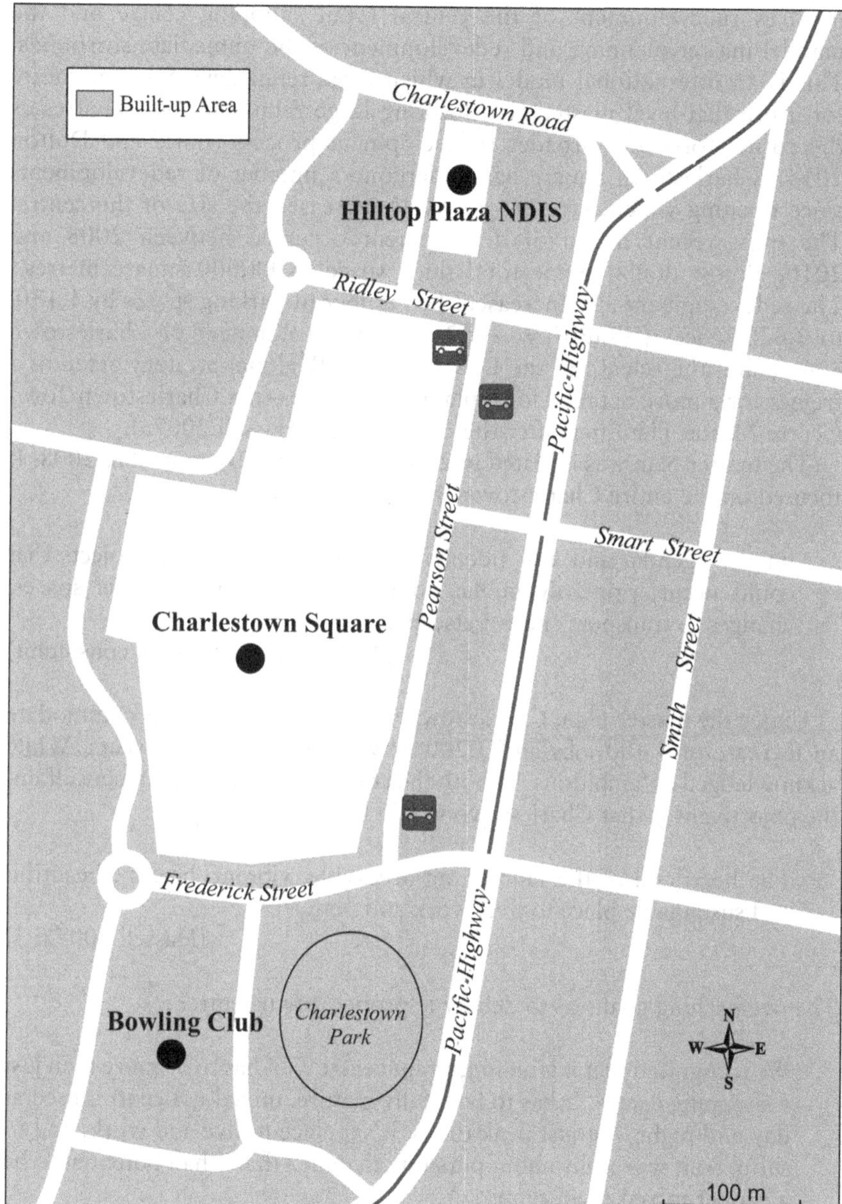

Figure 13.2 Charlestown Square (source: Olivier Rey-Lescure)

population growth and planning policies which encourage development around regional centres were important in providing the conditions for shopping centre-led regeneration in Charlestown. Charlestown's regeneration is characteristic of many middle-ring retail centres that involves redevelopment of the central retail shopping centre and the parallel master planning and redevelopment of the immediate surrounds. This is an international model in which large retail developers advocate for a precinct-level master plan locating large retail shopping centres as the cornerstone of a broader redevelopment process (Imrie and Dolton 2014). Charlestown Square has undergone a number of redevelopments since opening which have progressively increased the size of the centre. The most recent round of development occurred between 2008 and 2010, almost doubling the total floor space to 88,000 square metres.[7] The redevelopment also increased the number of parking spaces by 1,130, to 3,620 spaces (URaP-TTW 2007). The redevelopment of Charlestown Square and the release of the Lower Hunter Regional Strategy acted as a trigger for a more detailed local planning scheme – the Charlestown Town Centre Master Plan (hereafter the master plan) (Hassell 2007a).

The master plan was initiated in 2007 and adopted by council in 2008. It focused on the entire Charlestown area and aimed to:

> set the agenda, and also [identify] a number of catalyst projects that could occur, primarily public domain upgrades, new public spaces, changes to transport and streetscape.
>
> (Interview, master plan consultant)

Under the master plan, Charlestown's regeneration would accommodate an increase of 4,400 jobs and 3,200 new dwellings over 25 years. While acknowledged as 'ambitious' by both the council and master plan consultant, the objective was that Charlestown would be:

> The heartland of the Hunter, an accessible, vibrant, healthy, beautiful and sustainable place to live, work and play.
>
> (Hassell 2007a: 3)

The overarching goal was to deliver a 'proper' town centre:

> We recognised that it is an important centre ... it has to be more than just a shopping centre. It has to be a fully mature, mixed-use centre; used at day and night, not just peak times. It's a place to live and work, and to shop. So it was really about pursuing that idea that it had potential to be a proper mixed town centre.
>
> (Interview, master plan consultant)

This position as a regional centre was core in framing the built and commercial objectives of the master plan:

> As a regional commercial centre, it is expected that a greater number of taller buildings will need to be accommodated in Charlestown to

provide the floor space required for future job and housing targets in the Lower Hunter Regional Strategy.

(Hassell 2007a: 15)

Typical of many master plans, the strategy had a wide ranging set of generic objectives around improved open space, accessibility, community development, character, amenity, urban form and economic performance. The master plan divided Charlestown into 'Town Centre Core' and the 'Town Centre Periphery' (Figure 13.3). The core—containing Charlestown Square, the 'main street' (the Pacific Highway), the 'transit street' (Pearson Street), the bulk of community services and infrastructure, open space, commercial and retail land use, and, increasingly high-density housing—was the focus of the first wave of regeneration, led by the redevelopment of Charlestown Square and associated development contributions (including new or improved community and sporting infrastructure, negotiated under the voluntary planning agreement [VPA] between the owner, GPT, and the council). Like many middle-ring shopping centres, the redevelopment of Charlestown Square represented a process of compromise, where the redeveloped shopping centre had to manage both the historical shopping centre itself and an increased emphasis on interaction with the surrounding town centre. New development was limited to 3–4 storeys (except in designated 'gateways') and on the 'rejuvenation' of existing buildings. The master plan would:

ensure the provision of high-quality public domain, compatible building uses, form and scale, and the efficient and convenient movement of people and vehicles.

(Hassell 2007a: 8)

By contrast, the periphery would provide a transition from the largely detached single dwellings that dominate Charlestown, to the more high-density core. Its function would be primarily residential, with a significant increase in the number of 2- and 3-storey unit blocks. This strategy was designed to provide a high degree of housing choice while aiming to contain 40 per cent of all new dwelling construction in existing urban areas.

Since the adoption of the master plan, the strategic importance of Charlestown to the region's development has been maintained by council with the Town Centre Development Control Plan adopted in 2012 (LMCC 2012a). Running in parallel, the Draft Charlestown Streetscape Master Plan was released in 2012 and aimed to:

improve the aesthetic appearance of the town centres, promote the area's personality, increase opportunity, accessibility, functionality, economic effect and provide a modern and easily maintainable public domain

(LMCC 2012b: 2)

While a product of its unique context and history, the planning and regeneration of Charlestown offers a number of valuable insights into middle-ring regional centres, where retail and residential growth are targeted. The Charlestown case study reveals the synchronicities and challenges that arise from mixed-use shopping centre-led regeneration in middle-ring suburbs to develop suburban centres, encourage denser housing and reduce car dependence.

The town centre

Suburban shopping centres fit uneasily into traditional visions of a well-planned and conventionally aesthetically pleasing regenerated city. The redevelopment of shopping centres has sought to appropriate the notion of the traditional town centre. Drawing on the urban design principles of New Urbanism, the ideal town centre is seen as one which operates at a 'human scale', promotes walking, offers enclosed spaces and intimate plazas, and where the bulk of the built form does not overwhelm the public realm (Dudley 2003). Idealised town centres are designed to emphasise the unique

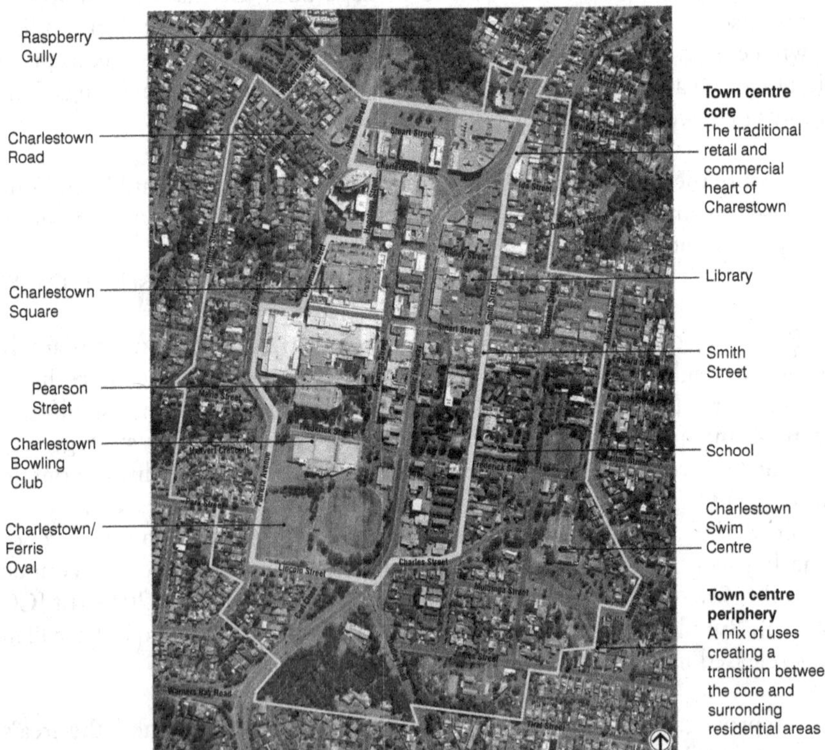

Figure 13.3 Master plan study area (indicating core and periphery town centre) (source: Hassell 2007a: 8)

characteristics of each location, promote a sense of place, and provide a mix of land uses. Yet, the inward orientation of large shopping centres has historically represented the antithesis of this, with their characteristic bulk, lack of legibility and generic 'blandscapes' (Staeheli and Mitchell 2016). This was echoed by many when discussing Charlestown Square:

> You can't have beautifully designed shopping centres. It's like box developments ... Charlestown was never sexy.
>
> (Interview, business chamber official)

Shopping centre-led regeneration projects increasingly look to emphasise a sense of place or identity and place-branding as a commercial strategy to increase patronage and better integrate the shopping centre with its surrounds (Catungal et al. 2009). For Charlestown Square, many of the public works associated with its redevelopment sought to emphasise the area's history. Walls and paths included images, key historical facts and motifs. Such improvements to the public domain support the objectives of the master plan:

> we all recognised, that Charlestown didn't have a clear character or identity beyond the shopping centre. That is why the Master Plan needed to articulate one.
>
> (Interview, master plan consultant)

For planners and developers, if regeneration is going to occur, town centres must be made appealing beyond the shopping centre itself. While the redevelopment process and associated development contributions might facilitate improvement of public spaces surrounding the shopping centre, at least in the short term, this improvement is limited. As residents indicated, this effectively defines the new and old in the regenerated town centre:

> when we drove down the Pacific Highway which is technically the main street of Charlestown, you kind of look at it the same way and go 'yeah, this is a little bit derelict'. But then you go one street over where the Square is and you go, 'oh no this is nice'.
>
> (Interview, Resident 1)

> I haven't seen the beautification in any other areas.
>
> (Interview, Resident 2)

The capacity of large shopping centres to act as 'town centres' servicing a local population relies heavily on design and built form. Charlestown Square reflects a series of issues arising from this reliance. First, as discussed below, the dependence on car travel and associated parking reduce opportunities for pedestrian and cycling movement. Even locals are likely to drive rather than walk or cycle to the shopping/town centre precinct. Second, the built form of the shopping centre itself potentially acts as a barrier to integration. This is, in part, due to the historical legacy of 'big box' style shopping centres constructs from the 1970s. The developer recognised this reality of contemporary middle-ring redevelopment:

> I would love to have started over again ... if you were [starting afresh], you would not have done it that way. But how do you start to take them and reshape them and remould them?
>
> (Interview, owner Charlestown Square)

There is only so much a redevelopment can do to facilitate integration with the local area. Unlike more traditional style town centres, large shopping centre complexes generally have a front and a rear. The rear of the shopping centre, often characterised by parking, loading bays, and garbage bins represents a major barrier to integration. To compound issues, the rear is associated with noisy truck movements at all hours. These barriers were well recognised in relation to Charlestown Square:

> [There is a] very big wall, facing that line of houses. You'd hate to wake up to that every day.
>
> (Interview, business chamber official)

The structure of the shopping centre can also act as a barrier to integrating other forms of transport into a regeneration strategy. For example, in describing the difficulties of accessing Charlestown Square by bicycle, a representative of the cycleways movement argued:

> A bicycle can, in most cases, get relatively close to the end destination. At Charlestown Square you've got these, almost shield-like car parks on the outside, shielding the activity that's happening on the inside.
>
> (Interview, Newcastle Cycleways Movement)

In contrast, the front of the shopping centre may be more successfully integrated with the surrounding area via plazas and piazzas, transport interchanges, improved public works and beautification initiatives. For Charlestown Square's developers, opening up and integrating the front of the shopping centre to the rest of the town centre was a central objective:

> It started with the really traditional shopping centre mentality of 'internalise everything, be totally inward refocusing'. You look at the Charlestown [Square] development ... it's trying to become outward focusing again but that is clearly a staged process and something that takes time.
>
> (Interview, owner Charlestown Square)

Yet the quality of the public space surrounding shopping centres is central to its capacity to catalyse town centre transformation. Drawing people out of the Charlestown Square was seen by council staff as central to achieving the vision set out in the master plan, in promoting other business and generating a sense of community:

> pulling people out of [Charlestown Square] is something we're trying to achieve ... The main issue we have is actually drawing people outside of the GPT development and getting into other areas. It's definitely not easy to do that.
>
> (Interview, council planner)

State intervention (local, state and federal) is often crucial to leveraging the drawing power of a shopping centre towards wider regeneration. In Charlestown, the local council has explored ways in which their assets (such as the local library) might be used to service locals and visitors alike and contribute to the vibrancy of the town centre. However, it was the federal government's decision to select Charlestown as a trial location for the National Disability Insurance Scheme (NDIS) that acted as a significant trigger for regeneration outside the shopping centre. The NDIS brought with it a major government agency (and associated employees) as well as a number of auxiliary medical services. The concentration of medical and government services outside the shopping centre was identified by all participants as a major driver of wider town centre regeneration. The medical services provided an alternative drawcard, beneficial to both local residents of Charlestown (with a number of residents moving to be close to these services) and the Newcastle and Hunter Valley more generally. Again residents registered the impact of the NDIS:

> those disability services, it's brightened the place up a bit.
>
> (Interview, Resident 3)

Drawing customers from large retail shopping centres is viewed by council, consultant planners and the wider business community as essential in ensuring the long-term commercial viability of the town centre. However, the commercial realities of being located close to a newly rejuvenated shopping centre pose a number of challenges. In Charlestown, these challenges led to a number of retailers moving: some to inside Charlestown Square, where rents and potential customer numbers were significantly higher, others to nearby town centres without large shopping centres, and others still to operate their businesses from their homes or via the internet. These processes led to high retail vacancy rates outside Charlestown Square. This was viewed as a major issue by council, consultant planners, the business chamber and local residents:

> Obviously when you put in an extra, I think, 40,000 [square metres] of retail floor space into a centre, you are going to have issues where

> the retail floor space outside of that actual building will suffer. That is inevitable ... Unfortunately, Charlestown does have a pretty high vacancy rate. I think we're operating – about 25 per cent of floor space outside of GPT is vacant.
>
> (Interview, council planner)

High vacancy rates reduced customer circulation outside the shopping centre, had a negative impact on visual appeal, and reduced scope for the community and social activities in the public domain. Importantly, the development of new mixed-use residential complexes was identified by many as contributing to this high vacancy rate, with retail and commercial spaces often remaining untenanted (irrespective of demand for residential units). The challenges faced by businesses outside the shopping centre and the importance of the serendipitous introduction of the NDIS to maintaining commercial viability outside Charlestown Square point to the importance of considering the implications of the local context of a shopping centre-led regeneration strategy.

Another cleavage in the logic of shopping centre-led regeneration centres on the extent to which the costs and benefits of the redevelopment flow to local residents or those traveling to the town centre from elsewhere. In Charlestown, significant divisions were observed between the opinions of business interests and local residents when it came to the appeal and use of the regenerated town centre (including Charlestown Square). For the business sector, Charlestown was now a 'lifestyle precinct' offering restaurants and entertainment which meant that local residents no longer needed to travel to other locations across the wider Newcastle area. Newer and younger residents were more likely to indicate that they frequented the town centre, in particular the various cafés and restaurants:

> I personally like it because for me Charlestown Square is such a central hub for me – socially as well. They're really making it into a social area as well.
>
> (Interview, Resident 1)

However, this perspective was not shared by all local residents. Many suggested that they actively avoided visiting the town centre and Charlestown Square in particular, except for specialised shopping trips. For this group of residents, these areas failed to provide the sense of place they desired and they were more likely travel to other, smaller centres for shopping and to the gentrified and regenerating centre of Newcastle (15 minutes away by car) for recreation:

> Charlestown has developed but we don't like it. We don't go up to Charlestown. We avoid it if we can.
>
> (Interview, Resident 4)

Nevertheless, the convenience (issues of traffic and parking aside) of having a major shopping centre was appreciated by most local residents. These findings suggest that Charlestown plays the regional role as outlined in strategic planning policy and desired by the owners of Charlestown Square who, in their commercial and marketing decisions, actively sought to attract customers from across the region:

> From a regional perspective, people come here from all over the Hunter... For young people, especially, the offer of the high level fashion, lifestyle now.
>
> (Interview, business chamber official)

The appeal of the shopping centre for customers travelling to the area raises a number of challenges for regeneration and management of the wider town centre. Once again, local specificity is important to how shopping centre-led regeneration might contribute to wider urban regeneration. In the case of Charlestown Square, the centre is seen as effectively providing shopping and entertainment services to attract a young population from across the region. Older local residents benefit from the retail access to the large-scale shopping activities centre, but are also attracted to smaller, less busy centres that offer easier parking and to the leisure and recreational activities elsewhere in Newcastle. The capacity of shopping centre-led regeneration to lead the regeneration of a middle-ring centre may thus depend on the specificity of other local opportunities within the town centre and the capacity of residents to access other opportunities across the city.

Funding and delivery

Achieving the aspiration of wider regeneration beyond the shopping centre itself demands wider public domain improvements. Both the master plan consultant and the council planner recognised the centrality of the public domain in achieving the master plan vision for Charlestown. However, funding such improvement remains a challenge for a financially constrained local government sector. Despite the tension between 'internal' shopping centre assets and 'external' spaces and retailers, the regeneration and retrofit of large shopping centres has the scope to directly and indirectly subsidise non-profit-generating facilities and wider public assets (Guy 2008). Legislated financial and in-kind contributions tied to planning approval offer one of the few large-scale injections of private funds into the revitalisation of public domain assets and infrastructure.

In NSW, compulsory developer contributions are managed legislatively or, for large developments, negotiated between the developer and the assessment agency (in this case, council) via a voluntary planning agreement. In Charlestown, funds and public works provided by the shopping centre developer were seen as vital in improving public domain and infrastructure:

> all of a sudden, council had to do a lot of documents on the pavers and what bike racks to put in because GPT were paying for it … Charlestown is like a pot of gold for council.
>
> (Interview, business chamber official)

The major contributions associated with Charlestown Square funded redevelopment of the local lawn bowling club, the mall and open space surrounding the shopping centre, and an adjacent sporting field. Unsurprisingly, the provision of these new spaces was supported by local council (see Dollery et al. 2013):

> This gave an opportunity to get a new building, a new sports field – probably one of our best sports fields we have … The outcomes for both parties seemed to be achieved. There was a little bit of a loss of public domain – public open space. Council did receive a bit of a funds to acquire some more land for that.
>
> (Interview, council planner)

Local residents generally approved of the asset upgrades packaged with Charlestown Square's redevelopment. Yet, after initial development contributions are spent, questions of how to fund new works and maintenance emerge as major issues facing local councils:

> Public domain is a little bit tricky in that it's difficult to fund. So at the moment, as development occurs, council tries to work with the developer to get some outcomes that suit, but what you end up is just getting a fragmented piecemeal approach, so it is difficult. Council doesn't really have the capacity to fund, say, a $20 million or $15 million in full redevelopment of the public domain.
>
> (Interview, council planner)

Despite the intention of the local council, the capacity to deliver and maintain improvement to public space remains unknown:

> Council is looking in the next five to ten years doing some substantial capital works in Charlestown – whether or not that actually comes to fruition, who knows.
>
> (Interview, council planner)

Nonetheless, development contributions allow major retail developers to exert considerable influence over planning decision-making and the form and function of the built environment as part of shopping centre-led regeneration both within and beyond the retail space (Imrie and Dolton 2014). The fiscal constraints of councils can see them become beholden to funds negotiated as part of retail redevelopment. How this works out in practice and the implications of the process in providing improved public domain assets for residents in middle-ring suburban locations is neither certain nor generalisable.

Access and transport

> All roads lead to Charlestown.
>
> (Interview, business chamber official)

The question of access and transport foregrounds another point of departure between the logic of shopping centre-led regeneration and its translation into development reality. One of the central tensions embedded in the logic of shopping centres as catalysts for urban regeneration relates to issues of access, transport and private motor vehicles. While the logic of TOD calls for increased travel via public transport to high-density residential and retail/commercial centres (Newman 2009), this is at odds with a shopping-centre model of redevelopment which depends upon a much broader retail catchment, accessed via private motor vehicles. Given that town centres (especially those with a large shopping centre) are embedded in the existing urban form and prone to traffic congestion, transport-related issues are often identified as the most challenging to address. In Charlestown, the transportation study initiated as part of the master plan process identified a number of major concerns to be dealt with as part of the regeneration, including: traffic congestion, high traffic speeds along the highway, need for better pedestrian crossings, and poor quality footpaths (URaP-TTW 2007).

The town centre's location at the intersection of two major arterial roads leads to traffic congestion. Moreover, the Pacific Highway, which runs through the middle of the town centre and is identified in the master plan as Charlestown's 'Urban Street', is a major arterial road to central Newcastle. This creates tensions between Transport NSW[8] and the Roads and Traffic Authority's[9] (RTA) priority to facilitate the fast movement of vehicles through Charlestown, and the master plan's 'town centre' redevelopment logic of transforming the Pacific Highway (Figures 13.4 and 13.5) into a more pedestrian-friendly environment via: reduced number of lanes on the highway; a dedicated bike lane; reduced parking; tree planting; and reduced speed limits. Many of these initiatives were opposed by the RTA. This suggests that the conflicting objectives of other state agencies can create significant barriers to middle-ring commercial centre regeneration, where the shift in transport requires both significant investment and a strategic realignment to the objectives of TOD.

A second issue relates to the impact on the local street network of redevelopment being designed to facilitate access to the shopping centre. The limits this places on achieving wider town centre regeneration was a major critique of council planners, the business chamber and residents alike:

Figure 13.4 Possible configuration of the Pacific Highway (source: Hassell 2007b: 45)

Figure 13.5 Pacific Highway, August 2013 (source: authors)

> The Charlestown road system was buggered – for want of a better word – because it's all leading into 'The Square'.
>
> (Interview, business chamber official)

> It's a very weird place to drive ... 'The Square' is like some kind of gravitational super-massive black hole and all of the traffic rotates around it.
>
> (Interview, Resident 5)

Thus, the local road network represents one area where large retail shopping centres can exert significant influence on the form and function of the town centre. While these road systems were approved (and maintained) by local council, they were constructed by the developer as a condition of planning approval (under the VPA). This type of road structure can prove to be a barrier to pedestrian and cycling access, undermining the capacity of shopping centre-led regeneration to reduce car dependence.

Likewise, parking becomes a flashpoint for conflicting logics of wider urban regeneration aspirations and those of shopping centre-led redevelopment. A long-term deficiency in car parking in Charlestown represented a major challenge for council. This challenge is three-fold. First, large shopping centre car parks provide convenient internal access to the shopping centre itself. As part of the redevelopment of Charlestown Square, the developers were required to provide additional parking spaces, with estimates of an oversupply of parking by about 1,000 spaces (except in peak retail periods of Easter and Christmas):

> When they developed their expansion, they ran at a surplus ... Those spaces are technically a public car park under their ownership, so there is no – while there's a limitation on time, people in the area can actually use it for short-term [parking]. They don't have to go into the [Charlestown Square] development.
>
> (Interview, council planner)

Yet this parking does not necessarily connect to surrounding retail, commercial or community spaces. In effect, shoppers are funnelled (via the local road system) into the shopping centre via convenient parking arrangements which potentially constrain access to locations outside the shopping centre itself.

Second, the emphasis on private motor vehicles access (and associated parking) calls into question the extent to which such centres can be pedestrian-oriented or 'walkable' with issues of traffic and pedestrian safety routinely raised (Newman 2009). Charlestown residents acknowledge that it was almost impossible to walk to Charlestown Square, no matter how close they lived:

> Those people live only a couple of hundred metres from Charlestown Square but because of the way they've made it, it's nigh on impossible for them to carry their groceries home ... Charlestown Square's not for people to walk to. It's built for people to drive to.
>
> (Interview, Resident 6)

Similarly, the Newcastle Cycleways movement argued that the construction of car parks and road access around Charlestown Square effectively shielded the square from cycling access and that bicycle racks were limited and poorly located further discouraging cycling. The redevelopment was seen as a missed opportunity to encourage alternatives to driving with activists critical of the view of cycling that was built into the redevelopment:

> a lack of seeing cycling as transport, as legitimate transport, it's still seen as recreation. So, for instance, getting to Charlestown Square [by bike] just wasn't considered. No one rides to a major shopping centre, they don't need to, so that was an opportunity lost. There was a good potential of having a fair network [of cycle paths] around the Square and there aren't any.
>
> (Interview, Newcastle Cycleways)

Third, the large car parks act as a barrier between the shopping centre and its surroundings (Dudley 2003). The emphasis on private motor vehicles potentially inhibits urban design and appropriate built form. In particular, large car parks become the primary entrance to the town and shopping centre. Other forms of access (such as public transport, cycling or walking) are, in effect, restricted:

> While car parking provides access to the town centre, it also has impacts on amenity and personal safety, as well as the visual character of the town centre.
>
> (Hassell 2007a: 19)

Thus, the extent to which wider regeneration aspirations towards TOD can be realised is limited.

Further, like many middle-ring retail centres, Charlestown does not have access to a rail line. Thus according to the Lower Hunter Transport Strategy Report 'the importance of a hierarchy of bus routes and services is greater than at most regional centres' (URaP-TTW 2007). The master plan proposed a transport interchange as a central element of a proposed revamped street running next to the Charlestown Square (Pearson Street). Pearson Street was defined as a 'transit street' providing 'an on-street interchange servicing local and regional bus, as well as taxi, services' (Hassell 2007b: 28). This space would also provide a new 'civic street', a pedestrian-focused space characterised by low traffic volumes, wide footpaths, gathering spaces, and outdoor cafes and restaurants. Pearson Street was strategically intended as a transformative project to produce a community oriented space to facilitate better interaction between the shopping centre and retail and other services located outside. Despite this, one of the most significant barriers to Charlestown's effective performance as a TOD town centre was the failure to secure funding from NSW transport agencies for a larger, more integrated transport interchange. Despite the best intentions of the TOD design, ongoing problems of access have limited its impact.

Indeed, council staff, business interests, cycling and public transport advocates and local residents alike, characterised the redevelopment of Charlestown Square and the surrounding precinct as a lost opportunity to achieve TOD and a more integrated and transformative transport interchange. The pivotal role of shopping centre-led (as opposed to simply retail-led) regeneration, with its focus on providing sufficient car parking spaces and roads designed for car access, was critical to this failure. Reliance on automobility left other transport options poorly considered and executed, or with funding shortfalls that developer contributions could not (or were not) used to overcome. The case study of Charlestown thus points to the importance of thinking about the possibilities and challenges of connecting shopping centre-led regeneration with other urban development imperatives, such as TOD.

Mixed-use and housing

> It will be mixed-use living but you are still in a suburban environment.
>
> (Interview, owner Charlestown Square)

Despite the invocation of mixed-use, shopping centre-led regeneration as a planning solution, redevelopments of existing shopping centres have largely not fitted neatly within the mixed-use discourse promoted in strategic planning documents or in academic debate. The retail components of such shopping centres have been spatially separated from residential land uses. However, recent redevelopment of shopping centres has begun to draw together these land uses in a number of ways.

First, a number of large shopping centre complexes have planned or begun construction of sizeable new residential components. Typically, these have been presented in the form of new 'residential towers' to be constructed on top of existing shopping complexes. NSW examples include Sydney's Top Ryde Shopping Centre and Macquarie Centre, and central Newcastle's Market Town. Many of these centres have benefited from planning policies which identified them as strategic centres, facilitating changes in land-use zonings and associated regulatory arrangements such as new permissible dwelling types, floor space ratios, dwelling heights and new parking regulations. To date, this option has not been pursued by Charlestown

Square, however, it remains an option for future centre redevelopment (given the tendency for major redevelopment every 10 years or so). To these ends, opportunities for expanded residential development represent a potential significant windfall for shopping centre owners located in strategic middle-ring suburbs. Residential development potential also presents local and state planning authorities with opportunities to pursue equity aims such as the provision of affordable or social housing and to leverage increased contributions for local services and infrastructure.

Second, redevelopment of major shopping centres, and associated changes to transport, public space and amenity (e.g., at Bondi Junction in Sydney), has offered opportunities for surrounding land-owners (commercial and residential) to develop mixed-use or housing projects. It is this form of development which has occurred at Charlestown. Importantly, the development has occurred at two scales: first, large development projects within the designated master plan precincts; and second, smaller-scale projects in surrounding areas. After testing a number of development models, the Charlestown Master Plan mobilised a strategy which promoted the establishment of two high-density 'gateways' at each end of the town centre. These gateways were cast as central to defining the town centre, creating a sense of arrival and departure (Hassell 2007a). In addition, an area in the middle of the town centre was also designated for high-density housing. Allowing buildings up to 15 storeys in height (very tall buildings for Lake Macquarie), these sites would, according to council:

> be viable and desirable from an urban design outcome as well as a viability point-of-view.
>
> (Interview, council planner)

Despite being identified in the master plan and allowable under altered planning regulations, high-density development has to date, only occurred in the northern gateway and the central location – the two locations closest to Charlestown Square. Thus, for Charlestown at least, proximity to the retail shopping centre and associated public space and infrastructure was vital in driving new housing construction. Indeed, a number of land owners in the southern gateway (including a number of large aged-care providers) indicated that they had no short-term intentions to develop their holdings.

Thus, in locations not serviced by railway stations, shopping centre-led regeneration offers one of few possible incentives for significant, high-rise development. High-rise development received support both within policy and from the broader public as an expression of Charlestown's status as a major centre:

> The placement of taller development at either end of the town centre will emphasise the natural landform of Charlestown, while creating landmarks or gateways that mark entry to the town centre. This profile

> of buildings creates a memorable identity for Charlestown, within the town centre itself and as it is viewed from the surrounding district.
>
> (Hassell 2007a: 15)

This planning aspiration was met with strong approval from Charlestown residents who approvingly likened high-rise Charlestown to Sydney's Chatswood:

> I quite like them because when you drive around Newcastle you can see them and you're 'oh, there's Charlestown'.
>
> (Interview, Resident 1)

In meeting the strategic vision for Charlestown, council-supported high-rise development and adopted a relatively flexible approach to design, planning and assessment:

> We just wanted to have that flexibility that, if a developer came along, had a specific site, that it was going to work for them financially, it was going to meet Council's criteria. We didn't want to have something there to block it. Any development in Charlestown of decent quality is what we're seeking.
>
> (Interview, council planner)

Nevertheless, issues of design and building quality were routinely raised by local residents as issues associated with high-rise apartments. Comments about design included 'eyesore' with a consensus that none of the high-rise apartments constructed to date represented an 'iconic building' for the area. Concerns around poor design and building quality were perceived as issues likely to affect the long-term regeneration of Charlestown as aesthetics and built form may deteriorate over time.

Away from the town centre core, Charlestown was viewed as simultaneously affordable within the context of the wider housing market, and as a high-growth location where profits could be made:

> Real estate is booming in Charlestown. It's a really good investment area.
>
> (Interview, business chamber official)

Dwellings were actively sought by young couples and families, driven by affordability concerns, opportunity for capital gains, access to local infrastructure and services and a central location which facilitated access to places of employment and education. To these ends, perceptions and experiences of young and middle-aged households who participated in this project supported the claims embedded in strategic planning documents, that middle-ring locations represent an ideal site for regeneration (in this case shopping centre-led regeneration), development and population growth.

However, housing redevelopment processes occurring in the largely suburban surrounds were limited. Here redevelopment consisted of primarily of small-scale renovation projects and a growing number of knock-down rebuilds (Legacy et al. 2013; Chapter 14). The past decade has also seen an increase in the number of duplexes and townhouses being constructed. Importantly, much of this construction is happening closest to the designated town centre (in the designated town centre periphery), much of it targeted towards seniors living (in part reflecting Lake Macquarie's ageing profile and attraction as a retirement destination). Thus, suburban shopping centre-led regeneration could be seen to offer a unique opportunity for housing Australia's ageing population. Many of the residents interviewed who were retired or approaching retirement age identified Charlestown as the ideal location to retire, primarily due to the proximity of a large shopping complex, access to health and medical services and central location which allows for easy movement across the urban area. Thus, shopping centre-led regeneration potentially drives housing solutions for an ageing cohort who might be attracted to the advantages of a middle-ring location (ACSA 2015). The local specificities of the Charlestown case are once again important, with an older population and relatively affordable housing combining with a concentration of medical services to help cement the link between shopping centre-led regeneration and the beginnings of the transformation of the Charlestown housing stock. More research on the ways shopping centre regeneration will play into the specificities of other local housing markets is urgently needed.

Conclusion

Middle-ring suburban shopping centres are increasingly identified within strategic planning documents as key strategic centres, ripe for regeneration and densification. Yet, the regeneration of shopping centre precincts in the middle of cities is not a panacea for the challenges of population growth, housing supply and affordability, services and infrastructure delivery or employment. Charlestown renders visible a number of both challenges and opportunities produced by middle-ring shopping-led regeneration.

First, the redevelopment of shopping centres offers an opportunity for financially constrained local governments to negotiate contributions for the wider public good, including contributions for social services and improvements to public spaces and amenities. However, the ability to negotiate sufficient contributions and ongoing maintenance costs represents challenges for the future town centre amenity. Equally, the geography at which these contributions are spent is also a challenge, as shopping centre owners push for improvements in the immediate area, while council and community priorities seek a more distributed expenditure.

Second, public works conducted as part of the redevelopment of shopping centres offers opportunities to improve local open/public space and amenity. However, these works should seek to promote interaction between the shopping centre and the wider retail, residential and public context. Capital works developed as part of shopping centre redevelopment should not reinforce the dominance of the shopping centre or funnel traffic (pedestrian or car-based) into the shopping centre with little regard for the surrounding area. In Charlestown, local road networks and the dominance of car parks are examples of how such works can strengthen the commercial performance of the shopping centre at the expense of those located outside. Together the provision of funds to council (via development contributions) and the direct provision of capital works, potentially place shopping centre owners in a position to exert significant influence over urban planning and built form, to their own ends.

Third, the commercial tensions of using large shopping centres as a catalyst for wider town centre regeneration remain largely unaddressed. The commercial logic of shopping centre regeneration is to increase the shops/services provided within, thereby increasing patronage and profit. This is in stark contrast to a wider town-centre urban regeneration model that also seeks to improve the economic and social performance of the area outside the shopping centre itself. The ability to regenerate locations outside the central shopping centre can rest upon the (sometimes serendipitous) actions of the state at all levels. In the case of Charlestown it was the introduction of the NDIS that acted as the catalyst for improvements outside Charlestown Square.

Fourth, while actions of the state can be catalytic, they can equally act as a barrier. In our case study, it was the inability to establish a clear, cross-government vision of transport that destabilised the TOD design principles of the town centre. Thus, where government department objectives do not align (or, in a worst case scenario, actively compete) coordinated regeneration of the town centre is unlikely. In the case of Charlestown, the TOD and pedestrian friendly vision of the town centre was opposed by the objectives of some departments to facilitate efficient traffic movement. Thus, both in terms of providing services and aligning objectives, the interaction and negotiations between state agencies is essential in regenerating middle-ring shopping centre precincts.

Fifth, shopping centre-led regeneration does offer opportunities for increased housing construction and densification. As Charlestown case illustrates, high-density residential development can be developed in the vicinity of a shopping centre. However, the influence of the town centre on housing construction diminishes quickly with distance from the shopping centre. Even locations designated within planning schemes as high density, are unlikely to materialise in the short term if distance from the shopping centre is too great. Overarching demand for higher-density dwellings and market conditions obviously also play a major role (Ruming et al. 2010).

Beyond the immediate shopping centre precinct, housing development is likely to remain unaffected with knock-down-rebuild, townhouse and duplex construction and renovations dominating.

Unlike brownfield regeneration or new greenfield sites where development might occur on a more or less blank canvas, the context and history of middle-ring shopping centre precincts is paramount. The contextual specificity and history of retail and residential development, transport and other infrastructure and local demographics, mediate the processes and outcomes of regeneration pressures. While the logics and mechanisms of master planning are largely ubiquitous, the influence of these contextual conditions will vary considerably, problematising any universal approach to shopping centre-led regeneration.

Notes

1 Research funded by an Australian Research Council Discover Grant (DP130100582). This chapter draws on interviews with seven key stakeholders and twenty-five residents.
2 http://profile.id.com.au/newcastle
3 http://www.lakemac.com.au/city/snapshot
4 http://haveyoursaylakemac.com.au/shapeyourfuture?tool=news_feed
5 http://www.lakemac.com.au/city/snapshot
6 http://haveyoursaylakemac.com.au/shapeyourfuture?tool=news_feed
7 http://www.charlestownsquare.com.au/centre-info/about-us
8 The state government agency responsible for delivering and managing public transport services.
9 The state government agency responsible for road infrastructure and traffic management. Abolished and replaced by Road and Maritime Services in 2011.

References

ACSA (Aged and Community Service Australia) (2015) *The Future of Housing for Older Australians*, Canberra: ACSA.

Catungal, J.P., Leslie, D., and Hii, Y. (2009) 'Geographies of displacement in the creative city: The case of Liberty Village, Toronto', *Urban Studies*, 46(5–6): 1095–1114.

Dollery, B.E., Kortt, M.A. and Grant, B. (2013) *Funding the Future: Financial Sustainability and Infrastructure Finance in Australian Local Government*, Sydney: Federation Press.

Dudley, S. (2003) 'The governance role of councils in urban development: lessons from the USA', Proceedings of the State of Australian Cities Conference, Sydney.

Emery, J. (2006) 'Bullring: A case study of retail-led urban renewal and its contribution to city centre regeneration', *Journal of Retail and Leisure Property*, 5(2): 121–133.

Espinosa, A. and Hernandez, T. (2016) 'A comparison of public and private partnership models for urban commercial revitalisation in Canada and Spain', *The Canadian Geographer*, 60(1): 107–122.

Fobker, S. and Grotz, R. (2006) 'Everyday mobility of elderly people in different urban settings: The example of the city of Bonn, Germany', *Urban Studies*, 43(1): 99–118.

Goodman, R. and Coiacetto, E. (2012) 'Shopping streets or malls: Changes in retail form in Melbourne and Brisbane', *Urban Policy and Research*, 30(3): 251–273.

Griffiths, S., Vaughan, L., Haklay, M., and Jones, C.E. (2008) 'The sustainable suburban high street: A review of themes and approaches', *Geography Compass*, 2(4): 1155–1188.

Guimaraes, P.P. (2014) 'The prospective impact of new shopping centres on the retail structure of Braga', *Bulletin of Geography Socio–Economic Series* 25: 167–180.

Guy, C. (2008) 'Retail-led regeneration: Assessing the property outcomes', *Journal of Urban Regeneration and Renewal*, 1(4): 378–388.

Hassell (2007a) *Charlestown Town Centre Master Plan Report, Prepared for Lake Macquarie City Council*, Sydney: Hassell.

Hassell (2007b) *Charlestown Town Centre Implementation Plan, Prepared for Lake Macquarie City Council*, Sydney: Hassell.

Imrie, R. and Dolton, M. (2014) 'From supermarkets to community building: Tesco plc, sustainable place making and urban regeneration', in R. Imrie, and L. Lees, (eds), *Sustainable London? The Future of a Global City,* Bristol: Policy Press.

Instone, P. and Roberts, G. (2006) 'Progress in retail led regeneration: Implications for decision-makers', *Journal of Retail Leisure Property*, 5: 148.

LMCC (Lake Macquarie City Council) (2012a) *Lake Macquarie Town Centres Development Control Plan 2012*, Boolaroo: LMCC.

LMCC (Lake Macquarie City Council) (2012b) *Draft Charlestown Streetscape Master Plan*, Boolaroo: LMCC.

Legacy, C., Pinnegar, S. and Wiesel, I. (2013) 'Under the strategic radar and outside planning's "spaces of interest": knockdown rebuild and the changing suburban form of Australia's cities', *Australian Planner*, 50(2): 117–122.

Lowe, M.S. (2000) 'Britain's regional shopping centres: New urban forms?', *Urban Studies*, 37(2): 261–274.

Matthews, H., Taylor, M., Percy-Smith, B. and Limb, M. (2000) 'The unacceptable flaneur: The shopping mall as a teenage hangout', *Childhood*, 7(3): 279–294.

McGreal, S. and Kupke, V. (2014) 'The spatial dynamic of retail planning and retail investment: Evidence from Australian cities, *Urban Policy and Research*, 32(3): 253–269.

McNeill, D. (2011) 'Fine grain, global city: Jan Gehl, public space and commercial culture in central Sydney, *Journal of Urban Design*, 16(2): 161–178.

Newman, P. (2009) 'Planning for transient oriented development: Strategic principles', in C. Curtis, J. L. Renne, and L. Bertolini, (eds), *Transit Oriented Development: Making it Happen*, Aldershot: Ashgate.

NSWDoP (New South Wales Department of Planning) (2006) *Lower Hunter Regional Strategy*, Sydney: NSW DoP.

Newton, P.W. (2010) 'Beyond greenfield and brownfield: The challenge of regenerating Australia's greyfield suburbs', *Built Environment*, 36(1): 81–104.

QLDDIP (Queensland Department of Infrastructure and Planning) (2009) *South East Queensland Regional Plan 2009–2031*, Brisbane: Queensland Government.

Raco, M. (2003) 'Remaking place and securitizing space: Urban regeneration and the strategies, tactics and practices of policing in the UK', *Urban Studies*, 40(9): 1869–1887

Randolph, B. and Freestone, R. (2008) *Problems and Prospects for Suburban Renewal: An Australian Perspective*, Research Paper no. 11. Sydney: City Futures Research Centre, UNSW.

Ruming, K., Randolph, B., Pinnegar, S. and Judd, B. (2010) 'Urban renewal and regeneration in Sydney, Australia: Council reflections of the planning and development process', *Journal of Urban Regeneration and Renewal*, 3(4): 357–369.

Southworth, M. (2005) 'Reinventing main street: from mall to townscape mall', *Journal of Urban Design*, 10(2): 151–170.

Staeheli, L. and Mitchell, D. (2016) *The People's Property? Power, Politics and the Public*, New York: Routledge.

Tallon, A. (2013) *Urban Regeneration in the UK*, London: Routledge.

Thomas, C.J., and Bromley, R.D.F. (2003) 'Retail revitalization and small town centres: the contribution of shopping linkages', *Applied Geography*, 23(1): 47–71.

Thomson, J., Benson, M., and McDonagh, P. (2015) 'The social and economic impact of improving a town centre: The case of Rotherham', *Local Economy*, 30(2): 231–248.

Tyndall, A. (2010) 'It's a public, I reckon: Publicness and a suburban shopping mall in Sydney's southwest', *Geographical Research*, 48(2): 123–46.

URaP-TTW (2007) *Charlestown Transportation Study for Lake Macquarie City Council*, Sydney: URaP-TTW.

Vanderbeck, R.M. and Johnson, J.H. (2000) 'That's the only place where you can hang out: Urban young people and the space of the mall', *Urban Geography*, 21(1), 5–25.

Voyce, M. (2006) 'Shopping malls in Australia: The end of public space and the rise of "consumerist citizenship"?', *Journal of Sociology*, 42(3): 269–286.

Wrigley, N., Guy, C. and Lowe, M. (2002) 'Urban regeneration, social inclusion and large store development: the Seacroft development in context', *Urban Studies*, 39(11): 2101–2114.

14 Incremental change with significant outcomes

Regenerating the suburbs through 'knockdown rebuild'

Simon Pinnegar, Robert Freestone and Ilan Wiesel

Introduction

Australia's suburban landscapes are continually evolving. This ongoing dynamism remains true to their early foundations in terms of how they were built out and financed: often in a patchwork and piecemeal way, through small-scale local builders and self-build. In the contemporary renewal of low-density suburbs, similar incremental processes are at play. Although essentially a myriad individual household decisions, the acts of reinvesting, reworking and replacing residential stock in existing neighbourhoods collectively represent a significant agent of change. While framed by both strategic and statutory urban planning agendas, rules and targets (Chapter 2), such processes arguably evolve outside the spaces of principal planning interest related to integrated precinct regeneration. This chapter focuses on one particular manifestation of incremental suburban change which has been a pervasive activity in Australian cities in recent decades, namely 'knockdown rebuild' (KDR), where homeowners and small developers replace individual older dwellings with new and often significantly larger houses. 'In-situ' reinvestment through KDR does not add to housing supply targets by virtue of its one-for-one replacement characteristic, but it does – as those individual acts coalesce – arguably bring about significant neighbourhood change over time, not least as older, tired and more affordable stock is swept aside to be replaced by typically larger housing floorplates with a commensurate step change in value.

The chapter draws upon extensive research on knockdown rebuild in Sydney by the authors (Legacy et al. 2013; Pinnegar et al. 2010; 2015; Wiesel et al. 2011; 2013a; 2013b), exploring its nature and extent; the motivations of those undertaking KDR activity; the built form that results; and how the process relates to wider planning frameworks and settings. From these elements, the discussion is developed in two ways. The first considers the drivers and outcomes of KDR within the emerging literature on 'soft' densification (Ministère d'Égalité des Terrritoires et du Logement 2014;

Touati 2012; Touati-Morel 2015) as a means of interpreting the contribution and significance of aggregated individual actions and events on broader urban change. Second, we reassess discussion to date (Legacy et al. 2013; Pinnegar et al. 2015) which has focused on the absences and deficiencies that KDR and other incremental renewal activity expose within contemporary spatial planning frameworks. While assertions that such activity remains 'off the strategic radar' largely hold, there are emerging signs that authorities are starting to engage with and encourage KDR as a means of strategically protecting the status of existing lower-density areas vis-à-vis more recent single-family housing developments in greenfield locations, on the one hand, and systematic medium-density infill in older neighbourhoods, on the other hand. The recent introduction of the Knockdown Rebuild Landscape Contribution Scheme by the Hills Shire Council in Sydney's northwestern suburbs offers a case study of a strategic direction making explicit use of KDR to help foster not only a traditional but updated built form, but with it socio-economic outcomes and political objectives (Hills Shire Council 2014; Petrinic 2014).

Knockdown rebuild in Sydney

Knockdown rebuild (KDR), involving the demolition of an existing property and its redevelopment on a one-for-one basis, is not a uniquely Australian phenomenon. However, it can be argued that it holds a particularly distinctive place within the wider narrative of urban change and renewal in Australia's major cities. It encapsulates many of the individualistic drivers that have shaped the development of Australian suburbia, particularly in the post-war years, in terms of the many actors and stakeholders involved and type of stock initially built. The prevalence of single, detached dwellings, coupled with statutory planning frameworks that both prescribe considerable 'rights' in terms of maximising the utilisation of lots through a tightly defined FSR (floor space ratio) within zoning controls, offer territories across which KDR activity can roll out relatively efficiently. In Sydney, annually updated land valuations (for the purposes of land tax) by the New South Wales (NSW) Valuer General also reflect a property market where land and dwelling each have intrinsic (and defined) values in their own right. As a wealthy city faced with population growth pressures and severe housing affordability constraints, the ability to fund KDR activity, and the financial attractiveness of doing so, can also be seen as a crucial driver of residential investment over the last 15 years. In the following section, we present a brief overview of key findings from our research on the KDR phenomenon in Sydney.

KDR: scale and extent

Capturing the scale and extent of KDR activity is notoriously difficult as might be expected with a process which by its very nature is instigated spontaneously by individual owners across the full breadth of neighbourhood typologies and separate local jurisdictions making up metropolitan Sydney. Nationally, the Housing Industry Association (HIA) tracks the relative importance of KDR output relative to all housing starts delivered by Australia's largest house builders. While many of these companies have developed a distinctive market position with KDR as a key part of their business, given that much of the redevelopment is driven by smaller local builders and developers, the HIA data inevitably falls short of providing a comprehensive picture of all activity. Nevertheless, it is instructive in terms of mapping trends over time and its relative prominence in terms of other housing types. For example, in 2003–04, KDR accounted for 25 per cent of all volume builder activity. While this dropped to around 10 per cent in 2013–14, in absolute terms this represented 4,000 more new dwellings than ten years earlier (Australian Construction Insights 2015).

Our research sought to build up a comprehensive picture of KDR activity across Sydney through systematically sifting through development application (DA) data at the local government area (LGAs) level over a period of five years (2004–08). There is no standard procedure by which councils record the KDR process. In some, the activity was clearly signalled and identifiable within DA information; in others, the need to link demolition DAs with subsequent redevelopment applications was a laborious process. In total, across 29 of Sydney's 42 LGAs (at the time of research), over 6800 units of KDR were identified against a strict definition of one-for-one replacement on a continuing single lot.[1] Activity is seen across different housing sub-markets. From our sample LGAs, it is most prevalent in the more affluent, lower-density LGAs, for example Ku-ring-gai on the upper North Shore, although it is also strong in the middle- and outer-ring of Sydney's west, where post-war communities in areas such as Bankstown and Ryde saw significant activity over our study period. The higher-density inner city (with fewer single dwellings to knockdown) alongside recently built-out greenfield areas (where the stock is relatively new) understandably saw more limited KDR activity. Table 14.1 highlights a representative sample of the councils included in our analysis. Although the overall numbers involved are relatively small, when considered as a proportion of new (if not additional) supply, the contribution in some cases – particularly middle-ring, lower-density LGAs – is significant.

The actors involved

A survey was sent out to all 6,800 addresses where recent KDR activity was identified with the intention of gaining insights from those directly involved in the process. The assumption that the instigator would remain living at the address following the rebuild – confirmed through responses returned – captures a distinctive characteristic of KDR activity. Over 1,200 surveys were completed, and follow-up in-depth interviews were held with around 30

Table 14.1 Incidence of KDR and impact on housing supply by selected local government areas, Sydney 2004–2008

Market typology	*Local government area (LGA)*	*Total development applications (DA) for KDR in LGA*	*KDR as a proportion of new supply in LGA*
Affluent low to medium density	Ku-ring-gai	616	19%
	Woollahra	147	19%
Middle ring post-war suburbs	Bankstown	627	25%
	Ryde	330	16%
Inner city, high density	City of Sydney	31	1%
Low density greenfield	Camden	21	1%

Source: Authors' research

of those respondents, allowing us to explore particular issues, experiences, motivations and behaviours in more depth.[2] Those responses help build up comprehensive pictures of both KDR activity and those undertaking the process. It is owner-led and locally driven, typically involving a property which they have held for some time or a property purchased with the intention of redeveloping for living in themselves. About half of those undertaking KDR had lived on the property in question for at least five years; amongst the more recent 'in-movers', two-thirds had moved from within the local or surrounding neighbourhood.

In contrast to the 'teardown' phenomenon in the United States where a much higher level of developer instigation can be seen (Charles 2013; PAS 2007), KDR is very much an owner-led homebuilding activity rather than an investment or house 'flipping' exercise. Due to the relatively high costs of KDR (including the costs of property ownership, reconstruction and temporary accommodation during the process), a greater proportion of activity is initiated by higher-income households, with over three-quarters of our respondents having annual household incomes in excess of $100,000, well above the Australian median. Although all types of household were represented across our survey respondents, the importance of growing families as a key driver stands out, with 65 per cent of our KDR initiators being younger to mid-life stage families with one or more children. Echoing the locally defined character of KDR activity, over two-thirds of respondents highlighted the importance of connection and desire to stay in the neighbourhood and retain established local networks, including schooling.

Recent media interest in those undertaking KDR in Sydney has become wrapped up with ongoing – anecdotal – assertions that Asian investors have played an instrumental role in the city's recent boom which has seen prices rise 50 per cent since the 2012 trough in the market cycle (Duke 2015; Law 2015; Rogers and Robertson 2015; West 2015). Inferring a correlation between parts of the city popular with Chinese investors where the market has been particularly strong such as Hurstville and Ryde, Conway (2015) suggests that hotspots of KDR activity are exposing loopholes within Foreign Investment Review Board (FIRB) rules. Typically, investment is restricted to new housing supply; however, where it can be demonstrated that a property has reached the end of its economic life, Conway (2015) infers that older existing homes are also being targeted:

> Federation-style homes and other older bungalows are vanishing as overseas investors gobble up swathes of suburbia and replace them with two-storey McMansions. An estimated 5,000 have sprung up in the past five years, worth about $10 billion.

Such loaded and polemic journalistic fervour should not distract from what the evidence behind KDR demonstrates: it is largely driven by local households and related to their own housing requirements and aspirations. Indeed, in contrast to the reshaping of the city at sites of major transformation by multinational companies driven by global finance and investment considerations, it is the importance of the 'local' in defining KDR significance as a driver of urban change which should be recognised.

Changing suburban landscapes

As an owner-led and uncoordinated activity, the resulting outcomes of KDR affect streetscapes and neighbourhoods gradually over time. KDR is both diffuse and pervasive. Concerns have been expressed primarily in terms of older housing stock lost and the nature, design and context sensitivity of what is built in its place. The many, and often complex interplay of, drivers shaping instances of KDR reflect multiple aspects of the 'use' value ascribed to the home: from wider cultural repositioning of comfort, material expectations, the shifting dynamics of household formation, and commensurate shifts in households' ability and options to respond to those dynamics (Wiesel et al. 2013a; 2013b). Some people undertaking KDR were clearly driven by the appeal of building their 'dream homes', often with the help of an architect-builder. Although not always the case, a significant increase in overall building footprint and total floor space is the norm. Reactions against 'Monster Homes' (Nasar et al. 2007) and the 'McMansion-isation' of neighbourhoods (Beal 2007; Stead, 2008) provide shorthand for such concerns, and our research did not dispute this trend (Wiesel et al. 2013a; 2013b). The typical KDR project described by survey respondents saw original three bedroom, one living room, one bathroom properties replaced by much larger floorplates with 4 to 5 bedrooms, 2 to 3 living rooms and

2 to 3 bathrooms. A tendency to greatly increase capitalisation of the lot – building toward FSR maxima and setback minima – can be seen across many KDR properties.

As might be expected, the key targets for replacement were properties built in the post-war years where the configuration of living space no longer suited modern lifestyles and where the costs of upgrading utilities and negotiating building materials such as asbestos meant that if there were no local heritage-conservation constraints, starting completely afresh rather than embarking on alterations and additions seemed financially more logical. Although changing streetscape and neighbourhood character caused by this lot-by-lot densification was acknowledged by many undertaking KDR, a number also explicitly expressed the view that their own patch of renewal was an indication of commitment to the locality, through modernisation and improvement of the housing stock, thereby adding to, rather than detracting from, the landscape.

Broader social, economic and environmental changes that accompany this physical change equally impact over a longer period of time. KDR activity often signals a generational shift, particularly in ageing 1950s–1960s neighbourhoods where younger families acquire properties being sold up by 'original' residents who have aged alongside their homes (Wiesel et al. 2011). Over the longer-term, replacement of ageing, smaller stock with modern larger properties inevitably alters the composite household profile of neighbourhoods undergoing extensive KDR. This upscaling will affect the status and affordability of the neighbourhood, and displacement over time, by incrementally crowding out choice within the housing stock and the price points at which it is available, raises concern. Alternatively, KDR might be seen to encapsulate a means of in-situ gentrification, enabling local households to 'step up' while remaining in their existing communities. The realities in many settings appeared midway between these extremes with reinvestment positioned as underpinning the ongoing sustainability or 'resilience' of ageing family neighbourhoods, revitalising their 'offer' relative to more recent master planned low-density subdivisions (Chapter 17).

Reconceptualising KDR: 'enabling' local coalitions and soft housing renewal

> The theory that a status quo in favour of preserving the living environment prevails in traditional suburbs, and the opposite growth coalition framework applied principally to central urban areas, does not reflect the full diversity of current regulatory dynamics in suburban regions.
>
> (Touati-Morel 2015: 610)

Our research highlights the significance of KDR as a change agent in Australian cities on a number of levels. We argue that the act and process of undertaking KDR, and the planning (and taxation) provisions that make it a relatively straightforward process, reflect distinctive characteristics of post-war suburban Australia. As originally built out, the resultant streetscape reflected a fragmented assemblage of decisions made by individual owners, and indeed 'owner-builders', at a time when materials and labour were scarce. As we have contended elsewhere (Legacy et al. 2013; Pinnegar et al. 2015), KDR has largely flown under the strategic planning policy radar as a result of not materially contributing to housing supply targets. Yet these individual uncoordinated processes are significant drivers of private-sector owner-led renewal with cumulatively significant impacts in terms of city form and function over time.

Major urban renewal 'activation' or 'transformation' precincts in Sydney, for example Green Square in the inner city, contribute only one part of the city's current urban renewal narrative, and – it might be argued – present the more obvious, less challenging element of that wider story. By contrast, KDR appears to operate within a richer, more variegated field of suburban renewal, involving households within existing neighbourhoods as part of a broader transitional shift toward more intensive habitation of settled areas. Our interest in these incremental forms of urban renewal echoes emerging debate regarding soft densification policy settings and supporting initiatives (Baker 2008; Ministère d'Égalité des Terrritoires et du Logement 2014; Touati 2012; Touati-Morel 2015). Different definitions as to what constitutes 'soft densification' arise in this literature, with the more inclusive definitions extending beyond single-lot-based changes (such as ancillary dwellings, granny flats and accessory apartments) to small-scale subdivision and modest levels of infill.

While KDR itself does not neatly accord with the concept of soft densification in that it does not add to overall housing supply numbers, we argue that – on the grounds of both its drivers and outcomes – it can be regarded as operating within the same 'soft' field of action. KDR reflects, albeit implicitly, planning policy settings conducive to the interests of an alliance principally involving individual land-owners, local builders and developers with a vested interest in increasing the value of suburban property (Touati 2012; Charles 2013; Pinnegar et al. 2015). This 'soft field' can be conceptualised as inhabited by coalitions of enablers (through strategic willingness and benign regulatory settings) and initiators (driven by individual, albeit small-scale, planning gain) of incremental change. Such coalitions, given the small-scale nature of the changes at play, present a modest and by inference more *acceptable* framework within which local renewal activity can be positioned and negotiated. While a certain extent of community backlash against KDR is inevitable, it is arguably less *systematic* in nature when compared to well-trodden arguments and campaigns against high density consolidation. With KDR, the protagonist is a neighbour and member of the community, rather than multinational developer driven by shareholder margins. The impetus for softer forms of densification reflects an alliance of interests which can accommodate other owners in the vicinity who in future may wish to undertake the same. Neither seeking to maximise uplift through re-zoning nor seeking to 'preserve' the historic status quo, this convergence of interests facilitates a seemingly mutually beneficial and community-accepted built form (Touati-Morel 2015).

As such, while KDR is characterised as predominantly an individual-led phenomenon of only limited interest within strategic planning frameworks, planning and urban renewal policy interests inevitably *do* facilitate, and help steward acceptable and encouraged forms of soft densification, even if this steer is typically rather more implicit than explicit. However, where local planning jurisdictions are starting to recognise that incremental activity is not only something that needs to be 'addressed' but in fact is an integral and potentially positive part of the wider narrative of urban change across LGAs and within particular neighbourhoods, the nature of policy adoption is instructive. In the following section, the Hills Shire Council offers an instructive case study exploring emerging engagement with KDR.

KDR and low density renewal: the Hills Shire Council, northwest Sydney

> Urban renewal and upgraded local infrastructure should not be confined to new densities and apartments ... Our Knock Down Rebuild (KDR) scheme reaffirms Council's support and commitment to single-lot housing and maintaining our Shire's low density character ... Although there are a lot of apartments on the way, I want to ensure that single-lot housing remains the dominant housing type in the Sydney Hills.
>
> (Councillor Andrew Jefferies, quoted in Hills Shire Council 2014)

The Hills Shire Council area has been a key site for housing growth and in particular lower-density single-dwelling provision in recent decades with metropolitan Sydney's North West (NW) Growth Area largely falling within its boundaries. The LGA extends from the northern edge of Parramatta, approximately 20km west of Sydney's Central Business District (CBD) in a north-westerly direction towards the Hawkesbury River and the metropolitan boundary. It primarily encompasses post-World War II suburbs such as North Rocks and Carlingford (in the south) and more recently built estates extending north along Windsor Road. It is a prosperous part of the wider metropolitan area with mid- to high-value family housing development in close proximity to higher-value employment areas such as Norwest Business Park and Macquarie Park as well as good connections to Parramatta CBD through the T-way dedicated bus link.

The construction of the NW metro rail link, scheduled for completion in 2019, presents a significant opportunity (and pressure) for uplift of residential densities in and around centres to be served by new railway stations. For an LGA in large part defined by, and which promotes itself, as low density in built form and residential character, the prospect of substantive medium-

and high-density development along the rail route represents a step change. It is within this context that the council's support for encouraging KDR in its existing, older neighbourhoods adds a further overlay in our understanding of the potential value of 'soft' forms of renewal within the growth management dynamics of a fast-growing suburban corridor. Council interest arose out of a strategic planning workshop involving councillors considering ways in which older suburbs and particularly those in the south could be supported in their ongoing contribution to the traditional family-oriented character of the shire. A number of considerations came into play here. First, as the earliest parts of the LGA to experience large-scale suburbanisation, the housing stock in these older suburbs is inevitably ageing and many homes are reaching the point where significant modernisation is required. Second, those neighbourhoods typically enjoy generous lot sizes by contemporary suburban land release standards, making them attractive not only as family housing areas but in terms of their potential for redevelopment if planning ordinances were to change. Third, taking a subregional perspective, these older suburbs are within the same housing market demand area as newer suburbs in the north of the LGA such as Rouse Hill. These contemporary master planned communities often enjoy higher public amenity in terms of modern parks and landscaping, and therefore some concern regarding ongoing appeal and competitiveness were voiced.

Knockdown rebuild was not the Council's assumed policy direction at the outset, with various alternative initiatives being explored. A survey of developers active in the shire was conducted in late 2014, exploring what they thought their clients were looking for and whether they felt there were impediments to development. The concerns expressed focused on issues tied to regulations and negative perceptions regarding council attitudes to urban renewal. Forums were also held with local community representatives that highlighted concerns regarding preservation of street character. In these forums, both developers and residents held the view that council presented barriers to achieving low-density renewal through KDR, particularly where it entailed the removal of mature trees or challenged existing set-back provisions.

An initiative around KDR thus emerged as a means whereby the council could signal to both developers and residents alike acceptance of incremental renewal and openness to a degree of flexibility in terms of low-density suburban form. Various policy options were explored, including dropping DA fees or offering rate relief through the construction process for those undertaking KDR. However, both were deemed too difficult procedurally and would have required formal council intervention on a case-by-case basis. Given community interest in streetscape character, the preferred policy response settled on providing a contribution to soft landscaping the front of rebuilt homes at the end of the process.

The resulting Hills Shire Knockdown Rebuild Landscape Contribution Scheme provides a simple grant administered through statutory planning mechanisms. The grant is for up to $2,000, payable on presentation to council of the property's occupancy certificate, a site inspection and receipts for landscaping work to the front of the KDR property. All single detached dwelling KDR activity in existing R2 zones (a standard zone comprising mainly low-density housing) is eligible, alongside any single dwelling undergoing extensive additions and alterations ($250,000+). Renewal involving subdivision is not included in the scheme, nor are alternative dwelling projects such as granny flats (Hills Shire Council 2015). Council has allocated $250,000 per year during its pilot phases (extended to 2017) and thus assumes a take-up of approximately 125 grants per year.[3] Alongside the financial contribution, the council has a case officer to assist and advise on KDR matters and be proactive in making people aware of the scheme.

In terms of its design, application and budget, the Contribution Scheme can be considered fairly modest. It does, nevertheless, encompass more significant strategic signals for suburban renewal activity which are of particular interest here. Promotion of KDR is positioned within broader strategic housing and urban policy directions for the LGA. While density pressures accompany the impending arrival of the NW rail link, it helps reinforce the council's commitment to single-dwelling residential environments and a particular focus on the shire as a good place to raise a family. And although available to KDR applicants across all existing neighbourhoods of the Hills District, the key areas where interest has emerged to date – in the south of the LGA in older suburbs such as North Rocks – align with the council's key geographies of concern. It is these suburbs which are looking tired relative to the new subdivisions to the north and also experiencing significant re-zoning pressures to allow dual occupancy or redevelopment at medium density. In such circumstances, proactive KDR policy constitutes an experimental mechanism to maintain investment in, and viability of, single-dwelling residential stock in ageing neighbourhoods.

The sums involved for the council are, within overall budgets, minimal, but its value is in providing a direct indication of the council's interest in private-sector low-density renewal, and 'reinforcing-through-revitalisation' low-density environments in R2 zones in parallel with negotiating significant increases in densification around key nodes activated by the NW rail link. Rather than KDR falling outside strategic planning's gaze as had been previously argued, here we have the process accommodated within the broader planning narrative for a rapidly growing part of the city. Importantly, its strategic positioning is framed in terms of reinvestment and longer-term neighbourhood sustainability rather than seeking 'preservation' of existing stock as a means of keeping density at bay.

Discussion

In recent years, urban consolidation debates in Australian cities have focused attention on densification around key activity centres, where large-scale – and increasingly high-rise – development has been at the heart of strategic planning interest. With the need to manage high population growth the paramount concern, discourse regarding urban futures has been first and foremost structured around numbers and targets, and strategic interest has essentially prioritised sites and localities which are seen as providing capacity to help meet those demands. This approach is evident in the suite of Sydney's metropolitan strategies prepared over the last decade, including the recently released draft Greater Sydney Region Plan A metropolis of three cities (Greater Sydney Commission, 2017). The renewal agenda has become both more prominent and assertive, with The UrbanGrowth NSW Development Corporation now leading the charge on a range of strategically located 'Growth Centres' (predominantly within the market-led, higher-value so-called Eastern Economic Corridor' within the new metropolitan strategy). Coupled with the residential construction boom engineered through accommodative fiscal policy since 2013 which has seen thousands of off-the-plan and investor-driven projects come on stream in and around many centres and neighbourhoods, the relative significance of incremental urbanism might appear somewhat muted in terms of its relative contribution of new stock.[4] KDR continues – in the background – to gradually reshape the 'other' geographies that are of less interest in strategic plans: the white spaces in the overall structure diagrams outside the zones shaded for transformation. However, we argue that the importance of KDR and other forms of gradual, 'softer' and incremental forms of redevelopment within the wider urban renewal narrative is not defined simply in terms of its modest contribution to the numbers game. Rather, the drivers and outcomes of KDR capture and highlight key characteristics of the dynamics of the Australian city which can be seen as pivotal in offering a more layered engagement with the changing shape and form of the suburban built environments in coming decades. Crucially its drivers, and the settings in place to facilitate lot-based renewal, reflect the historical trajectory of Australian post-war suburbia. As such, while its manifestation in the form of significant upsizing and upgrading of housing stock is contemporary (Wiesel et al. 2013a; 2013b), the ability for individual owners to undertake KDR can be seen to relate to the initial evolution, build-out and continuous modifications of housing stock in those suburbs. Australian suburbia has famously been lambasted for its ugliness (Boyd 1960), but the resulting urban landscape, resembling something of a patchwork quilt, reflects the myriad individual, fragmented decisions of lot owners as much as the structural overlay provided by strategic and statutory plans. Tomlinson (2012) has referred to the 'unintended' nature of our cities; in planning terms this might equate with a perceived weakness within those frameworks and regulatory systems. However, it is a landscape which foregoes visual consistency and 'order' for flexibility and, arguably, greater capacity for change and adaptation over time.

In terms of promoting both a strategic interest in, and approach to, the renewal of lower-density suburban areas, this historical context and trajectory is instructive. While urban consolidation debates often draw dichotomies between 'strategic spaces' for intense city change and those areas to be protected and preserved,[5] everyday processes of incremental urbanism disrupt the urban design-led pronouncements for the future city (Pinnegar et al. 2015). Far from being passive spaces within the wider development context, these ongoing activities, reflecting evolving household needs and demands and their willingness and capacity to invest in-situ, act cumulatively as a substantive transformative process within the evolving Australian city. As seen in the KDR policy of the Hills Shire Council, this one-by-one lot-based activity can become integral to and aligned with the broader strategic positioning and planning objectives of the local authority. As such, policies and initiatives facilitating incremental renewal in certain areas can be seen to work alongside (or more provocatively offset) more contentious high-density development in other parts of the LGA. Bringing these 'soft' processes into the strategic fold would help inform measures which seek to mediate and mitigate the externalities that result from such activity. For example, as previously discussed by the authors, both KDR and soft densification are likely to foster gentrification and displacement pressures, albeit gradually and over time (Pinnegar et al. 2015). These pressures may become more explicit within neighbourhood ageing and market contexts where KDR activity is seen to correlate with a generational shift (Wiesel et al. 2011; Wiesel 2012).

Insertion of 'soft' processes into the wider strategic planning context also reinforces the need for, and acknowledges the fundamental role played by, a complex overlay of alliances and coalitions bringing together initiators and enablers in the incremental development of cities. Sitting between developer-driven renewal, which tends to accrue benefits to those outside the communities most impacted, and community 'preservation' which resists any degree of change, KDR and soft densification mechanisms occupy a less explicitly contentious space. Nevertheless, while these individual renewal acts embed locally-derived reinvestment in the community and neighbourhood itself, they do not always transcend tensions arising through the potential externalities of change. Where the mutual benefits across stakeholders might be less apparent and tensions are heightened, consideration can be given to new policy mechanisms which may assist in encouraging acceptable levels of incremental renewal. For example, Baker (2008) proposes a model whereby incremental densification could be 'paced' in order to provide a degree of certainty regarding rate of change across the neighbourhood. The auctioning of densification rights – say ten properties in a suburb in any given year – would create a market mechanism which separates 'densification' rights from property rights. Value extracted through this process could be recycled for broader neighbourhood amenity and benefit. Baker notes that auctioning off densification permits hits upon a number of legal barriers – owners are, in effect, paying for the right to develop before the decision is made as to whether the ensuing proposal is appropriate or not – but such challenges are not insurmountable.

The question remains as to whether acknowledging the significance of KDR and other forms of incremental urbanism within broader strategic planning processes can happily translate into more effective urban policy in practice. It might be argued that part of the success of these fragmented, small-scale actions is *because* they have largely occurred without the burden of policy interest; a case where 'just letting things be' has not necessarily translated into wholly detrimental outcomes. The Hills KDR Landscape Contribution Scheme, modest as it is, nevertheless points towards the potential benefit of more coordinated thinking regarding housing provision, investment, need and demand.

The fragmented nature of individual lots, and the resulting uncoordinated nature of decisions taken associated with each of those lots over time, is often presented as a barrier: hindering attempts to amalgamate and produce sites economically viable for the developers of major transformation sites. One useful response, as seen in the work of Newton et al. (2012) is to develop progressive suburban renewal models that explore how such sites could be drawn together (if not necessarily physically) to facilitate synergies and enable the benefits of scale through more community-oriented or locally driven arrangements (also see Chapter 12). However it is contended here that the flexibility and relative efficiency of lot-by-lot 'soft' processes such as KDR can equally contribute to a more nuanced, more thoughtful, and more context-driven strategic planning process. It would signal a planning and urban policy process more in tune with the historic qualities of the Australian city, and its suburbs in particular. In this aim, we would echo and extend Dodson's recent clarion call for a suburban focus to national urban policy involving all levels of government in 'a preparedness to contribute to the task of suburban reconstruction' (Dodson 2016: 33).

Conclusion

Our argument here is not that 'hard', top-down renewal is bad and 'softer', bottom-up renewal is good. A context and market-driven approach to metropolitan wide issues will inevitably mean that – to use Touati's typology – a range of 'hard', 'soft' and 'mixed' densification approaches will be appropriate across different parts of the city (Touati-Morel 2015). Incorporating 'soft' processes such as KDR can help ensure that we are better placed to understand the full array of drivers shaping the city rather than privileging one and essentially ignoring the others. Extending the focus from areas identified as sites of transformation, to recognising that all areas experience transformation over time, provides a platform for more coherent and equitable 'city-building' policy frameworks. We can only meaningfully conceptualise what urban renewal should achieve in those major sites of transformation if they are positioned as part of, and contextualised within, the overall landscape of urban change and renewal – both large and small scale – throughout the city.

Urban renewal takes many shapes and forms, even though the primary policy and research foci have tended to concentrate on large-scale urban projects. We have explored the nature and ramifications of a 'softer' realm of urban change, one that is pervasive in middle-ring suburbs of not just Sydney but all Australian cities. While the KDR phenomenon has generally proceeded under the policy radar, our case study foreshadows the potential for a more nuanced consideration of the continual cycle of suburban renewal. In asserting this interest, we are not seeking to invoke a return to deterministic neighbourhood lifecycle models of the city, but to open up a conceptual space where the inevitability of urban change is progressively incorporated into broader, and hopefully more inclusive, engagement with the complexities of urban renewal (Pinnegar et al. 2015). This includes a more layered debate regarding densification, and how we are going to meet the continually evolving challenges of housing supply and demand (understood in terms of need, diversity, affordability and accessibility rather than bald targets) in our metropolitan landscapes.

Acknowledgment

The authors would like to thank The Hills Shire Council for their assistance in preparing this chapter. The original research drawn upon in this chapter was funded through an ARC-funded project DP0986122.

Notes

1 All Sydney metropolitan LGAs were approached by the research team to ascertain whether necessary data could be shared. The 29 LGAs able to supply or provide access to the required information constitutes a representative sample of all metropolitan area subregional geographies and housing markets. Since 2016, local government amalgamations in Sydney have taken place and the names of some LGAs have been superseded.

2 Survey respondents were asked whether they wished to take part in follow up interviews. Approximately two-thirds of respondents indicated that they were happy to do so. The thirty interviewees selected represented a good cross section of different geographies and market contexts across metropolitan Sydney.

3 In the year preceding introduction of the scheme (2013–2014), there were 60 development applications in R2 zones to undertake KDR and 27 for substantial alterations/renovations across the Hills Shire. As expected given the lack of first schemes coming through since launch of the Landscape Contribution Scheme in 2015, just eight grants (as of January 2016) had been approved.

4 This is a story more of the surge in multi-unit development in recent years claiming a bigger proportion of the pie (Australian Construction Insights, 2015).

5 For example City of Melbourne's Rob Adams proposes a '7.5 per cent city', where medium density development focused along tram corridors can accommodate projected residential growth while leaving low density neighbourhoods within the grid alone (Victorian Department of Transport and the City of Melbourne, 2010).

References

Australian Construction Insights (2015) 'Research snapshot: An insight into Knockdown-rebuild activity', http://aciresearch.com.au/-/media/HIA-Website/MiniSite/ACI/PDF/market-research/KDR_Feb2016.ashx?la=en&hash=8B4171373EA1DD22B52436F608FAF8B107D47C6E, 30 January 2015 (accessed 6 June 2017).

Baker, K. (2008) 'Incremental densification auctions: A politically viable method of producing infill housing in existing single-family neighborhoods'. Master's thesis, Massachusetts Institute of Technology.

Beal, W. (2007) 'McMansion', in R. Wier (ed), *Class in America: An Encyclopedia*, Westport, CT: Greenwood Press.

Boyd, R. (1960) *The Australian Ugliness*, Melbourne: Cheshire.

Charles, S. L. (2013) 'Understanding the determinants of single family residential redevelopment in the inner-ring suburbs of Chicago', *Urban Studies*, 50: 1505–1522.

Conway, D. (2015) Legal loophole lets foreign investors snap up huge numbers of older Sydney homes, 26 June, dailytelegraph.com.au, http://www.dailytelegraph.com.au/news/nsw/legal-loophole-letting-foreign-investors-snap-up-huge-numbers-of-sydney-homes/news-story/c0d41d82b8df393c603b06014608262a (accessed 6 June 2017).

Dodson, J. (2016) 'Suburbia in Australian urban policy', *Built Environment* 42(1): 22–36.

Duke, J. (2015) 'Chinese buyers account for 2 per cent of Australian home sales: new research, Domain', *Sydney Morning Herald*, 3 September, http://www.domain.com.au/news/foreign-buyers-account-for-1-per-cent-of-australian-home-sales-new-research-20150902-gjdycq/ (accessed 6 June 2017).

Greater Sydney Commission (2017) *Draft Greater Sydney Region Plan*, Parramatta: Greater Sydney Commission

Hills Shire Council (2014) 'Knock down and rejuvenate The Sydney Hills', http://www.thehills.nsw.gov.au/Building/Development-Approvals/Knock-Down-Rebuild-Landscape-Contribution-Scheme (accessed 6 June 2017).

Hills Shire Council (2015) 'Knock down rebuild landscape contribution scheme fact sheet', http://knock_down_rebuild_landscape_contribution_scheme_fact_sheet%20(1).pdf (accessed 6 June 2017).

Law, J. (2015) 'Chinese property investment through the roof: What it really means', http://www.news.com.au/finance/real-estate/buying/chinese-property-investment-through-the-roof-what-it-really-means/news-story/ba6df0a7bcdcae773677d72a4545a41d (accessed 6 June 2017).

Legacy, C., Pinnegar, S. and Wiesel, I. (2013) 'Under the strategic radar and outside planning's "spaces of interest": knockdown rebuild and the changing suburban form of Australia's cities', *Australian Planner*, 50: 117–122.

Ministère d'Égalité des Terrritoires et du Logement (2014) 'Towards soft public densification and intensification policies: Benefits, limitations and opportunities' A research and experimentation programme of the PUCA. http://www.territoires.gouv.fr/IMG/pdf/vers_des_politiques_publiques_de_densification_et_d_intensification_-_version_anglaise.pdf (accessed 6 June 2017).

Nasar, J., Evans-Cowley, J. and Mantero, V. (2007) 'McMansions: The extent and regulations of super-sized houses', *Journal of Urban Design*, 12: 339–358.

Newton, P., Newman, P., Glackin, S. and Trubka, R. (2012) 'Greening the greyfields: unlocking the redevelopment potential of the middle suburbs in Australian cities', *World Academy of Science, Engineering and Technology*, 71: 138–157.

PAS (Planning Advisory Service) (2007) *Teardowns*, PAS Quicknotes no. 9, Chicago, IL: American Planning Association.

Petrinic, I. (2014) 'Knock down rebuild incentive for Hills residents', *Hills News*, 26 November, http://hillsnews.com.au/story/2724621/knock-down-rebuild-incentive-for-hills-residents/ (accessed 6 June 2017).

Pinnegar, S., Freestone, R. and Randolph, B. (2010) 'Suburban reinvestment through knockdown rebuild in Sydney', *Research in Urban Sociology*, 10: 205–229.

Pinnegar, S., Freestone, R. and Randolph, B. (2015) 'Incremental urbanism: characteristics and implications of residential renewal through owner-driven renewal and rebuilding', *Town Planning Review*, 86(3): 279–301.

Rogers, D. and Robertson, S. (2015) 'Great wall of xenophobia makes for simplistic foreign investment debate', *The Conversation*, 14 October http://theconversation.com/great-wall-of-xenophobia-makes-for-simplistic-foreign-investment-debate-48973 (accessed 6 June 2017).

Stead, N. (2008)' Thoughts around the McMansion', *Reincarnated McMansion: Auditing, Dismantling, Rebuilding*, Sydney: Environa Studio.

Tomlinson, R. (ed) (2012) *Australia's Unintended Cities: The Impact of Housing on Urban Development*, Melbourne: CSIRO Publishing.

Touati, A. (2012) L'habitant maître d'ouvrage. Au cœur de la densification pavillonnaire [The inhabitant master of works. At the very heart of suburban densification], *Etudes Foncières*, 157: 34–39.

Touati-Morel, A. (2015) 'Hard and soft densification policies in the Paris City Region', *International Journal of Urban and Regional Research*, 39(3): 603–612.

Victorian Department of Transport and the City of Melbourne (2010) 'Transforming Australian cities for a more financially viable and sustainable future', http://www.transformingaustraliancities.com.au/wp-content/uploads/Transforming-Australian-Cities-Report.pdf (accessed 6 June 2017).

West, M. (2015) 'Wall of Chinese capital buying up Australian properties', *Sydney Morning Herald*, June 29, http://www.smh.com.au/business/comment-and-analysis/wall-of-chinese-capital-buying-up-australian-properties-2015068 (accessed 6 June 2017).

Wiesel, I. (2012) 'Can ageing improve neighbourhoods? Revisiting neighbourhood life-cycle theory', *Housing, Theory and Society*, 29: 145–156.

Wiesel, I., Freestone, R., Pinnegar, S. and Randolph, B. (2011) 'GEN-X-TRIFICATION? Generational shifts and the renewal of low-density housing in Sydney's suburbs', *Proceedings of the 2011 State of Australian Cities Conference*, Melbourne: State of Australian Cities Research Network.

Wiesel, I., Pinnegar, S. and Freestone, R. (2013a) 'Supersized Australian dream: investment, lifestyle and neighbourhood perceptions among "knockdown-rebuild" owners in Sydney', *Housing, Theory and Society* 30: 312–329.

Wiesel, I., Freestone, R. and Randolph, B. (2013b) 'Owner-driven suburban renewal: Motivations, risks, strategies in knockdown and rebuild processes in Sydney, Australia', *Housing Studies* 28: 701–719.

15 Regenerating Australia's public housing estates

Hal Pawson and Simon Pinnegar

Introduction

For the past 20 years interventions to redevelop or 'regenerate' rundown public housing estates have formed the main focus for Australian area-based initiatives (ABIs) relevant to urban poverty. Much the same could be said of the USA and the UK. In Australia, by the mid-1990s, most state and territory governments had come to recognise that the problematic social and physical condition of many large concentrations of public housing called for substantial re-investment (Arthurson 1998; Randolph and Judd 2000). More recently, growing awareness of the scope to 'unlock equity' tied up in public housing land has provided another estate renewal stimulant for estates in certain locations. The associated 'densified redevelopment' model connects with the broader 'compact city' agenda – the push to accommodate population growth largely by consolidating settlement within existing city boundaries (Troy 2012). The main purpose of this chapter is to identify, explore and briefly critique contemporary 'consensus approaches' to public housing estate renewal in Australia and to further illuminate these with reference to a large project undertaken by the NSW Government.

In varying degrees, state and territory estate-regeneration initiatives have typically encompassed asset improvement, stock demolition or disposal, community development, 'management-based strategies', and whole of government approaches (e.g. 'place management'). In the terms popularised in debates on ABIs in the UK, projects of this kind could be characterised as 'housing-led regeneration' (Kintrea 2007; Pinnegar 2009). As used in this chapter, the term 'estate renewal' refers to projects involving significant physical investment (e.g. large-scale housing demolition and replacement). 'Community renewal' interventions, by contrast, involve measures primarily concerned with promoting social and economic inclusion and enhanced quality of life for estate residents (Arthurson 2013). Contemporary 'estate renewal' projects tend to also incorporate community renewal components. However, not all community renewal programs also extend to large-scale physical investment. For the purposes of this chapter, 'regeneration' should be read as a generic or overarching term encompassing interventions involving both physical and social/economic investment.

The remainder of this chapter is structured as follows. First, since this crucially shapes the estate regeneration agenda, we highlight key aspects of Australia's public housing inheritance – the ways that historic patterns of development, management and funding have contributed to the social, financial and physical condition of public housing in 2016. Second, we explore and discuss the consensus principles which characterise the estate regeneration activity of the past 10–20 years. Third, focusing primarily on NSW, we exemplify estate regeneration policy and practice through reference to the NSW government's Bonnyrigg Estate project, arguably the largest and most ambitious initiative of this kind attempted anywhere in Australia over the past 10 years. Then, in the chapter's final section, we review some of the issues raised in the main body of the text and draw together some conclusions. Although focusing mainly on Australia – and, in particular, NSW – the chapter also makes reference to similarities and differences between practice in Australia and in the UK and the USA.

Australia's public housing legacy

As in many other developed countries, Australia saw a substantial public housing construction program running through the second half of the twentieth century. Initiated in earnest in the immediate post-war period (Troy 2012), this program saw at least 650,000 homes built by state and territory governments across the country in the following decades.[1] Even at the start of the new millennium, after extensive property sales and some demolitions, more than 400,000 (mainly) state-built low-rent homes remained in public ownership. Nevertheless, with public housebuilding virtually halted from 1996[2] and tenancy allocation increasingly targeted towards society's most disadvantaged groups, the early part of the twenty-first century sees state and territory governments managing an ageing property portfolio occupied by an increasingly needy and impoverished population.

Exemplifying the transformation of the public housing social profile, the proportion of NSW tenant households for whom wages are the main source of income, fell from 85 per cent in 1960 to just 5 per cent by 2013 (NSW Government 2014). As in other high-income countries, this shift has transformed public rental into a 'residual' tenure, restricted to those lacking the means to compete for market housing. This 'residualisation' largely results from a deliberate – and arguably quite justifiable – policy choice to 'align publicly supported housing provision with housing need'. Largely, this has been implemented through restricting tenancy allocations to those low-income households with the most complex problems – a 'safety net' model. Beyond this, a growing number of state and territory governments have aspired to transition further to an 'ambulance service' model where fixed-term tenancy frameworks enable residence to be ended for tenants significantly improving their economic position (Fitzpatrick and Pawson 2014).

An important result of the process described above has been the reduced rent-paying capacity of public housing tenants. Crucially, this change has come about without the creation of any counterbalancing social security entitlement or housing allowance.[3] Such a measure could have enabled state and territory governments to charge low-income tenants a 'realistic' rent (i.e. pitched at a level adequate to fund appropriate property upkeep) without pushing tenants into poverty.[4] Far from compensating for the above pressures, the value of Commonwealth government subsidies paid to state and territory governments to assist in funding public housing has also trended downwards over recent decades. For example, in the nine years to 2007–08, such funding paid through the prevailing Commonwealth State Housing Agreement fell by nearly a quarter in real terms (Productivity Commission 2009: 16.6).

In combination, the above factors have badly eroded the financial sustainability of Australia's entire public housing system (Hall and Berry 2007). In NSW, for example, this trend had progressed to the point where, by 2012–13, the state's housing annual revenue funding deficit approached $600 million (NSW Auditor General 2013). Since state and territory governments naturally wish to minimise the extent to which public housing needs to be subsidised from general taxation, it has become routine to defer non-emergency maintenance – with the inevitable result that stock condition has been gradually deteriorating. Again taking a NSW example, by 2013 the state government acknowledged that '30–40 per cent' of its public housing had fallen below an acceptable physical standard (NSW Auditor General 2013).

All of the social and financial considerations discussed above have been significant in shaping the context within which Australia's post-1990 public housing estate renewal initiatives have been developed. One additional set of 'public housing legacy' factors important to note, concerns the form and location of the public housing portfolio as it was developed through the latter half of the twentieth century. First, as in much of Europe and North America, a substantial proportion of Australia's public housing has been spatially concentrated. Hence, according to Arthurson and Darcy (2015) around half of all Australia's public housing is contained within 'estates' containing more than 100 dwellings. This factor is important when considering the socio-spatial impact of tenure-wide 'residualisation' as described above.

Second, in their built form, many of Australia's post-war public housing estates incorporated design features which would later come to be seen as problematic. The most notable example concerns the prevalence of the 'notorious "rabbit warren" Radburn layout' (Jones and Evans 2014: 153) common to numerous public housing estates developed across the country from the 1960s to the 1980s. The Radburn model, imported from the USA, aspired to separate cars from pedestrians. Key components include the 'back to front' arrangement of houses, the large areas of often unloved 'public' open space, and the prevalence of high fences impeding surveillance (Woodward

1997). While their typically very low density may present redevelopment opportunities where land values have risen, the other design features mentioned here have been considered as making Radburn estates especially prone to crime and antisocial behaviour, as well as to stigmatisation (Woodward 1997).

Third, the metropolitan fringe location of many public housing estates has turned out to be problematic (Peel 1995). Especially in Adelaide and Sydney, many large public housing developments of the 1960s and 1970s were sited in outer suburbs – in part reflecting the lower price of land in such places, but also in some instances intentionally consistent with the location of large-scale industrial employment, such as North Adelaide's car factories. Estates positioned in places of this kind have been vulnerable to the onset of de-industrialisation and the recent tendency towards the re-concentration of economic activity in well-connected central metropolitan locations (Rawnsley and Spiller 2012).

Principles underpinning Australian public housing estate-regeneration policy and practice

A number of largely consensus principles have characterised estate regeneration policy and practice in Australia as it has evolved over the past 20 years. Reflecting the policy transfer zeitgeist, these ideas draw substantially on urban policy thinking in the USA, the UK and in other high-income countries. As seen by certain critics, some of these ideas have been imported with far too little discrimination.

Poverty de-concentration

To the extent that Australia's public housing stock is configured as 'estates', the residualisation of the tenure has inevitably resulted in increasingly distinct spatial concentrations of disadvantage. Apparently in recognition of this reality, an ethic of 'poverty de-concentration' has formed a central pillar of estate regeneration thinking in Australia. Underlying this approach is the notion of 'neighbourhood effects', or 'places that disadvantage people'. Expressed most simply, this is the proposition that

> deprived people who live in deprived areas may have their life chances reduced compared to their counterparts in more socially mixed neighbourhoods ... living in a neighbourhood which is predominantly poor is itself a source of disadvantage.
>
> (Atkinson and Kintrea 2001: 3–4)

Phrased in more economic terms, neighbourhood effects are 'the consequences of spillovers and externalities that arise from the co-location or proximities of particular socio-economic groups or activities' (Maclennan 2013: 271).

The mechanisms through which neighbourhood effects (NE) are hypothesised as operating have been extensively elaborated and debated in the urban policy literature (Galster 2009). At one end of the spectrum are relatively uncontroversial notions such as stigmatisation by address – the idea that residence in a poorly reputed area can incur employment discrimination (Hastings and Dean 2003). At the other extreme are contentions that concentrated disadvantage results in social contagion and/or collective socialisation (e.g. a 'culture of worklessness'). Demonstrating official recognition of the NE thesis in the Australian context, 'agreed policy actions' included in the 2009 National Affordable Housing Agreement between the Commonwealth government and the states and territories was the aspiration for 'creating mixed communities that promote social and economic opportunities by reducing concentrations of disadvantage that exist in some social housing estates' (Council of Australian Governments 2009: 7). In the academic world, however, the NE thesis has sparked lively contestation, especially as linked with US government initiatives to 'de-concentrate poverty' through public housing reform initiatives such as the HOPE VI program. For critics such as Slater (2013), the NE concept is inherently suspect because of its association with 'underclass' ideas (Wilson 1987; Murray 1990) implying that individuals living in poverty are 'culpable for their own predicament' (Arthurson 2013: 254).

Specifically commenting on the Australian context, Darcy (2010: 11) contended that

> the 'problems' of poor households living in public housing have been reconstructed as problems of place in order to support a policy intervention [redevelopment] which is closely linked with dominant neo-liberal ideology and economic prescriptions.

Thus, as argued by Arthurson and Darcy (2015), public housing tenants with rights to security of tenure are viewed as obstructing opportunities for reinvestment and value capture. According to this critique, therefore, the main importance of the NE thesis in the policy arena is in legitimating rather than in inspiring contemporary estate renewal practice. Thus:

> the theory of cultural reproduction of poverty through 'neighbourhood effects' provides the necessary social justification for returning publicly owned land to the market by claiming that relocation of tenants is in their own interest.
>
> (Arthurson and Darcy 2015: 183)

As seen by Doney et al. (2013) complexities arising in the implementation of mixed tenure estate renewal (see below) stem in part from distinct and competing discourses of social exclusion within the broader problematisation of social housing. On the one hand there is the 'equity discourse' in which exclusion is interpreted as resulting from 'lack of access to quality housing, services and infrastructure' (Doney et al. 2013: 408). Thus, 'social mix policies could address structural gaps that create social housing tenants' lack of access to services and infrastructure'. On the other hand there is the 'culture of poverty' discourse which 'understands the underlying cause of disadvantage as the result of socio-cultural factors that relate to the need for the poor to change their behaviour'. Informed by this understanding of 'problematic' social housing estates, policy responses are likely to be preoccupied with 'modifying the behaviour of individuals' (Doney et al. 2013: 411).

Turning to the *consequences* of poverty de-concentration via 'demolition and replacement' estate renewal projects, concern has been expressed about instances where on-site public housing re-provision is at a reduced scale. Bearing this in mind it has been argued that the dispersal of disadvantaged populations has become a key focus of urban policy (Rogers 2012). Such policies are argued to be problematic partly because of their potential to undermine community voice where established spokespersons for a locality are rehoused elsewhere (August and Walks 2012; Rogers 2012). More broadly, such instances are characterised as 'breaking up communities'. For Darcy (2010), it remains in question whether people displaced from redeveloped Australian public housing estates in fact continue to experience disadvantage just as in their former location, but now – thanks to dispersal – in a less 'politically visible' way.

Certainly, as implemented through public housing reconfiguration in the USA, 'poverty de-concentration' indeed has tended to imply the scattering of disadvantaged former residents of demolished estates – e.g. through issuing vouchers to help low-income former public- housing tenants meet higher private rental costs in other localities (Popkin et al. 2009). This is part and parcel of a US approach to estate 'regeneration' in which the quantum of public housing stock has been substantially reduced. As reported by Goetz (2012) the 1993–2010 HOPE VI program resulted in the demolition of 150,000 public housing units, of which only 45 per cent were replaced as such (although other replacement dwellings included some less deeply subsidised 'affordable housing'). As reported by Schwartz (2015) only about a quarter of public housing tenants displaced by HOPE VI projects have ultimately been rehoused on site.

However, at least as implemented since the early 2000s, estate renewal practice in Australia tends to be somewhat different from this. First, there is generally a starting assumption that in an estate subject to redevelopment, political legitimacy demands that the quantum of social housing should be retained at its original level, with onsite re-provision strongly favoured. Recent and proposed NSW projects exemplifying this understanding include those at Bonnyrigg, Ivanhoe and Waterloo. Second, and associated with this, poverty de-concentration is to be achieved as far as possible by 'diluting' public housing through the additional on-site development of private-market housing – and through the consequential increase in local population. Squaring this circle is dependent on the capacity to redevelop the original site at substantially increased density; a factor generally crucial in underpinning project funding – as further discussed below.

None of this is to deny that estate renewal in Australia in some cases poses risks for the maintenance of cohesive communities. Even if pre-scheme public housing provision is numerically maintained, project implementation often necessitates 'temporary' off-site re-housing while works proceed, with displaced tenants often ultimately opting not to return even if that had been their originally stated intention (and 'entitlement' – i.e. where a 'right to return' commitment was volunteered by government at the outset). Nevertheless, while this may be a fair comment on US practice, we would question whether contemporary Australian estate renewal practice is actively aimed at fostering the off-estate displacement of disadvantaged people.

Mixed-tenure redevelopment

Inherent within the Australian interpretation of poverty de-concentration through the local dilution of the disadvantaged population (see above), another key consensus belief underlying policy-maker perspectives on public housing estate renewal is the need to redevelop former mono-tenure neighbourhoods as mixed-tenure localities. In NSW, for example, the default assumption is that public housing should account for no more than 30 per cent of the units of a redeveloped estate (Coates and Shepherd 2005). In passing, it is worth noting that despite its pivotal importance in contemporary NSW estate renewal project planning, this particular formula lacks any strong evidential basis. Rather than being an end in itself, mixed-tenure redevelopment is identified as a means to the end of achieving a 'socially mixed community'. Consistent with the precepts of the neighbourhood effects thesis, the associated rhetoric of 'socially balanced communities' is underpinned by the belief that importation of better-off households able to afford market-price housing will boost local purchasing power, offset area stigmatisation, provide positive role models, and reinforce (desirable) social control. More pointedly, as argued by Arthurson and Darcy (2015: 182):

> the social mix objective [of estate renewal projects] relies on the idea that owner occupiers, as mortgage holders, will be stable neighbours with a stake in the area and will underwrite an aspirational local culture based on employment.

This is part of the wider project which aims to 'turn public tenants into responsible citizens who work and pay their own way rather than relying on taxpayer assistance' (Arthurson and Darcy 2015: 184).

In practice, research evidence to validate key hypotheses of the pro-social mix policy agenda remains sparse. In particular, as emphasised by Galster

(2009), studies have demonstrated the often limited extent to which cross-tenure social contact takes place at the neighbourhood level (Atkinson and Kintrea 2000; Arthurson 2012). In any event, it may well be the case that neighbourhood 're-engineering' with the intention of drawing in middle-class owner occupiers will fail to turn out as such. Instead, as discussed by Hulse et al. (2004), market sale homes may prove more attractive to investor landlords. The resulting re-modelled estate would thus remain a community of renters—albeit divided between tenancies allocated through an administrative filtering process and those through the market. Indeed, focusing on a recent case-study example, Levin et al. (2014) report that investor landlord acquisitions accounted for 70 per cent of the market-sale units in Melbourne's redeveloped Carlton Estate (Phase 1). Given the typically high rates of tenancy turnover in the private rental sector, such an outcome is unlikely to contribute to the sense of community stability and resident responsibility to which mixed-tenure re-development projects often aspire.

As in relation to the associated issues of 'neighbourhood effects' and poverty de-concentration, 'mixed-tenure redevelopment' of former public housing estates excites some sharply critical perspectives in which such initiatives are portrayed as amounting to 'state-sponsored gentrification'. Advocates of this position contend that 'social mix' is little more than rhetoric that cloaks a gentrification strategy whereby the middle classes are invited into social and economically challenged neighbourhoods to 'save' them from permanent decline through consumption practices that boost the local tax base (Lees et al. 2012: 7). In her analysis of public housing estate renewal at Kensington, Melbourne, Shaw argues that the project will contribute to gentrification locally. In her view this was not government's primary aim in the privatisation of public land which formed a centrepiece for the project. Rather, the discourse of social mix was being used 'as a rationalisation for minimal government outlay on public works'. This 'political' interpretation rested on the argument that Victoria's ruling Labor Party administration viewed the project as enabling it to demonstrate the financially prudent credentials considered crucial to electoral success (Shaw 2012).

Over and above the *policy* issues connected with mixed-tenure redevelopment there are important *implementation* considerations which will need to be addressed in any such project. Linked with their evaluation of the Victoria state government's 2006–2017 redevelopment of Melbourne's North Carlton estate, Levin et al. (2014) identify four key questions. First, there is the matter of implementation scale. This refers to whether private and public housing should be mixed at the level of the building, the street or the locality.

Second, there is the often controversial question of to what degree tenure integration is desirable. When we speak of a 'mixed-tenure' project, is this an accurate label for a scheme where tenure is segmented at the block level? Or should the term be restricted to the model where public housing is entirely pepper-potted throughout a development? Relevant here is the understanding that 'pepper potting' is typically disliked by developers who consider that it devalues their for-sale properties. This was evidenced by Levin et al in the context of their North Carlton estate redevelopment case study where the pepper-potted layout for Phase 1 of the project, as originally proposed, was reportedly abandoned thanks to developer pressure. Even if private and public housing is contained within a single building this may involve separate entrances. The associated 'poor doors' – entrances specific to social renters – have sparked moral outrage (Osborne 2014). However, it has been pointed out that low-income tenants may well prefer not to share (via service charges or strata fees) the cost of providing and maintaining spacious and/or elaborate common areas designed to suit the taste of better-off residents of market-price units.

The third design/implementation question highlighted by Levin et al. (2014) concerns physical differentiation of tenures in terms of appearance. Some mixed-tenure proponents set great store by the 'tenure-blind' principle that private and public housing units should be built to an indistinguishable design to minimise stigmatisation. Finally, Levin et al. (2014) argue for the significance of building and site design in terms of the creation – or omission – of public spaces where cross-tenure interaction may occur. In terms of design features compatible with social integration, there appears to be consensus that mixed-tenure housing projects should contain a hierarchy of common spaces so that 'residents [can] feel free to stay with their own social group or mingle with people from other groups…' (Levin et al. 2014: 25). In the North Carlton case study, however, not only were housing tenures segregated by block, but an internal courtyard between the blocks was for the exclusive use of private housing residents.

Partnership working

Another consensus article of faith among those concerned with public housing estate regeneration is the project governance style termed 'partnership working'. As it originated in the UK, the partnership working vogue partly reflected an analysis of post-war planning failures – especially the large-scale industrialised approach to council housebuilding – as stemming from over-centralised or unilateral policymaking. While estate regeneration projects will generally happen only when initiated by a government agency, the success of such an enterprise is dependent on effective collaboration between that agency and other stakeholder entities and interests. Thus 'interagency working is central to creating holistic solutions to intractable social problems' (O'Malley 2004: 2). The partnership-working principle is closely related to the notion of network governance – decision-making involving 'self-organising, inter-organisational, networks characterised by interdependence, resource exchange, rules of the game and significant autonomy from the state' (Rhodes 1997: 15). This is about a shift from control exercised by hierarchical authority to influence exerted through a more diffuse distribution of power which involves 'various forms of co-production with other agencies and with citizens themselves.' (Newman et al, 2004: 204).

One requirement for genuine partnership working is that government recognises the necessity of sharing power. In the regeneration context, however, it is not just a matter of collaborating with non-state actors. It is a recognition that government is not a single undifferentiated entity. In the Australian context, the decision to redevelop a public housing estate may be the prime responsibility of the asset-holding state government agency ('housing authority'). However, other state government departments will need to play an active role. Local government, too, will be necessarily involved. Beyond this, other stakeholders who will be appropriately engaged may include civil-society organisations such as local community groups. Perhaps the most significant dimension of 'partnership working' involved in contemporary estate regeneration, however, is between public and private sectors. At least since the turn of the millennium it has been a universal assumption that Australian estate regeneration projects involving significant physical renewal will necessarily feature the central involvement of private developers. While this may be structured through traditional contractual agreements, there have been instances of more complex arrangements, such as joint ventures, and public–private partnership structures such as at Kensington, Victoria or Bonnyrigg, NSW.

AHURI research on innovative partnership models in the design and delivery of housing policy and urban renewal (Pinnegar et al. 2011) argued that a number of distinct issues and considerations shape the translation and operation of partnership arrangements in a housing and urban renewal context. First, the focus of partnership activity, certainly in the estate renewal context, is firmly grounded in *place*, and working together to ensure effective place-based outcomes alongside the communities' living day-to-day, in situ, through the process and outcomes of those partnership activities. With all entities necessarily focused on place, there is a distinct spatial dimension encapsulated within the partnership model. Second, where urban renewal activity is 'housing-led', as it inevitably is in the context of estate renewal, partnerships are working together on delivery of a particularly complex asset class – especially so where mixed-tenure housing is integral to schemes. Third, long-term renewal activity demands partnership arrangements that balance the need for certainty over many years but sufficient flexibility given the inherent reliance on changing housing market contexts. Fourth, and again especially in the context of mixed-tenure models where cross-subsidy is involved, there are significant risks to policy delivery and government accountability with increased interdependencies between public, private and not-for-profit sectors.

Recent partnership models have seen not-for-profit community housing providers (CHPs) starting to be cast in this 'development/investment partner' role. Examples include the Housing Tasmania Better Housing Futures program to upgrade the state's major 'broadacre estates' and the ill-fated Queensland state government 'Logan Renewal Initiative' planned to reshape a large body of public housing on the southeast fringe of Brisbane (Pawson et al. 2013; Pawson 2016a). Similarly, in the NSW government's 2015 Communities Plus estate renewal program, significant CHP involvement is likely (Pawson 2016b). In large flagship schemes, such as those currently planned for the Ivanhoe, Riverwood and Waterloo estates in Sydney, this is set to come about via partnerships with private developers where CHPs join in developer-led consortia bidding for project leadership via competitive tenders. Under this model it is understood that the participating CHP will be charged with managing the re-provided social housing to be constructed as an element of the rebuilt estate – albeit that asset ownership may remain with the NSW government, and the extent of CHP engagement with the development process itself remains to be seen. Beyond this, the CHP concerned could also potentially take on a post-redevelopment 'place management' role as in the case of Urban Communities, the NFP entity that performs such a function for the renewed Kensington estate in Melbourne (Social Traders Ltd 2009). The NSW Communities Plus program is also expected to involve smaller renewal projects in which CHPs will be lead players – as both scheme developers and post-renewal site managers.

A number of implications follow from these forms of CHP participation in estate renewal projects. First, from an institutional perspective, such involvement may in future form an important vehicle for building the operational scale and capacity of the not-for-profit affordable housing industry (Milligan et al. 2016). Second, from a socio-political perspective, some consequences flow from the fact that low-income residents of a redeveloped estate will be the tenants of a regulated non-government entity rather than government itself. Since CHPs are governed by 'skills-based boards' and are not directly answerable to elected ministers, landlord-accountability arrangements are different. Partly in connection with this observation some argue that effective replacement of public housing by community housing amounts to an unacceptable form of 'privatisation' (Save Public Housing 2016) although this view is contested (Pawson 2016a).

Community participation and placemaking

Associated with partnership working and the notion of 'power sharing' is the concept of community participation. The appeal of this concept is attributed to:

> a faith that involving 'the community' will give rise to more effective and sustainable solutions, help local people exert control over social problems, assist the improvement of mainstream services, as well as contribute to democratic renewal.
>
> (Atkinson and Kintrea 2001: 27)

By the early 2000s it was thus reported that, reflecting international thinking, 'A fundamental tenet of [the] new [Australian] consensus is that renewal work is unsustainable unless the community becomes actively engaged in ongoing arrangements' (Wood 2002: p. v). Subsequently, it has been recognised that community capacity-building investment is often a precondition for meaningful involvement (Judd and Coates 2008: 4–5).

Of course, the extent to which 'tenant participation' in the regeneration context in practice involves substantive government openness to resident influence is a debatable matter. This is partly a question of how government conceptualises 'participation'. Referring to the NSW government approach to resident involvement in estate renewal projects for example, Darcy and Rogers (2014) argue that this aspires to facilitate 'consensus building' rather than 'interest-based activism'. Even if tenant participation is substantive, this can be viewed in various ways. Thus, engaging local people in regeneration is a mode of exercising power although it brings with it both regulatory and liberatory possibilities for participants. The term 'placemaking' is increasingly being used in reference to the involvement of local residents in urban design (McGuirk et al. 2016). In the estate regeneration context, this is likely to call for 'community development' or capacity building activity.

Funding

While some early 1990s estate regeneration projects were part-funded through the federal administration's Building Better Cities program (Neilson 2008), direct financial assistance for such activities from this higher tier of government has been unusual. Although recent community renewal programs (e.g. Neighbourhood Renewal Victoria, Building Stronger Communities, NSW) have been largely or wholly state government funded, there has been an increasingly universal acceptance that estate renewal projects involving significant physical investment will happen only if largely or wholly self-financed. In practice, this means the targeting of activity in relation to (redevelopment) opportunity, rather than (socio-economic) need.

Self-financing, in this context generally involves the redevelopment of an estate at a much increased density, with the cost of re-providing demolished public housing met from the proceeds of land sales to developers for open-market housing construction. Key to the viability of this cross-subsidy model is the relationship between local land values and the scope for densification in terms of planning expectations and market capacity.

Highly relevant to financial considerations surrounding public housing estate renewal are the findings of analysis which sought to quantify the impact of estate renewal investment in terms of the uplift in property values enjoyed by house owners in surrounding areas (Wood and Cigdem 2012). This concluded that each \$1 invested generated an asset value gain of \$2.2. However, while this can be interpreted as demonstrating that such investment can be justified on 'efficiency' as well as 'equity' grounds, the study also emphasised that the absence of any 'value-capture' mechanism (e.g. land tax) meant that the vast bulk of the associated gains accrued to property owners rather than being shared by the wider community.

Exemplifying public housing estate regeneration: Bonnyrigg, Sydney

In order to explore how these principles have translated through policy development and then through actual implementation, we briefly focus on the high-profile comprehensive renewal of the Bonnyrigg estate in Sydney's west. Here we draw upon findings from the first wave of a longitudinal study conducted by the UNSW City Futures team (Pinnegar et al. 2013), established to track the expectations, experiences and outcomes of residents over the intended duration of renewal activity. The study involved establishing a panel which fully captured the diversity of the community, structured around talking to a representative sample of households experiencing different stages of the regeneration process. During the first wave, approximately one hundred households were interviewed (the large majority in their homes, all in their preferred language), including former tenants who had moved away from Bonnyrigg having made the decision (or felt they had little choice but) to be relocated, as well as a small number of both long-standing as well as recently arrived private residents.

Bonnyrigg was one of the last major public housing estates built by the then Housing Commission of New South Wales in the late 1970s and early 1980s as development of mono-tenure, low-density neighbourhoods was coming to an end. The government's impetus for Bonnyrigg's renewal was framed by a number of considerations. Asset-management issues were pervasive arguments for intervention. The low-density 'Radburn' design discussed above was seen as problematic (Figure 15.1). The proliferation of cul-de-sacs and limited 'eyes on the street' were seen as contributing, in good old environmentally deterministic terms, to anti-social behaviour within the neighbourhood. After a series of earlier unsuccessful attempts

Figure 15.1 Low-density Radburn layout of Bonnyrigg, pre-renewal (source: City Futures Research Centre Image Library)

to mitigate shortcomings in the design and layout of the estate, calls for a more comprehensive renewal approach gained momentum. The estate had become increasingly stigmatised, suffered high levels of crime, and the physical make-up of the estate was seen as compounding the multiple layers of social disadvantage shaping the lives of Bonnyrigg's residents. Equally, as incisively explored by Rogers (2014), the 'obsolescence' narratives produced to legitimise the need to intervene can be seen as a result of a desire to demobilise and decommission outmoded uses of space in the neoliberal, market-centric city.

Renewal imperatives were also driven by opportunity and perceived economic viability. Increasing density across the neighbourhood would enable a significant number of private properties to be introduced onto the original estate's footprint and facilitate pervasive policy interests to encourage social mix and to foster diversification in social housing provision and management. An ambition to deliver renewal through a public–private partnership (PPP) was also borne out of political commitment to the model – this had been applied to other forms of social infrastructure across NSW, and housing was seen by government as next in line. The desire to 'dilute' concentrations of social disadvantage further confirmed that renewal activity should no longer simply be treated as a matter for public funding alone: a 'whole of market' solution was required.

In 2007, the competitive tender to redevelop, manage tenancies and 'renew the community' was won by the Bonnyrigg Partnerships consortium, which comprised five partners: Westpac Bank (the financier); Becton Property Group (the developer); St George Community Housing (the tenancy manager); Spotless (the maintenance manager); and Bonnyrigg Management (subsequently branded 'Newleaf') established to manage partnership integration and community renewal activity. Their winning bid proposed the redevelopment of the estate to ultimately provide 2,500 new homes (from a little over 900 on the original estate), around 700 of which would be owned and managed by the community housing provider (and therefore broadly ascribing to the 70:30 private/social housing 'preferred' ratio).

The complexity of the ensuing contractual partnering arrangements manifested itself in intricately interwoven legal frameworks between the constituent agencies, arms of government and the myriad other stakeholders to be involved in the renewal process. Complexity also underpinned and shaped the spatial enactment of partnership arrangements and expected outcomes in the masterplan, with 18 spatially defined stages of redevelopment delivered over a period of 13 years. As well as 'packaging' up different sections of the estate, each stage encapsulated a specific bundle of debt in terms of the financing model, a certain number of residents to be rehoused at that time, and a certain number of homes for sale once that stage of redevelopment was completed. It translates into financing spread over the years, a capacity to manage the on-site 'shuffle' (Pinnegar et al. 2013) of residents around the estate rather than 'decant' all at once, and not flooding the housing market with hundreds of units all at the same time. It also meant that given the timeframes involved, the rollout of the renewal process would take almost a generation and therefore be exposed to the wax and wane of financial and housing markets as well as broader economic trends.

Actual commencement of renewal activity on site coincided with the onset of the global financial crisis in 2008/2009. Nevertheless, initial progress was positive, with the remodelling of Tarlington Reserve on the northern side of the estate, construction of the first tranche of new homes next to this enhanced green space, and residents moving into the their homes in the 'Newleaf' streets – both existing Bonnyrigg households as well as incoming owner-occupiers and private renters.

While consideration and expectations of households in relation to the renewal process are as diverse as the community itself, commonly observed views at the time of that research often captured a tension between the big picture of complex renewal and how it translates and impacts on a day-to-day level for residents. Arguments about tenure mix, increasing density and partnership working were understood by residents. However the renewal process which rolled out was quite rightly filtered through a more everyday lens. The complexities facing the renewal partnership tended to translate from the residents' perspective into concerns about reduced lot sizes or house layouts were as much about impacts on everyday spaces – the ability to keep pets, having somewhere to place their treasured garden shed, grow veggies and fruit in the garden, or having a dining table in the living area – as they were about more esoteric, academic debates about density and housing choice.

The necessary focus on 'place' arguably provided a shared imperative for Bonnyrigg Partnerships, the government and residents alike. Viability of later stages of the redevelopment process, dependent upon private purchasers buying into the evolving mixed-tenure 'Newleaf' neighbourhood, demanded a 'best for place' rather than simply 'best for project' approach. It meant that the pepper-potting of tenures had to be carefully considered, and the housing product built needed to appeal to both tenants and purchasers. Interestingly, one of the house types trialled in Stage 1, the 'quadplex' – a design which comprised four 'interlocking' apartments over two storeys intended for both social and market housing – was not rolled out in future development stages as a result of limited private purchaser appeal rather than tenant reaction about the loss of space and privacy.

For the residents living through change – at the time of the research – a general acceptance of (or least resignation to) the disruption and uncertainty tied to the gradual, staged approach was voiced. Commitment to the neighbourhood, and recognition that this arrangement enabled them to remain part of it, was typically the overriding concern: the decision to stay was, whether proactive or more resigned in nature, a commitment by residents to see the renewal process through. As such, the imposition of temporary relocation elsewhere on the estate while redevelopment of their part of the neighbourhood was undertaken was largely being matched with resilience, and households getting on with life as best they could.

However, the importance of place within partnership activity also exacerbates risk: having the residents remain on site, and redevelopment rolling out in stages over many years, ties in the community to the relative fortunes of the partnership in ways more transparent than otherwise would have been the case. Understandably, there is a heightened expectation regarding communication – not least when lives are placed in a degree of limbo while temporarily housed, or awaiting their turn in the staging process. With residents remaining in-situ, it becomes all the more apparent when things are not progressing to plan. It enrols residents and their everyday lives, into the complexity not only of the renewal process but also the partnership arrangements structuring that process, and indeed the myriad externalities shaping the operation and activities of the partnership. Renewal through the PPP structure exposes the community to the economic and market realities that define private-sector capacity and operation, and which – as eventuated in the case of Bonnyrigg – may ultimately undermine that contract and the partnership relationships it frames. The PPP was enabled, but also dependent upon, working with the 'grain of the market' (Cole 2007: 25).

When those dependent conditions falter – as seen shortly after this first wave of the longitudinal study with the collapse of the developer Becton in 2013 – it might have been hoped that having a 30-year contractual framework in place would offer a resilient template into which another developer could be slotted 'in'. However, notwithstanding that Becton's demise was unrelated to its involvement in Bonnyrigg, its departure from the project symbolised a fairly unattractive and risky venture – particularly since this coincided with a housing market upturn offering rather more straightforward developer return options. The result was stasis: just three stages out of the eighteen stages had been completed, other sections had been 'cleared' for redevelopment, but then no actual activity on site seen until mid-2016 when the state government (through UrbanGrowth NSW) had to return to fulfil its ultimate responsibilities to the community to complete those parts of the neighbourhood already cleared. As of late 2017, Stage 4 is under construction and pre-sales for Stage 5 have commenced. A Development Application DA for Stages 6 and 7 has also been submitted, although it is not clear whether original plans for redevelopment of the entire estate will now proceed.

Conclusions and future directions

The recently published ten-year strategy *Future Directions for Social Housing in NSW* (NSW Government 2015a) provides the guiding framework for the forward regeneration of public housing assets in NSW. The directions capture the complex contradictions inherent in the role of social housing in the twenty-first-century city, particularly those managing the challenges

– as well as opportunities – of strong population growth within neoliberal governance frameworks. In *Future Directions* the need for social and affordable housing is acknowledged as being as great as ever. Yet government continues to primarily frame its role as a 'social good' rather than more broadly fundamental to future economic productivity and efficiency. Despite reassuring developments in the form of the NSW premier's Innovation Initiative for social housing and the $1 billion Social and Affordable Housing Fund (Pawson and Milligan 2015), the base upon which any additional supply will materialise, is inadequately small.

More poignantly in the context of urban renewal, such investments are ever-more intertwined with the substantive redevelopment of the estates, particularly in higher value areas where the dividends attached to density uplift can be maximised. In the language of government this reflects the imperatives of 'an asset management framework that leverages the value of the existing portfolio to accelerate supply' (NSW Government 2015b). The geographical targeting that results from this kind of thinking is likely to contrast substantially with the needs-based prioritisation of area renewal projects under early 2000s initiatives such as Victoria's Neighbourhood Renewal Program (Department of Human Services 2005) and NSW's Building Stronger Communities (NSW Government 2007).

The justification for redeveloping existing estates ripe for 'asset value extraction' is clearly open to debate: while much of the stock is 'tired' and does not align with the demographic profile of those most in need of social housing, there are obviously different ways of addressing those apparent deficiencies rather than undertaking the highly complex, controversial and highly disruptive process of comprehensive renewal. As discussed above, the purported wider benefits of large-scale transformation are far from clear cut. The process is also accompanied by considerable risk.

Policy directions initially pursued in the low-density suburban estates of western Sydney have been recalibrated to place further emphasis on the NSW Land and Housing Corporation's asset holdings in places which align with broader metropolitan strategic planning objectives and where opportunities for market-led densification are greater. Communities Plus, a core element of the 10-year Land and Housing Corporation strategy enshrines asset-recycling as central to forward housing provision, and will see many of Sydney's key remaining concentrated sites of social housing 'offered to the market' (NSW Government 2015a: 9). Holdings across the portfolio will be identified, and developers will respond with proposals for renewal.

Although Communities Plus sites are located across NSW and include regional as well as metropolitan estates, it does encapsulate a more general shift in the focus of regeneration away from middle- and outer-ring Radburn estates to those located in higher-value parts of Sydney. One of the first of the next generation of Communities Plus partnership sites will see the eight-hectare Ivanhoe Estate at Macquarie Park in Sydney's northwest, currently home to a relatively modest 259 social housing properties, transformed into a mixed-use neighbourhood with a total of over 3,000 properties. Similarly, inner Sydney's Waterloo Estate is slated for a densified renewal involving on-site replacement of the existing 2,000 public housing units as part of a 7,000 home project (Pawson 2016b). Such levels of uplift reflect in part the economics required to facilitate renewal, but also capture a heady mix of dramatic shifts in relative land values, local housing demand tied to this jobs-rich and accessible location, and developer interest. The trade-off for the inevitable disruption to existing estate residents is an appeal to the wider 'public good': not only a big tick in terms of overall housing supply numbers in a city which needs to build an additional 664,000 housing units in the next 15 years, but – in the Ivanhoe case – a substantive increase in the number of social (at least 950) and affordable (128) homes provided, *on site*, as part of the redevelopment.

Policy decisions driven by developers' feasibility also risk exacerbating the spatial polarisation and systematic locational disadvantage which has become more pronounced in Sydney as in other Australian cities in recent decades (Randolph and Tice 2014; Pawson et al. 2012; 2015). For communities living in social housing located in high-value land markets, the ten-year strategy arguably reasserts their position as tenants with few rights associated with the actual spaces of their homes or 'place' of their neighbourhood: the spatial context of their lives is up for grabs in the name of maximising land utilisation for the greater good. At the same time, policy attention struggles in those areas less well geographically related to the economic and market drivers shaping Australia's cities: often those localities where need may well be greater. As with all market-centric approaches, such policies only 'work' where the market enables them to work.

Contemporary public housing renewal initiatives in NSW can be seen as consolidating the policy and political drivers of recent decades, caught up within the complex and contradictory tensions between seeking to manage (and increasingly exploit) the asset represented by social housing stock and the land on which it sits, while holding onto the vestiges of a value system which accepts responsibility in supporting some of the state's most vulnerable residents. Estate renewal activities will continue to be at the behest of this complexity, with investment dictated more by wider city demands and market imperatives rather than prioritising need. While pitched as a 'win–win', with additional social and affordable housing numbers provided overall, *somewhere* in the city, these settings necessarily demand interrogation, not least because they consolidate a bullish take by the NSW government on questions of equity, spatial justice, location and accessibility rights in a city where unfair, and unproductive, spatial polarisation appears to be not only an accepted but also justified cost.

Notes

1 Derived from Troy (2012) and ABS Table 8752.0. However, it is possible that data for earlier years may be less than comprehensive.
2 Albeit with the brief exception of the Social Housing Initiative which formed part of the counter-recessionary Nation Building Economic Stimulus Plan running from 2008–2011 (Ruming 2015).
3 While the Australian social security system includes a Rent Allowance payment available to low-income tenants in receipt of other means-tested benefits, public housing tenants are ineligible.
4 In New Zealand, for example, the government makes Income Related Rent Subsidy (IRRS) payments to social housing providers to bridge the gap between revenue derived from income-based rent charges and market rents for the properties concerned.

References

Arthurson, K. (1998) 'Redevelopment of public housing estates: The Australian experience', *Urban Policy and Research*; 16(1): 35–460

Arthurson, K. (2012) *Social Mix and the City*, Collingwood: CSIRO Publishing.

Arthurson, K. (2013) 'Neighbourhood effects and social cohesion: Exploring the evidence in Australian urban renewal policies', in D Manley, M. van Ham, N. Bailey, L. Simpson and D. Maclennan (eds), *Neighbourhood Effects or Neighbourhood Based Problems?* Dordrecht: Springer.

Arthurson, K. and Darcy, M. (2015) 'The historical construction of "the public housing problem" and deconcentration policies', in R. Dufty-Jones, and D. Rogers (eds), *Housing in 21st Century Australia,* Farnham: Ashgate.

August, M. and Walks, A. (2012) 'From social mix to political marginalisation: The redevelopment of Toronto's public housing and the dilution of tenant organisational power' in G. Bridge, T. Butler, and L. Lees (eds), *Mixed Communities: Gentrification by Stealth?* Bristol: Policy Press.

Atkinson, R. and Kintrea, K. (2000) 'Owner-occupation, social mix and neighbourhood impacts', *Policy and Politics*, 28(1): 93–108

Atkinson, R. and Kintrea, K. (2001) *Neighbourhoods and Social Exclusion: The Research and Policy Implications of Neighbourhood Effects*, Glasgow: University of Glasgow.

Coates, B. and Shepherd, M. (2005) 'Bonnyrigg Living Communities Project: A case study in social housing PPPs', paper presented to National Housing Conference, Perth: Western Australia, 28–29 October.

Cole, I. (2007) *Shaping or Shadowing? Understanding and Responding to Housing Market Change*, York: Joseph Rowntree Foundation.

Council of Australian Governments (2009) *National Affordable Housing Agreement*, Canberra: Council on Federal Financial Relations.

Darcy, M. (2010) 'De-concentration of disadvantage and mixed income housing: A critical discourse approach', *Housing, Theory and Society*, 27(1): 1–22

Darcy, M. and Rogers, D. (2014) 'Inhabitance, placemaking and the right to the city: Public housing redevelopment in Sydney', *International Journal of Housing Policy* 14(3): 236–256

Department of Human Services (2005) *Neighbourhood Renewal: An Interim Evaluation*, Melbourne: Government of Victoria.

Doney, R. H., McGuirk, P. M., and Mee, K. J. (2013) 'Social mix and the problematisation of social housing', *Australian Geographer* 44(4): 401–418

Fitzpatrick, S. and Pawson, H. (2014) 'Ending security of tenure for social renters: transitioning to "ambulance service" social housing?' *Housing Studies* 29(5): 597–615

Galster, G. (2009) 'Neighbourhood social mix: Theory, evidence and implications for policy and planning', paper presented at International Workshop at Technion University 'Planning For/with People', Haifa: Israel, June 2009

Goetz, E. (2012) The Transformation of Public Housing Policy, 1985–2011, *Journal of the American Planning Association*, 78(4): 452–463.

Hall, J. and Berry, M. (2007) *Operating Deficits and Public Housing: Policy Options for Reversing the Trend, 2005/06 Update*, Final Report no 106, Melbourne: AHURI.

Hastings, A. and Dean, J. (2003) 'Challenging images: tackling stigma through estate regeneration', *Policy and Politics*, 31(2): 171–184.

Hulse, K., Down, K. and Herbert, T. (2004) *Kensington Estate Redevelopment Social Impact Study*, Melbourne: Department of Human Services.

Jones, P. and Evans, J. (2014) *Urban Regeneration in the UK*, London: SAGE.

Judd, B. and Coates, B. (2008) 'Building community capacity in housing and urban design in a regenerating suburb: The Bonnyrigg experience', paper to European Network for Housing Research Conference, 6–9 July, Dublin, Ireland.

Kintrea, K. (2007) 'Policies and programmes for disadvantaged neighbourhoods: Recent English experience', *Housing Studies*, 22(2): 261–282

Lees, L., Butler, T. and Bridge, G. (2012) 'Introduction: gentrification, social mix/ing and mixed communities', in G. Bridge, T. Butler, and L. Lees, (eds), *Mixed Communities: Gentrification by Stealth?* Bristol: Policy Press.

Levin, I., Arthurson, K. and Ziersch, A. (2014) 'Social mix and the role of design: Competing interests in the Carlton Public Housing Estate redevelopment, *Cities*, 40(Part A): 23–31

Maclennan, D (2013) 'Neighbourhoods: Evolving ideas, evidence and changing policies', in D. Manley, M. van Ham, N. Bailey, L. Simpson, and D. Maclennan, (eds), *Neighbourhood Effects or Neighbourhood Based Problems?* Dordrecht: Springer.

McGuirk, P.M., Mee, K.J., and Ruming, K.J. (2016) 'Assembling urban regeneration? Resourcing critical generative accounts of urban regeneration through assemblage', *Geography Compass*, 10: 128–141

Milligan, V., Martin, C., Phillips, R., Liu, E., Pawson, H. and Spinney, A. (2016) *Profiling Australia's Affordable Housing Industry*, Final Report no 268, Melbourne: AHURI.

Murray, C. (1990) *The Emerging British Underclass*, London: Institute for Economic Affairs.

Neilson, L. (2008) 'The Building Better Cities program 1991–96: A nation-building initiative of the Commonwealth government', in J. Butcher, (ed), *Australia Under Construction*, Canberra: ANU Press.

NSW Auditor General (2013) *Performance Audit: Making the Best Use of Public Housing*, Sydney: NSW Auditor General's Department.

NSW Government (2007) *A New Direction in Building Stronger Communities 2007–2010*, Sydney: NSW Government http://www.housing.nsw.gov.au/__data/assets/pdf_file/0010/330499/BookletNewDirectionStrongerCommunities.pdf (accessed 13 February 2017).

NSW Government (2014) *Social Housing in New South Wales*, a paper for input and comment, Sydney: NSW Government.

NSW Government (2015a) *Future Directions for Social Housing in NSW*, Sydney: NSW Government http://www.socialhousing.nsw.gov.au/?a=348442 (accessed 13 February 2017).

NSW Government (2015b) 'Communities Plus – Overview' http://www.communitiesplus.com.au/

Newman, J., Barnes, M., Sullivan, H. and Knops, A. (2004) 'Public participation and collaborative governance, *Journal of Social Policy* 33(2): 203–223.

Osborne, H (2014) 'Poor doors: the segregation of London's inner-city flat dwellers', *The Guardian*, 25 July, https://www.theguardian.com/society/2014/jul/25/poor-doors-segregation-london-flats (accessed 13 February 2017).

O'Malley, L. (2004) 'Working in partnership for regeneration: The effect of organisational norms on community groups, *Environment and Planning A*, 36(5): 841–857.

Pawson, H. (2016a) 'Ditching Logan public-private regeneration sets Queensland back on social and affordable housing', City Futures Research Centre Blog, 25 July http://blogs.unsw.edu.au/cityfutures/blog/2016/07/ditching-logans-public-private-regeneration-sets-queensland-back-on-social-and-affordable-housing/ (accessed 13 February 2017).

Pawson, H. (2016b) 'Is community housing set to feature in the Waterloo Estate rebuild?' City Futures Research Centre Blog, 2 June http://blogs.unsw.edu.au/cityfutures/blog/2016/06/is-community-housing-set-to-feature-in-the-waterloo-estate-rebuild/ (accessed 13 February 2017).

Pawson, H. and Milligan, V. (2015) 'Are we seeing a new dawn for affordable housing?' *City* Futures Research Centre Blog, 17 December http://blogs.unsw.edu.au/cityfutures/blog/2015/12/are-we-seeing-a-new-dawn-for-affordable-housing/ (accessed 13 February 2017).

Pawson, H., Davison, G. and Wiesel, I. (2012) *Addressing Concentrations of Disadvantage: Policy, Practice and Literature Review*, Final Report 190, Melbourne: AHURI.

Pawson, H., Milligan, V., Wiesel, I. and Hulse, K. (2013) *Public Housing Transfers in Australia: Past, Present and Prospective*, Final Report no 215, Melbourne: AHURI.

Pawson, H., Hulse, K. and Cheshire, L. (2015) *Addressing Concentrations of Disadvantage in Urban Australia*, Final Report no 247, Melbourne: AHURI.

Peel, M. (1995) *Good Times, Hard Times: The Past and the Future in Elizabeth*, Melbourne: Melbourne University.

Pinnegar, S. (2009) 'The question of scale in housing-led regeneration: Tied to the neighbourhood?', *Environment and Planning A*, 41(12): 2911–2928.

Pinnegar, S., Wiesel, I., Liu, E., Gilmour, T., Loosemore, M. and Judd, B. (2011) *Partnership Working in the Design and Delivery of Housing Policy and Programs*, Final Report no 163, Melbourne: AHURI.

Pinnegar, S., Liu, E. and Randolph, B. (2013) *Bonnyrigg Longitudinal Panel Study First Wave: 2012*, Sydney: City Futures Research Centre, UNSW.

Popkin, S., Levy, D., and Buron, L. (2009) 'Has HOPE VI transformed residents' lives? New evidence from the HOPE VI panel study', *Housing Studies*, 24(4): 477–502.

Productivity Commission (2009) *Report on Government Services*, Melbourne: Productivity Commission.

Randolph, B. and Judd, B. (2000) 'Community renewal and large public housing estates', *Urban Policy and Research*, 18(1): 91–104.

Randolph, B. and Tice, A. (2014) 'Suburbanizing disadvantage in Australian cities: Sociospatial change in an era of neoliberalism', *Journal of Urban Affairs*, 36(1): 384–399.

Rawnsley, T. and Spiller, M. (2012) 'Housing and urban form: a new productivity agenda', in R. Tomlinson, (ed), *Australia's Unintended Cities*, Collingwood: CSIRO.

Rhodes, R. (1997) *Understanding Governance: Policy Networks, Governance, Reflexivity and Accountability*, Buckingham: Open University Press.

Rogers, D. (2012) 'Citizenship, concentrations of disadvantage and the manipulated mobility of low-income citizens: The role of urban policy in NSW', paper presented at Australasian Housing Researchers' Conference, 8–10 February, University of Adelaide: South Australia.

Rogers, D. (2014) 'The Sydney Metropolitan Strategy as a zoning technology: analysing the spatial and temporal dimensions of obsolescence', *Environment and Planning D: Society and Space*, 32: 108–127.

Ruming, K. (2015) 'Reviewing the Social Housing Initiative: Unpacking opportunities and challenges for community housing provision in Australia', in R. Dufty-Jones, and D. Rogers, (eds), *Housing in 21st Century Australia*, Farnham: Ashgate.

Save Public Housing (2016) '5,000 homes saved from privatisation! Huge win for public housing in Queensland!!' http://savepublichousing.blogspot.com.au/2016/08/5000-homes-saved-from-privatisation.html (accessed 13 February 2017).

Schwartz, A. (2015) *Housing Policy in the United States*, 3rd edn, New York: Routledge.

Shaw, K. (2012) 'Beware the Trojan horse: Social mix conceptions in Melbourne' in G. Bridge, T. Butler and L. Lees (eds), *Mixed Communities: Gentrification by Stealth?* Bristol: Policy Press.

Slater, T. (2013) 'Capitalist urbanisation affects your life chance: Exorcising ghosts of "neighbourhood effects"', in D. Manley, M. van Ham, N. Bailey, L. Simpson, and D. Maclennan (eds), *Neighbourhood Effects or Neighbourhood Based Problems?* Dordrecht: Springer.

Social Traders Ltd (2009) 'Social enterprise case study: Urban Communities Ltd', http://cdn.socialtraders.com.au/app/uploads/2016/05/Urban-Communities-Case-Study.pdf (accessed 13 February 2017).

Troy, P. (2012) *Accommodating Australians,* Sydney: Federation Press.

Wilson, W. J. (1987) *The Truly Disadvantaged: The Inner City, the Underclass, and Public Policy*, Chicago, IL: University of Chicago Press.

Wood, M. (2002) *Resident Participation in Urban and Community Renewal*, Final Report no 23, Melbourne: AHURI.

Wood, G. and Cigdem, M. (2012) *Cost-effective Methods for Evaluation of Neighbourhood Renewal Programs*, Final Report No.198, Melbourne: AHURI.

Woodward, R. (1997) 'Paradise lost', *Australian Planner*, 34(1): 25–29.

16 Resisting regeneration

Community opposition and the politicisation of transport-led regeneration in Australian cities

Crystal Legacy and Elizabeth Taylor

Introduction

While metropolitan strategies across Australian capital cities articulate the importance of land use and transport integration in planning to help manage present and future growth (WAPC 2004; Government of Victoria 2014; NSW Government 2014), research has examined the planning and market-related challenges associated with its delivery in practice (Curtis et al. 2009; Searle et al. 2014). However, research has rarely examined land use and transport integration as a form of urban regeneration, and the distinctive character of resident and community resistance it can attract. Such resistance can add to implementation-related challenges. In this chapter we aim to extend the Australian research examining resistance to urban regeneration, to which there have been many important contributions made over the past two decades (Burgmann and Burgmann 1998; Davison 2004; Porter and Shaw 2009; Ruming 2010; Iveson 2013; Howe et al. 2014; Darcy and Rogers 2015), by bringing into dialogue the role that transport-led urban regeneration has played in igniting recent experiences of citizen opposition and resistance. Australian cities recently undergoing investment in both light and heavy rail, and freeway infrastructure – including Melbourne's East West Link and Sky Rail projects, Sydney's Westconnex and Anzac Parade Light Rail projects, the Gold Coast's Light Rail project, Canberra's Light Rail and Perth's Freight Link projects – are encountering intense local resident resistance to some of the land-use impacts associated with these projects. In the case of some of these projects (Canberra's Light Rail and Melbourne's level crossing removal or 'Sky Rail' as two examples) proposed investments into public transport have been framed as having significant and region-wide transport-led opportunities for increased mobility and accessibility, as well as offering adjacent urban regeneration possibilities.

Transport proposals have a long history of attracting resistance, but this resistance has often been directed towards freeway construction that serves to move people through neighbourhoods (Davison 2004; Legacy 2016). In contrast to freeway construction, investment in public transport seeks to produce opportunities for non-car-based mobility, local accessibility, as well as regeneration of adjacent urban spaces – the latter a feature of light-rail investment (Cervero 1984; Cao and Porter-Nelson 2016). Yet in the face of this investment, some affected residents and traders fear that investment in public transport in their neighbourhood can bring about unwanted land-use changes.

In this chapter we show how the renewal (maintenance and upgrade) of public-transport infrastructure attracts local community resistance. Our aim is to highlight the spatial and scale-based nature of resistance (from the local to the non-local) to transport-led urban regeneration. In doing so, we join Ruming et al. (2012) in highlighting the dimensions and various scales of resistance and the multiple publics through which resistance is mounted. However, we aim to extend this debate by considering the political spaces through which the materiality of local resistance confronts the inclusive and seemingly more progressive public policy agendas of accessibility and inclusivity. We seek to depart slightly in this chapter from the important critical scholarship that exposes the injustices of capitalist production driving urban reproduction (Lees et al. 2013; Harvey 2012; Porter and Shaw 2009), by emphasising the unique complexities involved in transport-led urban regeneration. We do this with a view to garnering new understandings of urban resistance in cities experiencing a shortfall of critical social infrastructure (e.g. affordable housing, high-quality public transport), but where neighbourhood resistance forces are strong.

This chapter focuses on one project, Melbourne's number 96 Tram accessibility upgrades and the resistance mounted to it at the project proposal stage. In this case the transport-led regeneration project was part of a larger urban agenda to invest in urban transport to address broader mobility challenges faced across metropolitan Melbourne. The case shows a failure to manage how the strategic imperatives of accessibility gain expression through targeted transport investment but yet attract different levels of acceptance across different scales of association with a proposed project (Ruming et al. 2012). Our analysis is drawn from a desktop review of policy documents, council minutes and media analysis on this controversial project. The findings offer insight into the complex interplay between local resistance and policy agendas aligned with a social inclusion and non-car agenda orientation.

Resisting urban regeneration in the Australian city

Cities and their urban spaces are produced through struggle and contestation. Struggles to overcome and challenge the extraction of surplus value have characterised forms of urban reproduction, whether through the urban revitalisation of neighbourhoods in the 1960s and 1970s, to forms of multi-sectoral (e.g. housing and transport) urban regeneration (Wilson

1963; Palen and London 1984; Smith 2002), to metropolitan-wide policies of urban consolidation and intensification (Randolph 2006). All of these characterise the urban reproductive efforts of contemporary Western cities. Likening regeneration to a 'natural process', Smith (2002) criticises the suggestions that the gentrification produced is 'natural' which renders invisible the 'the politics of winners and losers out of which such policies emerge' (Smith 2002: 445). Lees (2008) also warned that the suggestion of positive gentrification that mediates dislocation by introducing social mixing, often through housing provision, simply 'conceal the inequalities of fortune and economic circumstances that are produced through the process of gentrification' (Lees 2008: 2463). As the social challenges and inequalities of urban regeneration processes become evident, and in order to manage possible citizen opposition, attention has been given to governance. This includes partnership formation in coordinating public- and private-sector activities and in legitimating these programmes (Raco 2000). Citizen contestation associated with urban reproduction, and how to use participation in these processes so as to manage the local politics, has a long association with urban renewal programmes dating as far back as the 1960s (Wilson 1963).

Within the Australian literature, books by Burgmann and Burgmann (1998), Davison (2004) and more recently Howe et al. (2014), provide important historical accounts of the public protests that mounted in some inner-city Australian neighbourhoods against the negative effects of state-sponsored growth. These texts document the force of the union movements prevalent in Australia in the 1960s and 1970s, providing the backbone to some instances of community campaigns against urban renewal. One of the better-known and documented accounts of urban resistance is that associated with the Green Bans movement[1] in Sydney in the 1970s (Iveson 2013; Burgmann and Burgmann 1998). What marked the Green Bans campaign was the connection to union organising led by a concern for the widespread state-enabled capitalist dispossession of people's livelihoods and communities which challenged conceptions of 'the people'. Iveson (2013: 1008) described the community action as 'not simply the *products* of a prior process of alliance building and subjectification as "the people", they were also a significant *constitutive element* of that process'. Going further, Iveson (2013) emphasised the extent to which conceptions of the 'public interest', in this case the right to the city, were framing the substantive content and were acting as a mobilising agent to strengthen resistance: 'The green ban activists asserted a right to the city by invoking the needs of "the people" before profits that sought to *politicise* the forms of inequality and injustice that are naturalised in capital cities' (Burgmann and Burgmann 1998).

Reminiscent of the concerns which catalysed the resistance observed four decades earlier, the recent sell-off and redevelopment of public housing in Millers Point, resulting in the dislocation of long-time public-housing tenants, provides another example of the hotly contested nature of forms of urban regeneration in Australian cities and of the community resistance it can attract (Darcy and Rogers 2015). Resistance to urban regeneration also forges platforms from which concerned and affected citizens may challenge the economic rationalisation that drives urban redevelopment. In these spaces, more equitable social outcomes can be demanded (Shaw 2009). Resistance to urban regeneration calls into question those development processes that adhere predominantly to economic growth objectives over those of equity, environmental sustainability, diversity and inclusivity. In addition, in those cities investing in urban public transport this investment is directed to deliver land-use regeneration goals. In the case of transport-led urban regeneration, the framing of resistance, as that against the 'bad' forces of economic development, becomes fraught and too simplistic to fully capture the complexity involved in community resistance to urban infrastructure investment and the multiple publics through which this resistance transpires (Ruming et al. 2012).

The connection between transport planning and urban regeneration has a long history in Australian cities. The proposed freeway expansions associated with the urban land clearance of the early 1970s was a concentrated period of resident agitation and community resistance to protect inner-city neighbourhoods from wide-scale demolition (Ruming et al. 2010). Recently in Melbourne, the proposed East West Link inner-city road tunnel harked back to the proposed inner-city urban-renewal programs associated with modernist freeway construction that Davison (2004) and more recently Howe et al. (2014) described in their historical accounts of inner-city urban transformation in Melbourne. This time, the proposed freeway traversing the inner city suburbs of Melbourne, would be built as a six-kilometre underground tunnel so as to deliver increased public amenity by putting the proposed freeway underground. Yet the project still attracted significant broader community and localised NIMBY-related resistance over the impact to an inner-city park, practical challenges related to acquisition of homes and the location of exhaust stacks, as well as deep-founded concerns about building an inner-city road at the expense of other critical urban transport infrastructure projects (Legacy 2016).

When the East West Link project contracts were cancelled following the election of a Labor government in late 2014, Anthony Main, one of the most vocal activists, produced a self-published book celebrating the defeat of the East West Link and the residents' role in that defeat. The book was titled *Beating the Big End of Town* (Main 2015) and chronicled the work of the direct action and NIMBY-related protests, but it would be remiss to disregard the reach of the opposition to this project (which Main did not document) to parts of Melbourne that extended well beyond the project corridor (Legacy 2016). The examples of resistance to urban-regeneration projects and other forms of urban change continue to illustrate the constant struggle and resistance (emanating from affected localities as well as extending to other non-local places where concerned residents and community-based groups also reside) to preserve community assets in the face of growth and development interests (Ruming et al. 2012).

The opposition to change mounted by affected residents underscores what many have been described pejoratively as NIMBY (not-in-my-backyard) campaigns. Originally attached to proposed affordable housing projects (Dear 1992; Iglesias 2002) and waste facilities or locally unwanted land uses (LULUs) (Wolsink 1994), NIMBY tends to be associated with projects with wide benefits but with real or perceived localised costs. Affected residents may purportedly agree with the broad aims associated with these projects in principle (e.g. support for inclusive and diverse housing choices in strategic planning is a classic example) but will deflect when the associated projects are proposed for their own area (Legacy et al. 2014). NIMBY can imply an irrational and selfish response to 'higher-order' projects; or to the rational preference of residents, particularly homeowners, to relocate the externalised costs of development. Critical scholarship warns that the language of NIMBY is limiting and pejorative, and cautions against a tendency to categorise all opposition as NIMBY (Burningham 2000; Devine-Wright 2009; Wolsink 2006), a language which continues to be used broadly to invalidate opposition (Burningham 2000; Devine-Wright 2009; Wolsink 2006).

As in other locational conflicts, the concerns raised by residents against transport-regeneration projects are often multiple, combining local objections with 'higher-order' urban-planning issues (e.g. loss of public open space, compromising heritage and character). Robinson (1999) argued that opposition to specific proposals can form bridges between parochial, defensive concerns and broader 'moral' interests which compose the higher-order issues that motivate action of some groups (Not On Planet Earth – or NOPE). Research into the substantive content of community campaigns highlight several critical issues being exposed including the failure of planning process to adequately account for community concerns (Mouat et al. 2013), the dispossession of land from vulnerable community (Porter and Shaw 2009), and housing unaffordability. Certainly not exclusive, but taken together, these resident concerns reveal the continuous privileging of short-sighted economic imperatives that lack creative design and inclusionary solutions to redevelopment (Woodcock 2016). On the other hand, resistance to urban redevelopment can also identify more practical challenges related to local amenity including car parking – as the case in this chapter will show – as well as concerns about property values and disruption to local character.

In the following sections of the chapter we examine the revamp of the 96 Tram in Melbourne where a clear 'public interest' agenda to upgrade the tramway's accessibility, safety and speed was combined with efforts by local councils and the state government to explore ways to reduce car dependency along Melbourne's longest and busiest tram route. In Australian cities, as public transport infrastructure buckles from years of under-investment (Currie et al. 2009; Victorian Auditor General 2013) and as population pressures continue to mount, we are observing considerable resident resistance directed at public-transport infrastructure proposals, including Sydney's Anzac Parade Light Rail, Canberra's proposed light rail, and Melbourne's tram Route 96 upgrades, that focus upon both local-level practical problems (such as parking, loss of trees) and also broader urban and social policy questions (such as accessibility and gentrification). We examine how this public investment into the 96 Tram Route was marketed as supporting an improved transport experience for passengers, regenerating sections of the route through the introduction of tram stop upgrades in the form of 'super stops', and creating distributed impacts on mobility and access to popular 'villages' that punctuate the route. In this context, examining local concerns focused on car parking loss, provides insights into land use and transport integration as a form of urban regeneration, and the distinctive character of resident and community resistance it attracts.

Contesting transport-led urban regeneration

Australian cities undergoing investment in transport infrastructure are encountering intense local-resident resistance to some of the land-use impacts associated with these projects. Melbourne's 96 Tram upgrade project is one example.

In 2012, Public Transport Victoria (PTV), the public transport government agency in Victoria, announced plans to upgrade the Number 96 Tram route, with a set of infrastructure works seeking to improve tram network accessibility, reliability, and travel speeds; all long-held urban-policy commitments as evidenced in a succession of metropolitan strategic plans in Melbourne. The $800 million upgrade project – with funding from PTV, Yarra Trams, and VicRoads – comprised a set of new low-floor trams, tram stop accessibility upgrades, and changes to road space allocations giving priority to trams. Promotional material for the upgrade stressed that Melbourne's trams shared large amounts of their routes with road traffic, with significant slowing during peak hours (PTV Victoria 2012; 2013). The Number 96 Project was presented as an opportunity to demonstrate the advantages of prioritising trams over road traffic (Table 16.1). Principles used to justify the project were social (accessibility), environmental (increasing reliability and speeds for public transport); and, more broadly, economic demand (citing increased congestion and patron numbers, the prioritising of trams over congested road traffic promised to be 'reliably moving large numbers of passengers more efficiently'). The project was thus justified by asserted 'higher-order' public and economic benefits at the individual and metropolitan scales.

In June 2013, PTV released conceptual plans for the project. A formal 'community and stakeholder engagement process' was undertaken by PTV between June and August 2013 (PTV Victoria 2013), with a goal that 'the Route 96 Project will work closely with local councils, business groups and

Table 16.1 96 tram upgrade: comparison of debates framed 'inside' and 'outside' formal participation

	'Inside'	*'Outside'*
Focus	• "Tram freeway"	• "Neighbourhood spot fires"
Justification	• Social – accessibility, safety • Environmental – priority to public transport • Economic – growing demand for route	• Economic (local businesses)
Scale	• Individual – accessibility for disabled patrons • Metropolitan - public transport speed and service reliability	• Local 'villages' along route: Acland Street, Nicholson Street, East Brunswick • Local businesses
Transport mode	• Public transport over car traffic	• Cars over public transport
Assumptions	• Tram upgrades have broad benefits • Increased tram speed and size is positive	• Loss of car parking damages businesses • Tram speed and size is divisive for street
Scope of consultation	• Design and placement of stops • "Key stakeholders" – councils, business groups and residents	• Project and justification • Loss of car parking • Prioritisation of trams over cars
Forums	• Surveys (1,333 completed) • Drop-in sessions (400 attendees) • Local council meetings	• Newspaper complaints • Demonstrations • Protest signs • Local council meetings
View on consultation	• Consultation showed near-consensus support for project and principles • Minority opposition to specific sites	• Inadequate consultation • Inadequate consideration of impacts on local areas from loss of parking

Source: Authors

residents along the route'. The scope of public consultation reflected the project justifications and assumptions. The stated purpose of the engagement was to 'seek feedback on the project's early design concepts, and generate awareness of the project's purpose and benefits'. Public input was thus effectively limited to the specifics of stop placements and designs; always framed in terms of meeting project goals for accessibility and for tram prioritisation. The proponents expressed openness to 'alternative designs' but not to revision of the project's prioritisation of trams over cars (News Ltd 2013b). Formal consultation on the project included 1,333 feedback forms completed and returned; 17 submissions from key stakeholders; and 400 attendees at drop-in sessions; in addition to online and mailed correspondence.

A summary document of the formal consultation results states that the significant majority of feedback showed 'overall public support for the proposed designs across all three key areas' and 'support for the key project objectives and principles, particularly those which aim to improve access, safety and tram service reliability'. However, not all consultation feedback was supportive. As a comparatively minor but, as it turned out, crucial point, the scope of formal consultation included to 'identify locations where a reduction of on-street parking, median opening and tram stops may be required'. Some tension around parking loss was anticipated in the process, but typically subsumed by metropolitan-level goals – as in: 'an important part of the development process will be working with key stakeholders to understand their views about delivering improved tram priority and reliability', which 'may result in a reduction in on-street parking'. Similarly, in the design consultation phase of the project there was an optimistic summary that while the tram route upgrades 'may result in some changes for road users, improved priority for trams will help ensure local communities continue to have access to efficient and reliable tram services into the future'.

Despite this optimism, car-parking loss was raised as a concern: as in 'the most significant concern raised during the engagement process related to the project's proposal to reconfigure Acland Street to create a pedestrian mall/plaza'. Consultation documents referred to how 'although this option was widely supported by the community and the City of Port Phillip, there were strong concerns raised (in particular by the St Kilda Village Traders Group) that this would impact local businesses in Acland Street' (see Chapter 13 for similar discussion of shopping centre-led regeneration). The strongest views put forward in the formal feedback process were those concerned with car parking, and specifically its loss in the process of prioritising trams. Few if any submissions criticised the project overall. The conflict was largely in terms of the scale of effects.

Politicising the 'neighbourhood spot fires'

The 96 Tram upgrade project was justified by larger-scale or 'higher-order' policy goals emphasising public transport and accessibility. These justifications came into conflict, however, at the level of individual communities or 'villages' (strip shopping areas along the tram route) (Figure 16.1). The fact that Australian cities have instilled regimented and highly formalised processes of community engagement, which restrict the form and content of engagement, means that any opposition needs to be mounted beyond these rigid formal settings (Legacy et al. 2014; Mouat et al. 2013). In the case of the 96 Tram upgrades, the chief point of conflict observed was mounted 'outside' the formal participatory process, highlighting the local impact of the 96 Tram upgrade on existing allocations of road space to private vehicles, and parked vehicles in particular. This 'outside' opposition bought those concerns out into a public dialogue (Lord 2013).

The effect of tram upgrades on car parking in strip shopping centres along the route emerged as a substantial stumbling block for the 96 Tram project. Traders' groups at locations at both ends of the route fought the proposed upgrades on the basis of economic concerns about loss of car parking. Traders argued that the upgrades would destroy small businesses, and by association local services, by deterring patrons wishing to park their cars. To trader groups, the formal consultation process for the project was perceived to be tokenistic and undemocratic in its prioritisation of tram upgrades over local vehicular rights. Most of their concerns sat 'outside' of the formal planning process, as shown in the right-hand side of Table 16.1. Table 16.2 illustrates the timeline of how the community input offered through the formal planning process was quickly matched by protests in other forums, as seen in other contentious transport proposals such as the East West Link (Legacy 2016).

The initial promotion of the 96 Tram upgrade project saw support from councils and public-transport users (Hingston 2012a; 2012b). However, by late 2013, when the first of the new tram stock was trialled, a schism was emerging between how the project was presented for tram users – including disabled patrons – and how it impacted local trader groups on both ends of the route (Acland Street St Kilda, Nicholson Street, and Brunswick). In October 2014, an *Age* report surmised that the 96 Tram route upgrade was struggling with the 'grassroots campaign' and 'series of neighbourhood spot fires' against it. The key issue was the loss of car-parking space in retail areas – hence 'a plan to turn Melbourne's busiest tram route into a flagship light railway is struggling to launch as residents and traders along the line fight moves to allocate road space at the expense of parking and car lanes' (Carey 2014).

Traders in Acland Street St Kilda opposed the tram upgrades and associated removal of car parks – saying it would 'sound the death knell for their iconic thoroughfare', and arguing that there had been 'no consideration of how the absence of parking spaces would influence visitors' (News Ltd 2013a; Carey 2013). In a description evoking, perhaps accidentally, the famously fascist prioritisation of transport running times by Italian dictator Mussolini, a representative for the Traders Association spoke of the feared effects of parking space removal, suggesting that even if the upgrades were beneficial on

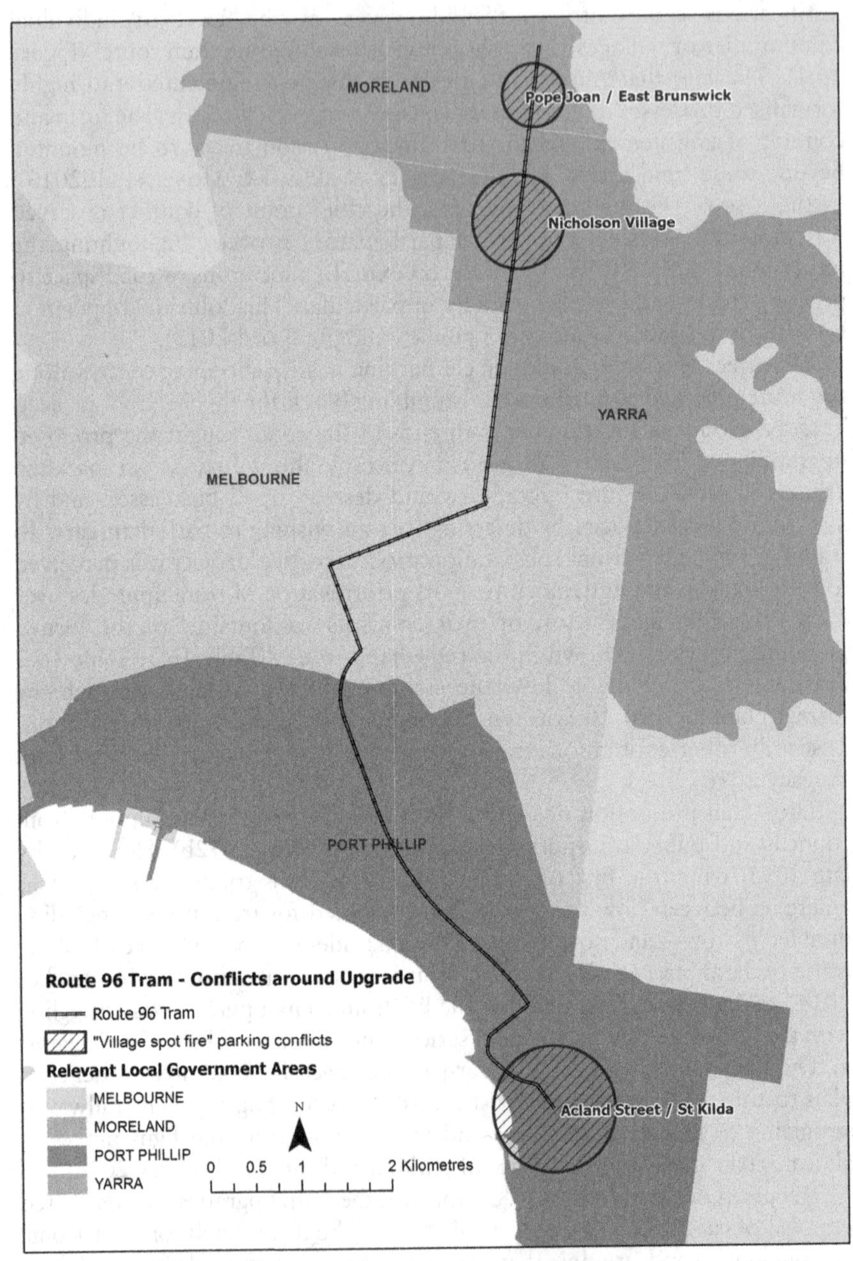

Figure 16.1 Route 96 Tram and 'village spot fire' conflict sites (source: authors)

Table 16.2 Timeline of events 'inside' and 'outside' the 96 tram consultation process

Timeline	*Events 'inside'*	*Events 'outside'*
April 2012	• PTV, Yarra Trams and VicRoads announce plans to upgrade route 96 with accessible stops	
June–August 2013	• Conceptual plans released • Formal 'community and stakeholder engagement process': feedback forms, submissions, drop-in sessions.	
November 2013	• Transport minister emphasises consultation with disability groups	• Acland Street traders opposing car park removal. Criticise impacts on street. Stage 'mock funeral' for Acland Street • Nicholson Village traders urge council not to approve tram upgrades with parking removal
June 2014	• Yarra Council meetings on Nicholson Village impacts	
August 2014		• Newspaper editorial voices opposition in Acland Street to tram size and frequency – 'will Acland Street be eaten?' • Criticism of pedestrian access in Acland Street
September 2014	• Revision of Nicholson Village plans – removal of 20 instead of 66 parking spaces • Yarra Council revision of plans • Yarra Councillor states 'near-consensus' support for upgrade project • Stalling of Acland Street works	
October 2014	• Parts of the route upgrade progressing	• Grassroots campaigns and spot fires – key issue is car parking loss and businesses • Deferral of Nicholson Village plan • Petition to 'save Nicholson Village' • East Brunswick campaign against 'parking purge'
November 2015	• Most stop upgrades except Acland Street completed • Acland Street council (Port Phillip) votes to support combined tram / pedestrian mall	• Traders submit alternative plan for Acland Street

Source: Authors

a broader scale that did not justify a destructive imposition on a local shopping area: 'we think these new E-class trams, while they're absolute monsters they're probably a good thing for Melbourne. But let's not destroy St Kilda village just because we want the frequency of trams to be improved' (Lord 2013).

At the Brunswick end of the route (Figure 16.1) traders similarly criticised the extent and intent of the consultation process over the upgrades, with the owner of a Brunswick café accusing PTV of wanting 'to turn route 96 into Melbourne's flagship tram route' and hence being 'not genuinely interested in traders' concerns' (car parking). Also at the northern end of the route, although the local council (Yarra) had given in-principle support to new stops in late 2013, traders in the 'Nicholson Village' urged Yarra Council to not approve the tram upgrades, due to loss of parking spaces along their shopping strip. Yarra Council deferred its decision, pending meeting with traders. As with Acland Street, the Nicholson Village traders were concerned about loss of visitors through parking loss and an associated impact on business revenue (News Ltd 2014c).

At the launch of the first trams, the transport minister emphasised consultation with disability groups rather than traders (Harris 2013). PTV likewise continued to emphasise the accessibility benefits of the project. In 2014 parts of the route upgrade were progressing, but trader objections to the upgrades were resulting in changes to parking removal. As shown in Table 16.2, these objections were voiced through newspapers, public protests (including a 'mock funeral'), and through contact with local government councillors.

In a bid to balance trader feedback with the broader project momentum, in September 2014 Yarra Council revised plans for the upgrade down to two instead of three stops, to remove 20 instead of 66 car parks (News Ltd 2014a). A Greens councillor in Yarra Council claimed that there was near-consensus support for the upgrades and that the project would support both traders and passengers, which, she emphasised, outnumbered drivers in the area (News Ltd 2014b). This consensus claim jarred with how, as the report noted, 'a petition to "Save Nicholson Village" and its parking spaces still looks out from shopfront windows in North Carlton'. In June 2014 'Nicholson Village traders continue[d] to fight proposed new tram stops that would remove car parking from the strip'. Further north in Brunswick, a campaign led by café Pope Joan against the 'parking purge' in that strip still argued it 'would badly hurt small businesses', and was pushing for consideration of an option to preserve all parking in the area.

Opposition to tram upgrades in St Kilda continued in 2014 and 2015. In an opinion piece, a St Kilda newspaper adopted the accessibility and inclusiveness language of the tram upgrade proponents to question the project's local impacts. The piece positioned the street itself as a victim and suggested that at the street level for pedestrians and drivers, the new tram stops would 'create hardship for people with a disability and those with mobility issues, such as parents with children in prams'. Asking whether Acland street would be 'eaten' by trams, the piece emphasised the size and frequency of the new trams (a key selling point used for the project) as an impact on local streets, dominating and dividing through their feted capacity and frequency – pointing out that 'the sheer length of these super-sized trams means they will dominate as well as divide the street. This will be highlighted by the frequency of their service – anywhere from every 4 to 10 minutes'.

As of November 2015, many of the upgrades planned for route 96 had been completed – albeit in a 'stuttering' fashion. However, Acland Street St Kilda remained unresolved. Following the abandonment of plans due to trader hostility, a revised plan prepared by Port Phillip Council, PTV and Yarra Trams for the Acland Street end of the route was formally considered – and approved – by the council in December 2015. The latest plan was based on turning the end of the street into 'an open air plaza', with 330 square metres of public open space intended to (using regeneration language) 'make the popular tourist strip more walkable, with more space for outdoor eating, drinking and events'. The catch was still that the plan meant 'the loss of about 50 on-street parking spaces'. A report noted succinctly that 'an alternative proposal for the street, put forward by traders who fear the loss of on-street parking will discourage visitors and damage business, has been rejected'. Traders again staged protests in favour of retaining parking on the street, with Acland Street Village Business Association representative warning that the proposal would 'grind [vehicle] access to a halt', and that there had not been enough community consultation. The council is strongly supportive of the plan and its benefits for the strip, with the mayor stating that 'to move people around efficiently, the best way to do that is public transport, walk and bike and it's the way our community has told us they want to enjoy Acland Street' (Johnson 2013; Carey 2015).

The use of statistics from the consultation process placed the strong criticisms from traders into a minority category. This being said, the supportive submissions – the majority of the formal feedback – did not necessarily present any view on local economic concerns such as the loss of parking. Support and momentum for the project came from broader social, environmental and economic justifications. Opposition was comparatively parochial, focused on local business concerns. Concerns from traders about parking, and later (in Brunswick) from local residents concerned about parking rights, were set aside in the face of broader support for and justifications of public transport improvements. This is not to say that the trader concerns about parking were successfully allayed. Parking impacts were reduced in the Nicholson Village and in Brunswick, whereas in Port Phillip, the loss of general parking is still proving contentious, while it seems the PTV and project team have limited patience for the interminable parking debates.

Discussion: resisting urban regeneration and questions of scale

As one lens through which to examine resistance to transport-led urban regeneration, we call upon Purcell's (2006) theorisation of the 'local trap' to consider the ways that locally situated politicisation of urban regeneration fail to show if or how resistance connects with wider discussions of urban structure and change. Purcell (2016: 394) describes the local and non-local relationship as: 'how a particular community (however defined) should relate to other communities, and more specifically, how a particular community might relate to a wider "public"'. In this instance, Purcell offers two challenges. The first is to consider 'how a given community might better take into account, and even value, the interest of people beyond their community'. The second challenge Purcell offers is to think about the political relationship between the state (proposing the project and directing the investment) and the public to form a robust understanding of the public interest.

The question of public benefit in transport-led urban-regeneration projects is complex. In our Acland Street tram versus parking case study described above, the wider public benefit achieved through accessibility upgrades was conceived by the Acland Street traders to be in direct conflict with serving what was perceived to be a valuable (to the traders) car-based clientele. However, the pedestrian amenity that would accompany the introduction of superstops following the December 2015 council approval for an open air plaza showed how this infrastructure investment could serve to further renew (in the case of the already thriving Acland Street) existing land uses to serve low-carbon forms of mobility and reducing car-based travel further. Despite these promised benefits, local traders remained distraught – arguing that failed public engagement had served to marginalise their dissenting voices. For some groups who resist forms of urban regeneration, their opposition is founded on concerns that the forthcoming development would displace existing residents. Without leveraging affordable housing outcomes for existing residents to enable them to stay, lifts in property value from the ensuing development would undermine housing affordability. Conversely, others worry about loss of property values. Other groups express concern about the distributional benefits of investment – questioning why one part of the city is being prioritised as beneficiaries of investment over another. It is clear that many redevelopment decisions stem from path-dependent ideas of greatest economic benefit – without privileging social, equity or indeed accessibility benefits.

What does this case reveal for community resistance to transport-led urban regeneration more broadly? It is common to observe that once governments determine infrastructure investment decisions, including the form of technology (e.g. bus rapid transport versus light rail) and route selection based upon the development potential that can be leveraged, resistance to these decisions mount (Ruming et al. 2016). This oppositional positioning and resistance to state decision-making forms the cornerstone of the post-political urban literature (Allmendinger and Haughton 2012). This body of literature casts a new light onto the political context in which investment decisions and citizen engagement are structured by and the various ways in which citizen engagement can be a force of political action against a project.

This is where the 96 Tram case departs from some other contested case studies including the highly controversial East West Link in Melbourne which has been described as illustrative of a post-political decision-making landscape (Legacy 2016). In the case of the 96 Tram upgrade the project aligned with an accessibility and pubic mobility agenda – to move a greater volume of people along this route through public transport. The East West Link proposal was another manifestation of a deep-seated, and path-dependent commitment to serve car-based travel and where community engagement was seen to be restricted to consensus formation (rather than inviting citizens to comment on the efficacy of the proposed project) removing opportunities for community dissent on the grounds of social and environmental impacts of road building. The 96 Tram case saw an inverse, in that the substantive focus of the opposition was founded on maintaining an economic logic as the basis of decision-making.

The resistance observed in transport-led urban regeneration exposes the framing of the injustices that beset those affected by planning decisions, whether those decisions are steered by market rationality or by socially inclusive policy frameworks. But it also highlights the limitations on a planning system's ability to fully accommodate the breadth of interests at multiple scales; to engage with the complexity of considerations; and to also assert a visionary mandate of change (Purcell 2006). Beyond what the formal planning system can readily respond too, citizen opposition continues to show how urban regeneration is engaged with as a political project. Resident action groups seek to influence urban-regeneration projects by politicising the proposal in question despite the public interest claims that are attached, forcing the planning system to 'open-up' to respond to the very resistance it sought to suppress and exclude (e.g. reducing parking concerns to marginal status).

The local protests against transport-led regeneration, illustrate how tensions between different scales of 'public goods' as described by Purcell (2006) take shape. Local controls can be exclusionary and a mechanism for strengthening the power of already powerful groups and areas (Parvin 2009). Local interests also retain broader pejorative meanings of NIMBY irrationality and selfishness; sympathetic if cynical 'homo economicus' (Matthews et al 2015) or 'homevoter' (Fischel 2004) versions of local opposition notwithstanding. Yet the local is also valorised as the most direct and meaningful form of political engagement. Clarke and Cochrane (2013) highlight its inherent contradictions: that local engagement always has a tension between being seen as problematically parochial, fragmented, disjointed from broader possible impacts and visions, and its potential to build significant progressive movements with deeper roots.

Deas (2013) argues, in the context of localism, that the relationships between local concerns and centralised or globalised politics is more complicated than a dichotomy between neoliberalism and progressive politics. Deas also argues:

> the complexity and heterogeneity of urban policy can be easily forgotten as researchers attempt to demonstrate the degree to which neoliberalism dominates policy concerns – especially when they set out to 'fit' the empirical realities of different policies to a wider storyline about the rise of neoliberal thinking.
>
> (Deas 2013: 78)

Our case study demonstrates that the local can disrupt, broadly, both sides of politics. And, alongside the long history of local activism against roads, that local politics can disrupt public transport projects.

Conclusions

While research examining forms of urban resistance is plentiful, in this chapter we have tried to show through the case of Melbourne's 96 Tram Route upgrades, how transport-led urban regeneration is a form of urban regeneration that attracts resident resistance that is disctinctive. In this case, urban-transport investment was positioned as a means to meet growing demands for public-transport services; increase mobility; and enhance accessibility. Yet the tram upgrade was met by fierce resistance from local traders fearing economic impacts through parking loss: thus offering something of an inversion in the positioning of opposition to transport (freeway) projects in Australian cities.

Modernising and building transport infrastructure shares similar profiles to what we might associate with urban regeneration more broadly across Australia. A firm economic agenda will typically overlay any public benefit that might accrue. Like most urban-regeneration projects, larger urban-infrastructure projects such as freeways and urban rail involve a complex mix of public and private agencies, some of which hail from places well beyond the local context (e.g. higher-orders of government, including the Commonwealth government) in which the proposed development will be built. A focus on transport-led regeneration renders visible questions about how resistance is expressed, and how it is seen by others (planners, other residents, politicians). Opposition to the 96 Tram upgrade – as with other light rail debates in Australian cities, for example in Canberra and in Sydney's Anzac Parade – reveals distinct aspects of transport-led urban regeneration given that the developments are underscored by transport and land-use goals, including new pathways for mobility and accessibility (Curtis and Scheurer 2010) which seek to serve broader social and environmental imperatives.

Local resistance to transport – whether road or transit based – is, as with resistance to other 'NIMBY' changes such as higher density housing or waste facilities, not 'only' parochial. Yet it is also not 'only' a process of holding economic interests to the broader public interest to account. Local resistance, as in the 96 Tram upgrade, exposes and politicises the myriad decisions, beneficiaries, and costs that underlay any transport and regeneration project. In the case of car parking lost to a tram route, the 'loss' is one of unsettling previous prioritisations of car-based infrastructure and space over other transport and land-use options. Despite concerns over the disjointed nature of transport planning in Australian cities from strategic land-use planning (Dodson 2009), there is still a renewing and a reshaping of the urban landscape occurring which is driven by transport investments. Yet there is a politics of transport-led urban regeneration that is foregrounding questions about the kind of transport infrastructure and investment desired and which ought to be built to serve the 'public interest'. These questions form the foundation to then discuss and consider the positive and negative impacts of a proposed investment, how these impacts are distributed across the local and non-local scales, and what public benefits are being achieved (Young and Keil 2014: 1602). In most cases, these questions are political questions because of the land use, transport and strategic impact they embody, and the impacts, be they positive or negative, will be understood differently by the multiple and diverse publics impacted by transport-led urban regeneration.

Note

1 A series of construction bans, initiated by the Builder Labourers Federation, on projects deemed to be environmentally or socially undesirable.

References

Allmendinger, P. and Haughton, G. (2012) 'Postpolitical spatial planning in England: a crisis of consensus?' *Transactions of the Institute of British Geographers*, 37(1): 89–103.

Burgmann, M. and V. Burgmann (1998) *Green Bans, Red Union: Environmental Activism and the New South Wales Labourers' Federation*. Sydney, UNSW Press.

Burningham, K. (2000) 'Using the language of NIMBY: a topic for research, not an activity for researchers'. *Local Environment*, 5(1): 55–67.

Cao, X. J. and Porter-Nelson, D. (2016) 'Real estate development in anticipation of the Green Line light rail transit in St. Paul'. *Transport Policy* 51:24–32.

Carey, A. (2013) 'Traders plan mock funeral for "death of Acland Street"'. *The Age*, 4 November. http://www.theage.com.au/victoria/traders-plan-mock-funeral-for-death-of-acland-street-20131104-2wwf1.html (accessed 12 March 2017).

Carey, A. (2014) 'Tram route 96 freeway struggles as opposition mounts', *The Age*, 20 October. http://www.theage.com.au/victoria/tram-route-96-freeway-struggles-as-opposition-mounts-20141019-118arp.html (accessed 12 March 2017).

Carey, A. (2015) 'Trams and pedestrians win, motorists lose in planned Acland Street makeover', *The Age*, 28 October. http://www.theage.com.au/victoria/trams-and-pedestrians-win-motorists-lose-in-planned-acland-street-makeover-20151027-gkjudk.html (accessed 12 March 2017).

Cervero, R. (1984) 'Journal report: light rail transit and urban development'. *Journal of the American Planning Association*, 50(2): 133–147.

Clarke, N. and Cochrane, A. (2013) 'Geographies and politics of localism: the localism of the United Kingdom's coalition government'. *Political Geography*, 34: 10–23.

Currie, G., Richardson, T., Smyth, P., Vella-Brodrick, D., Hine, J. et al. (2009) 'Investigating links between transport disadvantage, social exclusion and well-being in Melbourne: Preliminary results'. *Transport Policy,* 16(3): 97–105.

Curtis, C. and Scheurer, J. (2010) 'Planning for sustainable accessibility: developing tools to aid discussion and decision-making'. *Progress in Planning,* 74, 53–106.

Curtis, C., Renne, J., and Bertolini, L. (eds.) (2009) *Transit Oriented Development: Making It Happen*. Aldershot: Ashgate.

Darcy, M. and Rogers, D. (2015) 'Place, political culture and post-Green ban resistance: Public housing in Millers Point'. *Cities*, 57:47–54.

Davison, G. (2004) *Car Wars: How the Car Won our Hearts and Conquered our Cities*. Crows Nest, NSW: Allen and Unwin.

Dear, M. (1992) 'Understanding and overcoming the NIMBY syndrome'. *Journal of the American Planning Association*, 58(3): 288–300.

Deas, I. (2013) 'Towards post-political consensus in urban policy? Localism and the emerging agenda for regeneration under the Cameron government'. *Planning Practice and Research*, 28(1): 65–82.

Devine-Wright, P. (2009) 'Rethinking NIMBYism: The role of place attachment and place identity in explaining place-protective action'. *Journal of Community & Applied Social Psychology*, 19(6): 426–441.

Dodson, J. (2009) 'The "infrastructure turn" in Australian metropolitan spatial planning'. *International Planning Studies,* 14(2): 109–123.

Fischel, W. A. (2004) 'An economic history of zoning and a cure for its exclusionary effects'. *Urban Studies*, 41(2): 317–340.

Government of Victoria (2014): *Plan Melbourne: Metropolitan Planning Strategy*, Melbourne: Department of Transport, Planning and Local Infrastructure.

Harris, A. (2013) 'Super-tram hits streets at last', *Herald Sun*, 5 November.

Harvey, D. (2012) *Rebel Cities: From the Right to the City to the Urban Revolution*. London: Verso.

Hingston, C. (2012a) 'Thumbs up to 96 boost', *Melbourne Times Weekly*, 25 April.

Hingston, C. (2012b) 'Terminus upgrade just the ticket', *Melbourne Times Weekly*, 5 September.

Howe, R., Nichols, D., and Davison, G. (2014) *Trendyville: The Battle for Australia's inner Cities*. Melbourne: Monash University Press.

Iglesias, T. (2002) 'Managing local opposition to affordable housing: A new approach to NIMBY'. *Journal of Affordable Housing & Community Development Law*, 12(1): 78–122.

Iveson, K. (2013) 'Building a city for "The People": The politics of alliance-building in the Sydney Green Ban Movement', *Antipode* 46(4): 1–22.

Johnson, S. (2013) 'Makeover for Iconic St Kilda Strip', *ArchitectureAU,*11 November. http://architectureau.com/articles/makeoverforiconicmelbournestrip/ (accessed 12 March 2017).

Lees, L. (2008) 'Gentrification and social mixing: Towards an inclusive urban renaissance?' *Urban Studies,* 45(12): 2449–2470.

Lees, L., Slater, T. and Wyly, E. (2013) *Gentrification*. London: Routledge.

Legacy, C. (2016) 'Transforming transport planning in the postpolitical era'. *Urban Studies*, 53(14): 3108–3124.

Legacy, C., March, A., and Mouat, C. (2014) 'Limits and potentials to deliberative engagement in highly regulated planning systems: Norm development within fixed rules'. *Planning Theory and Practice*, 15(1): 26–40.

Lord, K. (2013) 'St Kilda traders concerned about loss of parking spots to make way for supersized trams'. *ABC News*, 5 November. http://www.abc.net.au/news/2013-11-05/st-kilda-traders-concerned-about-loss-of-parking-spots/5070034 (accessed 12 March 2017).

Main, A. (2015) *Beating the Big End of Town: How a Community Defeated the East-West Toll Road*. Melbourne: Lighting Source, Melbourne.

Matthews, P., Bramley, G. and Hastings, A. (2015) 'Homo economicus in a big society: understanding middle-class activism and NIMBYism towards new housing developments', *Housing, Theory and Society*, 32(1): 54–72.

Mouat, C., Legacy, C., and March, A. (2013) 'The problem in the solution: Testing agonistic theory's potential to recast intractable planning disputes'. *Urban Policy and Research*, 31(2): 150–166.

News Ltd (2013a) 'Groups not on board with tram proposal', *Bayside Leader*, 12 November.

News Ltd (2013b) 'City's busiest tram line set to undergo work', *Melbourne Leader Yarra Edition*, 18 November.

News Ltd (2014a) 'Brakes put on plans for 96 tram', *Melbourne Leader*, 8 September.

News Ltd (2014b) 'Route 96 is our future', *Melbourne Leader*, 15 September.

News Ltd (2014c) 'Route 96 stop upgrade plan deferred', *Melbourne Leader*, 30 June.

NSW Government (2014) *A Plan for Growing Sydney*, Sydney: Department of Planning and Environment.

Palen, J. and London, B. (eds.) (1984) *Gentrification, Displacements, and Neighborhood Revitalization*. New York: SUNY Press.

Parvin, P. (2009) 'Against localism: does decentralising power to communities fail minorities?' *The Political Quarterly*, 80(3): 351–360.

Porter, L. 2009, 'Whose urban renaissance', in L. Porter and K. Shaw (eds): *Whose Urban Renaissance? An International Comparison of Urban Regeneration Strategies*, Routledge, London.

Porter, L. and Shaw, K. (2009) *Whose Urban Renaissance? An International Comparison of Urban Regeneration Strategies*, London: Routledge.

PTV Victoria (2012) 'Route 96 Project Overview', http://ptv.vic.gov.au/assets/Trams/Route-96-Project-Overview.pdf (accessed 18 November 2015).

PTV Victoria (2013) 'Route 96 Project: Community and stakeholder feedback summary report June – August 2013, http://ptv.vic.gov.au/projects/trams/route-96-project/ (accessed 18 November 2015).

Purcell, M. (2006) 'Urban democracy and the local trap'. *Urban Studies*, 43(11): 1921–1941.

Purcell, M. (2016) 'For democracy: Planning and publics without the state'. *Planning Theory*, 15(4): 386–401.

Raco, M. (2000) 'Assessing community participation in local economic development: Lessons for the new urban policy'. *Political Geography*, 19: 573–599.

Randolph, B. (2006) 'Delivering the compact city in Australia: Current trends and future implications'. *Urban Policy and Research*, 24(4): 473–490.

Robinson, A. (1999) 'From NIMBY to NOPE: Building eco-bridges'. *Contemporary Politics*, 5(4): 339–364.

Ruming, K. J. (2010) 'Developer typologies in urban renewal in Sydney: Recognising the role of informal associations between developers and local government'. *Urban policy and Research*, 28(1): 65–83.

Ruming, K., Tice, A. and Freestone, R. (2010) 'Commonwealth urban policy in Australia: The case of inner urban regeneration in Sydney, 1973–1975'. *Australian Geographer*, 41(4): 447–467.

Ruming, K., Houston, D., and Amati, M. (2012) 'Multiple suburban publics: Rethinking community opposition to consolidation in Sydney'. *Geographical Research*, 50(4): 421–435.

Ruming, K., Mee, K. and McGuirk, P. (2016) 'Planned derailment for new urban futures? An actant network analysis of the "great [light] rail debate" in Newcastle, Australia', in Y. Rydin and L. Tate (eds), *Actor Networks of Planning: Exploring the Influence of Actors Network Theory*, London: Routledge.

Searle, G., Darchen, S., and Huston, S. (2014) 'Positive and negative factors for transit oriented development: Case studies from Brisbane, Melbourne and Sydney'. *Urban Policy and Research*, 32(4): 437–457.

Shaw, K. (2009) 'Rising to a challenge', in L. Porter and K. Shaw (eds): *Whose Urban Renaissance? An International Comparison of Urban Regeneration Strategies*, London: Routledge.

Smith, N. (2002) 'New globalism, new urbanism: Gentrification as global urban strategy'. *Antipode*, 34: 427–450.

Victorian Auditor General (2013) *Developing Transport Infrastructure and Services for Population Growth Areas*. Melbourne: Victorian Auditor General.

WAPC (2004) *Network City: Community Planning Strategy for Perth and Peel*. Perth: Western Australian Planning Commission.

Wilson, J. Q. (1963) 'Planning and politics: Citizen participation in urban renewal'. *Journal of the American Institute of Planners*, 29(4): 242–249.

Wolsink, M. (1994) 'Entanglement of interests and motives: Assumptions behind the NIMBY-theory on facility siting'. *Urban studies*, 31(6): 851–866.

Wolsink, M. (2006) 'Invalid theory impedes our understanding: A critique on the persistence of the language of NIMBY'. *Transactions of the Institute of British Geographers*, 31(1): 85–91.

Woodcock, I. (2016) 'The design speculation and action research assemblage: "Transit for all" and the transformation of Melbourne's passenger rail system'. *Australian Planner* 53(1): 15–27.

Young, D. and Keil, R. (2014) 'Locating the urban in-between: Tracking the urban politics of infrastructure in Toronto'. *International Journal of Urban and Regional Research*, 38(5): 1589–1608.

17 On the fringe of regeneration

What role for greenfield development and innovative urban futures?

Kristian Ruming, Kathy Mee, Pauline McGuirk and Jill Sweeney

Introduction

Australia is a suburban nation. An estimated 77 per cent of the population of the 16 largest cities live in suburban neighbourhoods and 78 per cent of population growth 2006–2011 occurred in suburban locations (Gordon et al. 2015). Fringe development continues at a rapid rate, despite decades of explicit consolidation policies (Burton 2015; Dodson 2010). Given its continued importance, exploring how fringe suburban growth happens is vital to understanding how contemporary Australian cities are being regenerated. Regeneration, as we understand it in this chapter, is not just the renewal of an individual site or broader existing built form, but rather the ongoing renewal of the entire urban form. Within this wider process of urban change, ongoing fringe development is important alongside regeneration processes happening in inner and middle-ring suburbs and landmark urban regeneration projects occurring on brownfield sites.

In this chapter we explore urban regeneration in Australian cities through the lens of the transformation of suburban development. We draw on a case study of Huntlee, NSW, to highlight new trends in reshaping the urban fringe. As the chapter illustrates, contemporary suburban fringe development both connects with and departs from the ideals which drove suburban development in the twentieth century, to aspire towards more varied, better serviced and sustainable models. In this regard, it has much in common with regeneration in Australia's inner cities and middle-ring suburbs.

The template of twentieth century Australian suburban development has begun to fray on the basis of shifting economic, social, political, environment, and cultural conditions of urban regions (Burton 2015). Suburbs have been increasingly characterised as unsustainable, unjust and expensive due to high infrastructure and travel costs, and segregated socio-spatial structures (Keil 2015). In response, inner- and middle-ring urban regeneration and densification is now the primary logic underlying contemporary strategic planning across Australia's cities (Chapter 2). The goal is to constrain development within the existing urban footprint, limit sprawl and consolidate growth around designated centres, connected by efficient transport systems (Forster 2006). As a result, low-density, energy-intensive forms which traditionally characterised fringe development are being tempered by principles of consolidation and New Urbanism, subject to more stringent strategic intent, and restricted to a series of designated locations, promoted by specialised strategic zonings/urban centres located on the edge of cities. For example, since *Cities of Cities: A Plan for Sydney's future* (NSWDoP 2005), North West and South West Growth Centres have been identified in Sydney's metropolitan planning strategies as central in delivering new housing. The current *Plan for Growing Sydney* (NSWDPE 2014) reinforces the need for fringe developments to provide a broad range of services to new communities.

Despite these aspirations, fringe development has often been driven by large residential developers, leading to an entrenched conservatism and development style that maximises profits over other concerns such as the provision of community services (Buxton and Scheurer 2007). For Newton et al. (2012) and others (Burton 2015; Dodson and Sipe 2008), this is the antithesis of good planning. It perpetuates the unsustainable city-building process of at least the last half a century promoting car-based, low-density sprawl in an era of resource depletion and increasing petrol prices (Burton 2015; Dodson and Sipe 2008). Gleeson (2006) provocatively highlights how fringe development inflames ongoing affordability and equity concerns by shaping 'urban welfare camps' where transport deficits entrench social divisions. Yet for others, fringe development is essential to maintaining the centrality of the detached family home to Australian urbanism (Gibson 2012) and the popular belief that an idealised suburbanism provides a 'template for the good life for most Australians' (Burton 2015: 504). Importantly, Anderson (2006) argues that both critics and defenders of suburbia are potentially blinkered and fail to acknowledge the true diversity and heterogeneity of Australian suburbs.

In ongoing debates on the value of Australian suburbanism, one constant is the notion that suburbs represent something different to inner and middle-ring locations. In this chapter, rather than positioning suburban development as something that happens 'out there', we position fringe development as a central aspect of the urban regeneration story (see also Beattie and Haarhoff 2014). Only by considering how fringe locations are planned and developed can we capture holistically the diverse ways cities are being shaped and reshaped. In addition, this view captures how the form of fringe development now occurring across Australian cities departs from the suburban development model rolled out in the post-World War II period, via an observable transition to more complex, dense and diverse suburban landscapes (Charmers and Keil 2015). Fringe developments have increasingly drawn on established design elements of New Urbanism and a recent history of master planned residential estates (McGuirk and Dowling 2009; Johnson

2010). Yet they are also potential sites of innovation and experimentation through which more diverse forms of suburban development are emerging. The new build fringe suburb is, therefore, a central site of this evolution, where new ideals and established traditions of suburban development combine to regenerate the city.

In what follows we explore this complex emergence. We begin by discussing the evolution of Australian suburban forms before exploring the ideas and influence of New Urbanism, master planning and the 'post-suburb'. We then focus on Huntlee in the Hunter Valley, NSW, 50 km northwest of the city of Newcastle: a New Urbanism inflected, master planned 'post-suburb' (Keil 2015) with its own greenfield town centre development, increasingly typical of new fringe developments in strategic release growth centres in Australian cities. Drawing on interviews, planning documents, marketing documents, social media sites and media reports spanning over a decade,[1] we trace Huntlee's planning and development, its departures from traditional suburban forms through the innovation and experimentation to which it aspires, and its parallels with regeneration processes found elsewhere in the city.

Australian suburbs

In Australia, as across most Western economies, fringe residential development represented a key element of the post-war Fordist–Keynesian regime of accumulation, as land developers and house builders sought to maximise and protect profits via the roll-out of relatively standard, mass-produced housing (Keil 2015). Australian suburbs expanded rapidly as owner-builders took advantage of the availability of cheap land to favour single-storey detached residential dwellings, and developers, speculative builders and state housing agencies embarked on an unprecedented level of fringe housing construction (Spearitt and DeMarco 1988; Davison 2013). However, minimal regulatory controls resulted in a low-quality built environment, much of which was unserviced (Dodson 2016). And while the initial post-war fringe expansion was predicated on the roll-out of new transport infrastructure and services, the 1960s saw a reorientation of infrastructure provision towards automobility and freeway construction which saw the 'in-fill' of spaces between rail infrastructure and rapid residential expansion beyond the rail system (Burke 2016; Dodson 2016). Although this was also the period when the first coherent metropolitan strategic planning documents were released, the capacity to manage and service new development at the local scale was severely constrained (Forster 2006).

Mirroring international trends, fringe development was 'housing-led not infrastructure-led' (Cochrane et al. 2015: 574), and the provision of important infrastructure (sewerage, roads and parks) was often delayed (Morison 2000). While this backlog was progressively addressed by state and federal government initiatives, the mode of suburban development shifted in the late 1970s and early 1980s as land developers were progressively required to provide infrastructure directly or pay development contributions to state or local governments (Dodson 2016). From the 1970s, the continued fringe expansion by private land developers led to a sparse and disjointed urban landscape, where there was little integration or connection between land releases. Developers provided the necessary infrastructure within their estates, but paid little attention to the broader urban fabric. By the 1980s and 1990s, the growing spatial mismatch between housing and employment locations was a major concern (Dodson 2016).

Furthermore, concerns around built form and aesthetics were raised as suburbs were judged to be homogenous, devoid of character or a unique sense of place (Gleeson and Low 2000). Suburban homes have been variously labelled the 'great Australian dream', the 'great Australian ugliness', and almost everything in-between (Boyd 2010; May 2009) and Australian suburbs popularly characterised as bland and uninspiring: the antithesis of the exciting, innovative and culturally superior inner city (May 2009; Dowling and Mee 2000). Yet, even as media, academic and policy critique increased from the 1970s, the demand for detached dwellings with a yard and garden on the urban fringe remains strong (Burton 2015). In defence of suburbs, authors such as Davison (2013), Gibson (2012), Mee (2002) and Troy (1996) have championed their virtues as sites of opportunity, community, and a rich cultural diversity.

Contemporary fringe development represents something of a departure. The push for infrastructure, increased density, a mixture of dwelling types, affordable housing provision and a more diverse demand base has led to the reconfiguration of suburban form and function. This has led Dodson (2016: 10) to argue that "Australian suburbia now finds itself in a place of ambiguity". It is no longer the promised land of the quarter acre lot with a detached dwelling in owner occupation (or at least being purchased), but is increasingly positioned in public discourse and characterised in urban policy by "social, cultural and political complexity, the interwoven contradictions and combinations of opportunity and vulnerability, the environmental costs and the infrastructure needs" (Dodson 2016: 10). This ambiguity may have objective dimensions but rather than re-enforcing the division it implies between the outer- and the inner- and middle-ring suburbs, we suggest it offers an opportunity to reconceptualise fringe development as a form of urban regeneration that connects to the planning and design logics of urban consolidation and also contributes to the ongoing regeneration of the city as a whole. We suggest that the suburban promise itself may have changed, as we show in the case of Huntlee, to combine some of the key elements of a more traditional suburbia (e.g. home ownership, private space, access to a garden) with benefits (such as improved access to services and recreational opportunities) thought to accrue through the regeneration of inner and middle-ring urban locations. In the next section we turn to the growing literature on the post-suburb, New Urbanism and master planning.

This literature comes closest to reimagining fringe development as part of a broader regenerative process of urban change.

The post-suburb and urban regeneration on the fringe?

The ideal of the post-suburb – densification of new fringe developments compared to traditional suburban subdivisions (Charmers and Keil 2015; Touati-Morel 2015) – draws on debate around the role and value of suburbs in North America. Cochrane et al. (2015: 571), discussing fringe development in London, similarly characterise the post-suburb as an 'archipelago of settlements rather than a continuous urban space', with each island of settlement centred on a town centre. For Keil (2015: 582) post-suburban cities can be characterised as a:

> new dispersed, multi-focus, discontinuous, variably dense, and multi-centred city, not centred towards the traditional core.

Increased densification on the fringe is delivered through a shift in the type of dwellings constructed and a decrease in the land size allocated to detached dwellings (Burton 2015). In Australia, this form of suburban development has evolved alongside strategic planning attempts to promote the virtues of consolidated built form around designated centres distributed across the city and, increasingly, on the fringe (Bunker 2012).

The 'new' form of post-suburban development draws heavily from the design logic of New Urbanism (albeit often unacknowledged). Keil (2015: 583) quips that New Urbanism has 'carried the torch' of densification in outer metropolitan regions, while Foster et al. (2015: 150) argue it:

> advocates for the design of the compact, pedestrian-friendly, mixed-use developments thought to promote walking, minimise car dependence and enhance sense of community.
>
> (Foster et al. 2015)

In the Australian context, the logic for fringe urban consolidation mirrors that of inner and middle-ring urban areas, promoting the benefits for community ties, access to transport, employment, education and other services (McDougall and Maharaj 2011). Implementation of development based on New Urbanist logic has taken shape via the increasingly popular 'master planned' estate (Johnson 2010). Consolidation and town centre development on the fringe differs from the deployment of these forms of regeneration in other parts of the city because they are not required to deal with the legacy of previous urban development (although, as we show in this chapter, they may have to deal with other legacies). These post-suburban developments differ equally from traditional suburbs in that ideas of place uniqueness and a sense of community (both supposedly absent from traditional suburbs) are centred in their planning, development and marketing (Cheshire et al. 2013). In the next section we explore new fringe suburban development at Huntlee. In doing so we trace the connections to and parallels with inner and middle-ring regeneration to that occurring via new suburban development on the fringe.

Huntlee, New South Wales: moorings and departures in regeneration

Huntlee's 1,700 hectare site spans the borders of two local-government areas, City of Cessnock and the Singleton Council in the Lower Hunter Valley, approximately 50 km from Newcastle, the second-largest city in NSW with a population of approximately 552,000 (Figure 17.1). First proposed in 2006, Huntlee is planned to be the biggest single housing development in NSW. Approval for the project was granted in 2009 by then NSW minister for planning Kristina Keneally, but was put on hold amidst a series of community protests and legal challenges which cited concerns about adverse environmental impacts, potential negative impacts on nearby commercial districts in Cessnock and Singleton, and political corruption of the decision-making process (Walsh 2009; Nichols 2013; Page 2013). In 2010 the project was deemed 'state significant' and assessed under Part 3A of the NSW Planning and Assessment Act, with developers, LWP Property Group, receiving final

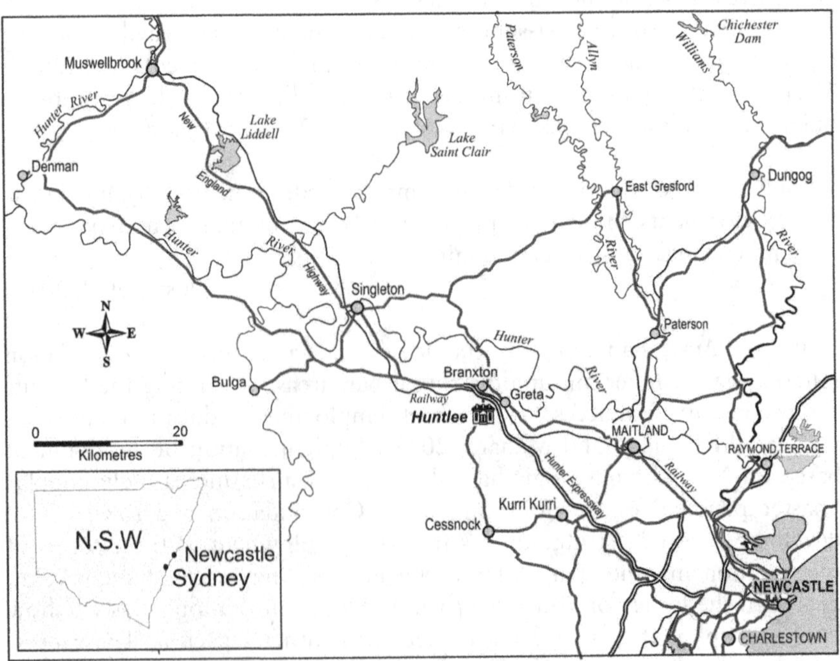

Figure 17.1 Huntlee, NSW (source: Olivier Rey-Lescure)

approval for their amended plans in April 2013. LWP Property Group is a 'land developer specialising in medium- to large-scale urban projects. LWP's work encompasses residential, commercial, retail, industrial and community developments nationwide' (LWP 2016). Development at Huntlee commenced the following February. The site will provide up to 7,500 dwellings for 20,000 people, and spaces for commercial properties, community facilities and conservation land. In addition to a designated town centre, Huntlee will consist of four distinct villages/land releases (Figure 17.2).

The Huntlee site has a complex history. It forms part of the traditional lands of the Indigenous Wonnarua Nation, yet prior to its acquisition for residential development, the site was owned by a coal company (Kirkwood 2013), who used the land for mine works and coal preparation. As a result, parts of the land are heavily undermined and contaminated by coal waste. LWP have used the site's contamination to promote the necessity of the development, with managing director Danny Murphy stating, 'It won't be fixed unless we come in and fix it' (Page 2013).

Public discussion, promotion and marketing of Huntlee regularly emphasised the connection between Newcastle's city-centre urban regeneration and the fringe. For example, Stephen Leathley from the Planning Institute of NSW argued:

> with Newcastle's rejuvenation well underway, the focus is now shifting to areas such as ... Huntlee, a new town, I think it's going to be very exciting to see what happens there.
>
> (ABC News 2015a)

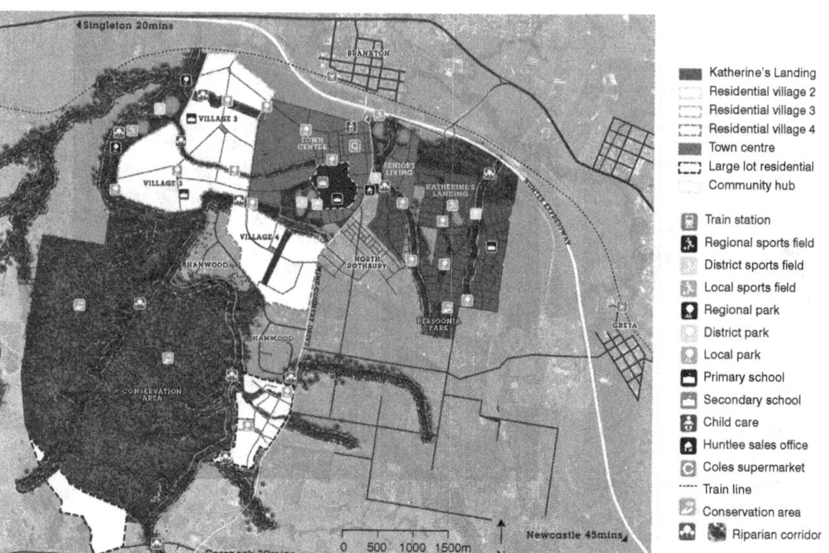

Figure 17.2 Huntlee master plan (source: LWP Property Group)

Similarly Cessnock local mayor Bob Pynsent argued:

> It's going to be a real trailblazer in terms of urban renewal.
>
> (Page 2013)

In many ways Huntlee already reflects the multi-centre city. Marketing material cites its location in terms of proximity to the nearby town of Branxton, as well as the Hunter Valley centres of Cessnock, Maitland, Singleton and Muswellbrook, and the urban centres of Newcastle and even Sydney which is 184 km away (Figure 17.1). Rather than being understood as being on the fringe of a single urban centre, Huntlee is imagined as part of a complex integrated constellation of (sub)urban centres.

At the time of writing,[2] Huntlee is in the early stages of construction with a planned final completion date of 2035. The first residents moved into the development in late March 2016, and the display village opened the following June. We focus, therefore, on Huntlee's planning and early development stages. Our purpose here is to chart Huntlee's demonstration of emergent fringe development tendencies that mobilise similar policy and ideological tendencies to urban regeneration elsewhere and which act to regenerate Australian cities in a way different from the traditional form of suburban development.

Regenerating the city: innovations on the edge

The enactment of New Urbanism on the fringe offers a series of innovative opportunities for regenerating the form and function Australia's future

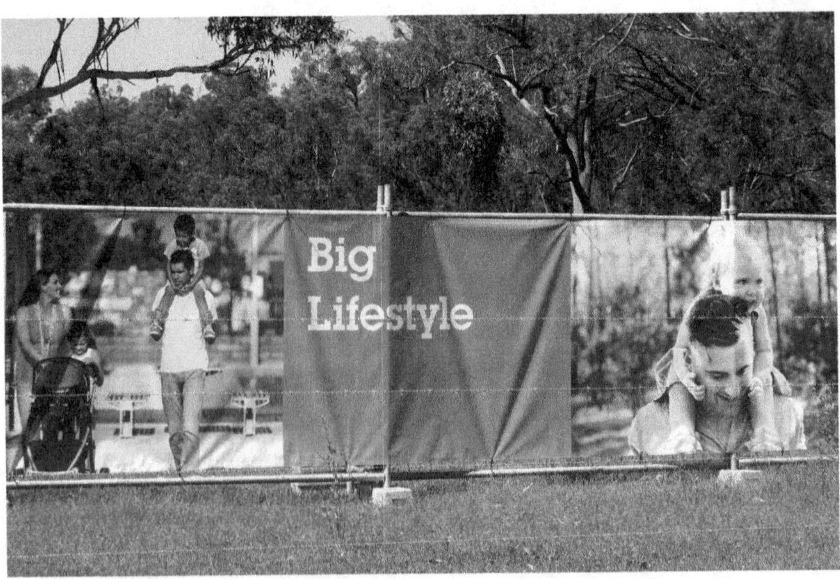

Figure 17.3 Advertising at Huntlee (source: authors)

cities. Drawing on Huntlee, four arenas of innovation emerge: (i) nurturing community, (ii) housing and built form, (iii) infrastructure and service provision, and, (iv) urban sustainability.

Nurturing community

> Welcome to your new town in the heart of the Hunter. Like never before, Huntlee is opening up to reward residents with an unprecedented location, rich with community spirit and outstanding facilities. This is a vibrant, new and uplifting town with everything you'll need to call it home, forever.
>
> (LWP, n.d.a.)

Contemporary fringe developments like Huntlee draw heavily on New Urbanism planning and design logics which, like the Garden City and City Beautiful movements before it, advocate a built form that promotes a stronger 'sense of community' and civic engagement achieved via a diversity of built form, a walkable scale, the clear demarcation of the 'centre', and limiting the dominance of private motor vehicles (Sander 2002). Huntlee's New Urbanist design principles are championed by the developers. As one project manager espoused on a Huntlee website blog post:

> New urbanism is an urban planning philosophy ... that it is 'people' who make places great ... That's what we believe at LWP ... Developments that fail, fail because they focus on chopping up land as quickly and cheaply as possible, rather than building the social and cultural layers that make a thriving community.
>
> (Thompson 2014)

Similarly the Huntlee project manager argued:

> We want people to think of Huntlee as a community ... It's not just 'bought a block at Huntlee just because it was cheap'. We really want to get that community vibe happening.
>
> (Interview, Huntlee project manager)

LWP have been attempting to nurture the development of community through their consultation strategy which has included engaging with the Indigenous community through the Huntlee Aboriginal Reference Group (HARG) and working on issues of cultural heritage protection and opportunities for Indigenous training and employment (Wonnarua Nation Aboriginal Corporation 2015a; 2015b). Huntlee's community vision document foresees annual community forums, a community reference group, regular community newsletters and an online forum. Thus far, LWP have sought community input about public spaces and facilities, including a competition to design public open spaces. In discussing the competition, the Huntlee development manager argued:

> (We) wanted to give the public the opportunity to contribute to the development of the town ... Huntlee will be a town designed around its residents and, right from the start, we want those residents to be a part of how the town will evolve.
>
> (*Cessnock Advertiser* 2014)

Design competitions and community input have been regeneration planning staples for many years (Atkinson 1999; Foley and Martin 2000; Maginn 2007). Also mirroring tactics employed in inner urban regeneration in Newcastle, Huntlee developers are deploying social media techniques (albeit in limited and controlled ways) to connect new and potential residents with the developer, and nurture community development. LWP's blog enables company members to answer questions and reflect upon aspects of estate development. The Huntlee Facebook site provides a venue to post developer news, while the My Huntlee Facebook site claims to be 'the place to be for all members of the Huntlee Community!', where members are encouraged to 'post on this page to communicate with your neighbours and LWP.'[3]

In seeking to create a new community in Huntlee, LWP have actively sought to mobilise a vision of community in Huntlee. Of course, this is not to suggest that community was absent from earlier forms of suburbia (indeed much research shows the opposite), but rather that the 'creation of a community' is vital to experiencing (and selling) the new suburbia. Certainly LWP promotion claims to be building community into Huntlee. Just as residents buy into the 'feel' of a community or location in inner and middle-ring suburbs, this is increasingly positioned as an essential element of the 'lifestyle' appeal of new fringe communities, even for initial residents with very few neighbours. Consultation and early community initiatives seek to position new development as different from and superior to 'traditional' suburbia.

Importantly, the lived experiences of Huntlee residents might be very different to the community ideal promoted through planning and marketing material. Moreover, involvement via planning processes and online forums does not, in itself, ensure that community has any real say in the planning process. Rather, involving purchasers (and potential purchasers) in the process at the early stage forms part of a marketing/development toolkit which seeks to 'pre-form' community association where residents might feel as though they have connections to the development and other residents upon arrival. This consultation seeks to create a 'history' for residents which is often absent from new build suburbs, but is often an important component of housing market choice in inner and middle-ring l ocations.

Housing and built form

One of the major critiques of Australian suburban development is the lack of variation in built form (Gleeson 2006). Contemporary fringe town centre developments have sought to respond to this critique via the provision of a diversity of lot sizes and dwelling types. In Huntlee, developers argued the project represented something new for fringe development:

> On paper you put a mix of lots down but we're trying to change market perceptions on what size lots people want to live on, especially out this way on the fringe.
>
> (Interview, Huntlee project manager)

Fringe development is often positioned as an affordable alternative (at least in terms of housing costs) compared to middle- and inner-city housing (Burton 2015; Mee 2002; MacKillop 2013). Unsurprisingly, housing affordability and diversity were central to the justification of Huntlee as it played out via various planning processes and legal challenges (Brunton 2011). However, even in fringe locations, the cost of housing challenges many prospective home purchasers, especially first-home buyers (Burton 2015). In Huntlee, smaller lots and dwellings are positioned as an affordable housing option:

> Most people still can't really afford [800 square metre blocks] to be honest so we're trying to provide a mix. We've still got that product but we're trying to also provide a mix of the 300 to 400 square metres.
>
> (Interview, Huntlee project manager)

Importantly, however, the town centre model also seeks to create the amenity, infrastructure and services of inner-city locations, so often seen to be missing on the fringe. In justifying this model, the Huntlee project manager repeats much of the planning logic embedded in current inner- and middle-ring regeneration projects, advocating proximity to services and open space as essential design elements:

> The parks are within walking distances of those smaller lots so that people don't think they need a big backyard because there's parks and recreation areas that provide that requirement for their family.
>
> (Interview, Huntlee project manager)

Diversity in lots sizes and housing design (ironically within the context of strict developer-defined design guidelines) is also seen to respond to lifecycle changes: small lots for first-time buyers (and investors) looking to enter the housing market; larger lots that these households might aspire to move to in the future; and smaller lots as options for households looking to downsize:

> We're going to have everything here so you can move five streets away into a bigger house in four years when you can afford it. Vice versa, you can move back downwards if you're here for 20 years and want a smaller house.
>
> (Interview, Huntlee project manager)

In supporting the development, the mayor of Cessnock, Bob Pynsent, argued:

> Being able to move across one community as you age and not lose your friends is completely different and exciting.
>
> (Page 2013)

As LWP has found in its other developments (e.g. Ellenbrook), for projects with a long timeframe, appealing to multiple housing markets presents a business model that allows multiple sales to single households as new stages are released. The model reflects regeneration pressures operating across Australian cities and is one feasible response to lifecycle pressure for knock-down-rebuild redevelopments as households have the opportunity to stay in their location and get a new home.

As unashamedly New Urbanist, the Huntlee developers claim the diversity of lots/dwellings provides for a mixture of households akin to more established neighbourhoods:

> Because of the lot and product mix we'll have the full spectrum of purchasers so your first home owners, your downsizers, big families, everyone really.
>
> (Interview, Huntlee project manager)

Of course delivering this diversity relies on the capacity and willingness of the relevant planning authority to negotiate planning conditions (as well as on market demand). In Huntlee, local councils (especially Cessnock) have supported the project and recognise the development as central to meeting local housing demand as well as increasing the quantity and quality of local services and infrastructure. Whether in the inner-city, a middle-ring suburb or on the suburban fringe, planning and consent authorities' responses to housing propositions are fundamental to the way regeneration unfolds across the city.

Infrastructure and services

The provision of transport infrastructure was a major objective of Huntlee's developers. The Huntlee site sits adjacent to both the Hunter Valley heavy rail corridor and the Hunter Expressway, completed in March 2014, and access to both are acknowledged by Huntlee developers and local council staff as vital for the future viability of the site. But while Huntlee appears to align with the integrated form of fringe development advocated in strategic planning documents across Australia (Chapter 2) – due to its location close to a rail line – it is likely that transport will remain car-based at least in the mid-term. Huntlee has no train station and coal trains dominate the rail line. Passenger trains to Branxton station (a 20 minute walk/4 minute drive away) stop too infrequently to suit working commuting needs. Likewise, bus services are currently limited. Nevertheless, the developers claim that population growth will catalyse improved transport services and infrastructure (such as a train station) and have proposed that Branxton railway station be developed into an integrated transport interchange. However, ensuring delivery of such interchanges has been a major challenge for other fringe town centres (Cook and Ruming 2008; Ruming 2005).

In response to concerns around transport infrastructure and the promotion of car-based transport, the Huntlee development promotes a more inward-looking urban future. Huntlee is indicative of fringe residential developments across Australia that claim that the bulk of services and infrastructure will be provided onsite:

> It's a new town so obviously we're trying to contain employment and those sorts of trips. If we were just building 7,000 lots without any employment, any retail, any inward destinations then I guess I would have to accept that argument that to do anything everyone has to leave Huntlee, but that's not going to be the case.
>
> (Interview, Huntlee project manager)

It is claimed that Huntlee will provide approximately 3,000 jobs. Other contributions (totalling $39 million) to services on the development include Stage 1 of the Huntlee Community Hub (a combination of library and youth centre), a community hall, a child-care centre, recreational facilities and parks, as well as contribution to regional infrastructure such as the Branxton Aquatic Centre and regional sporting fields (LWP n.d.b.). Innovation, therefore, emerges from the capacity of large greenfield developers to negotiate the onsite provision of services and infrastructure early in the development processes. Table 17.1 outlines expected facilities by project completion. In addition to these services, LWP are also responsible for providing land for community facilities: according to the voluntary planning agreement with Cessnock Council, LWP must provide Education Contribution Land appropriately serviced with sealed road access, power, water and telephone. Huntlee's Development Control Plan (NSWDPI 2013) states that land should be provided for a community health centre, emergency services, government primary schools, a government high school, and tertiary (vocational) education. Access to these facilities for residents is an important part of the development's marketing strategy. However, they are equally a wish list included in planning documents, the delivery of which rests upon negotiation with and commitments from a series of state and non-state agencies and organisations (Cook and Ruming 2008; Ruming 2005). The failure to secure future commitments from other agencies/

Table 17.1 Huntlee community facilities

Facility	*Number*
Primary schools (public and private)	5–6
Secondary schools (public and private)	2–3
Pre schools	6
Library	1
TAFE (tertiary and further education)	1
Aged care (nursing home)	1–2
Youth centres	2
Neighbourhood centres	4

Source: LWP (n.d.a.), Huntlee project brochure

organisations represents a potentially major obstacle to delivering Huntlee's town centre vision.

Claims that services and employment will be provided in the early stages of the development are strengthened by the fact that a major shopping retailer (Coles) has agreed to locate in the first stage of the town centre. Equally the supermarket played a major role in project marketing:

> Generally they would wait until there's a critical mass and see what's going on. Normally the shops are almost the last thing to go in or 10 years into a project. For us that's a massive coup to get them signed up and basically try and get them in as soon as possible.
>
> (Interview, Huntlee project manager)

The town centre is positioned as something that draws residents together, though this is dependent on whether it is possible to deliver the planned services, jobs and infrastructure to Huntlee. The extent to which this will be a success depends on the ability of Huntlee to disrupt existing suburban planning and development norms.

Urban sustainability

Fringe developments have long been critiqued as an unsustainable extension of Australian cities which promote private motor vehicle usage, high electricity consumption and the destruction of natural habitats (Burton 2015; Dodson 2010). One of the justifications for an increased policy push for urban consolidation has been the claim that a more dense urban form is a more sustainable one (Newman 2014). Advances in technology and negotiated planning schemes offer other opportunities for fringe development to contribute to sustainability advances through environmental innovation. Plans for Huntlee include three such forms of innovation around electricity, water and environmental conservation.

Notwithstanding regulatory challenges around off-grid energy supply, Huntlee is proposed to become 'Australia's first off-grid town' (ABC News 2015b) via onsite generation. In late 2015 a study into renewable energy options was jointly funded by the developer and the Australian Renewable Energy Agency. The study examined a combination of solar, gas and geothermal energy to power dwellings on the development. When news of the study broke, it was ongoing but, preliminary results have purportedly been promising:

> We've progressed far enough already that we think there is a financial and technological case for it ... If there aren't any regulatory barriers, by the end of this year Huntlee could become Australia's first new town near existing infrastructure that relies entirely on renewable energy.
>
> (Kane 2016)

The Huntlee developer has described Huntlee as a 'blank canvas' and 'an example which can be applied across all greenfield projects throughout Australia' (ABC News 2015b). Huntlee would work as a 'micro-grid' model with energy generated and stored locally (via new battery technology) and could potentially be 'community owned' (Kane 2016).

With respect to water, while potable water would still be drawn from the central Hunter water network, it is claimed that up to 70 per cent of daily water needs could be generated onsite via a membrane-bioreactor recycled-water plan, rainwater harvesting and air-to-water generators (Kane 2016):

> A refined water network purifies rainwater, storm water and wastewater to the highest Australian standards at Huntlee Water Centre, and it will be used for washing machines, toilets, irrigation, air conditioning and community amenity.
>
> (LWP n.d.c.)

Importantly, it is expected that the cost of onsite electricity and water generation and storage would be included within the land purchase price (Kane 2016). For the developer, this is important on a number of fronts. First, it adds an additional facet to the marketing strategy for the estate, although the developers acknowledge that green infrastructure and sustainability do not appeal to all purchasers:

> If you look at your whole market to some people sustainability is an important consideration ... but it might be 15 per cent.
>
> (Interview, Huntlee project manager)

Second, this allows the developer to further its claim about housing affordability. Here the primary argument relates to the fact that dwellings will not require the onsite water tanks mandated for new dwellings in NSW planning legislation (BASIX), a particularly important consideration for smaller (more affordable) lots:

> Using Huntlee Water means that you do NOT have to install a rainwater tank at your new property, which is a considerable saving at a time when you want to put the finishing touches to your new home. Plus, ongoing you will be paying less for your water requirements.
>
> (LWP n.d.c.)

In addition, environmental design at Huntlee includes water-sensitive urban design and innovations such as a carbon-neutral target for the development. Thus, Huntlee's environmental and sustainability measures suggest the potential for fringe suburban development to experiment with sustainability technologies and provide new suburban responses to resource depletion and climate change. While similar responses are evident in inner and middle-ring regeneration initiatives, greenfield fringe developments are unfettered by the constraints of retrofitting existing built environments.

One of the consistent criticisms of the unsustainability and environmentally damaging impact of greenfield development, arises from its broadacre land clearing to provide space for new homes and associated infrastructure (Dodson 2016). That fringe development removes vegetation and species habitat cannot be denied. However, it also has the potential to play a productive role in the conservation of endangered species. Greenfield developments commonly encounter endangered flora and fauna species and, indeed, these encounters have regularly reshaped development configurations on the urban fringe (Ruming 2009). In the case of Huntlee, the endangered flowering shrub the North Rothbury persoonia (*Persoonia pauciflora*) – identified *as* critically endangered under both the Commonwealth Environment Protection and Biodiversity Act 1995 and the NSW Threatened Species Conservation Act 1995 – has influenced project implementation. The species is only found in the North Rothbury area, surrounding Huntlee. Twenty-two of the 354 recorded plants are located within Huntlee's borders, and habitat loss due to residential land clearing constitutes a recognised threat to its survival.[4] The presence of persoonia was mobilised by a local community group, the Sweetwater Action Group as one of the reasons for opposing Huntlee. In response to persoonia, local opposition and a request from NSW State government, LWP increased the amount of dedicated conservation land, including a new 'Persoonia Park' on land originally designated for development:

> We've dedicated thousands of hectares of land for conservation purposes ... Our offset ratio is almost 10 to one which is the high end of the spectrum. That, I'd say, is a lot higher than what most other developers around here would have to provide.
>
> (Interview, Huntlee project manager)

> We also have Persoonia Park that we have set aside to keep some of those persoonias safe in perpetuity.
>
> (Project director in ABC News, 2015c)

LWP argue that the scale of Huntlee's development makes the protection of threatened species possible due to the halting of activities that threaten species, funds supplied by the developer and their capacity to leverage action from other organisations (including government and residents). As the Huntlee project manager stated:

> The development provides funds for those activities which may not have happened if farmers were still here grazing, cutting down trees. Who knows in another 30 years if there'd be any left ... If you drive around most of our areas, people that were here before have already really thinned out the vegetation. They've had cattle grazing on it which really, the cattle grazing is what stops these plants from growing.
>
> (Interview, Huntlee project manager)

It is impossible to test these claims, as the potential management arrangements implemented under alternative land ownership will never be known. It is also important to acknowledge that the level of protection came in response to local opposition, rather than being a core element of the original development proposal. However, due to the size of the development, a higher degree of protection is now ensured, a fact now mobilised by the developer as a positive element of the development and used as part of the marketing campaign.

In addition to setting aside conservation lands, the Huntlee development has appointed a Huntlee conservation manager, prepared a Huntlee Conservation Management Plan (RPS 2014) and provided funds to the Office of Environment and Heritage to assist in the implementation of the National Recovery Plan North Rothbury Persoonia. Additional innovations initiated by LWP include a partnership with the Royal Botanical Gardens to establish a translocation program for persoonia. Moreover LWP have attempted to engage community involvement and connections via the My Huntlee Facebook page which encouraged participation in the North Rothbury persoonia planting day in May 2016, where residents (and future residents) would work with the Office of Environment and Heritage to plant 100 persoonia cultivated by the Botanical Gardens. Residents were also able to take plants to plant in their gardens or in nearby bushland. Whether persoonia protection is enhanced by Huntlee's development or not, the resources and land dedicated to its protection is considerable and outweighs that likely to have been available had the site remained in multiple ownership and used for agricultural purposes. The history of unsustainable suburban development, and the reaction against it has, over time, resulted in a regulatory framework for large-scale fringe development aimed to ensure more environmentally sensitive development is a higher priority and to enable innovations to achieve this.

The nascent development of Huntlee points to the potential for large-scale and somewhat innovative fringe developments that contribute to regenerating the city in less unsustainable ways. Huntlee's building in new ways of generating electricity, processing water and protecting threatened species at least plans to instantiate more sustainable approaches to urban resources. And while Cessnock Council has largely been supportive of the development, some Cessnock councillors have been highly critical of Huntlee. When Huntlee was officially opened, councillor James Ryan (The Greens) stated:

> Huntlee would put thousands of people in an environmentally sensitive area with no local jobs and few facilities, making everyone reliant on the motor vehicle.
>
> (Kirkwood 2014)

Ryan's criticisms connect developments such as Huntlee to continued concerns about the environmental and social impact of suburban development on the fringe. While parallels remain between Huntlee and the environment pitfalls of past suburban developments, particularly in terms of car dependency, innovation around accounting for environmental impacts through the project's design, build and resource provisioning, and the role imagined for residents in environmental protection, is somewhat encouraging. It suggests the potential for greenfield fringe developments to contribute to more sustainable urban regeneration that mirrors that of inner-urban brownfield regeneration (see Chapter 10) and distinguishes itself from prior patterns of suburban development.

Conclusion

As Australia's population has grown over the last decade or so, much of this growth has taken place in outer suburban areas across the main metropolitan regions (Gordon et al. 2015). In some respects this growth runs counter to longstanding strategic planning objectives that have sought to contain suburban growth. Notably, however, any conceptualisation of urban regeneration needs to acknowledge that fringe development plays a central role in the evolution of cities. Three decades from now (the planning horizon for most of our strategic planning documents, Chapter 2), cities' inner- and middle-rings will look vastly different, but so will the outer suburbs. Future fringe developments are likely to be inherently post-suburban and polycentric in nature (Keil 2015). More specifically, new fringe development will depart from past processes which resulted in a built form that was widely critiqued in terms of aesthetics, cultural and social isolation, equity and social justice, car dependence and unsustainability (Dodson 2016; Burton 2015; Gleeson

2006). Contemporary and future developments posit different promises and challenges. The Lower Hunter's Huntlee development offers insights into emergent trends that maintain some of the perceived benefits of traditional suburbia (e.g. home ownership and single detached family homes), while simultaneously departing from this model and advancing additional forms of development that are essentially the product of urban regeneration processes typically associated with inner and middle-ring areas. To these ends, new fringe suburban development offers a series of potential innovations, much like large brownfield regeneration in the inner and middle-ring.

New Urbanism has been critiqued over decades as a utopian, exclusionary and bland planning ideal, representing the last great expression of modernist planning principles (Grant 2006). Yet on the city fringe, New Urbanist ideals, logics and justifications are mobilised as a means to recreate the qualities of inner and middle-ring regeneration sites. Gleeson (2006) advocated for a form of 'new urbanist' fringe development in Australia to promote: openness and diversity by opposing exclusionary planning principles and maintaining the public domain; larger release scales of the town centre-focused variety aimed to promote social inclusion; and greater opportunity for democratic design. Huntlee attempts to address some of these issues and offers innovative ideas and lessons on four key fronts.

First, new fringe town centre developments offer opportunities for early and innovative *community consultation and participation*. This consultation potentially facilitates 'buy-in' from residents, fostering a level of community connection which might otherwise take an extended period to develop. This is a distinct point of departure from traditional suburban development and an effort to promote a development more akin to inner and middle-ring regeneration characterised by an in situ community. At Huntlee there have been some attempts to foster social inclusion via community consultation. However, it must be recognised that orchestrated efforts at community engagement may delimit space for alternative forms of community expression and activity.

Second, compared to traditional suburban development, new fringe town centre developments promote a more *diverse housing and built form*. Conforming to New Urbanist ideals, the diversity of built form is to: provide affordable housing options; promote community interaction; prompt active transport and use of public open space; and, drive design innovation. This diversity is a response to wider urban regeneration processes according across Australia. It marks new fringe developments as different to the historical norm of suburban development and appeals more to the diverse urban form more typical of inner and middle-ring locations.

Third, the up-front provision of infrastructure and services represents an opportunity for *innovative fringe regeneration*. One of the appealing aspects of inner and middle-ring locations, especially as sites of regeneration or designated strategic centres, is the presence of existing infrastructure and services. Likewise, a persistent critique of traditional suburban development has been the relative absence of similar infrastructure. Thus, contemporary town centre development on the urban fringe seeks to position itself as well serviced from the outset. Achieving this has represented a challenge for fringe town centre developments. Nevertheless, the size of new fringe developments offers an opportunity to negotiate commitments for new infrastructure and service delivery that have been lacking in traditional suburban developments. In Huntlee, the commitment of a large shopping centre to the early stage of the development is one example.

Finally, and perhaps most innovatively, new fringe town centres offer opportunities for *advances in the environmental sustainability*. On one hand, large fringe developments offer opportunities for a more coordinated and strategic assessment of endangered species and the implementation of innovative management responses. On the other hand, fringe development offers important testing grounds for new technologies which have the potential to improve environmental sustainability of urban growth. While much attention goes to new technologies being implemented in new large brownfield developments, such as Central Park in Sydney (Chapter 10), similar advances can be made in our suburbs. In Huntlee, onsite electricity generation and recycled-water systems may represent advances on the traditional form of suburban development. Indeed, given that fringe development will remain an important element of urban growth in Australia, the potential for new technologies rolled out across new fringe developments to reduce the environmental impact of growth is significant.

Ultimately, of course, caution is needed in relation to the gaps that can open up between developers' rhetoric and how these 'innovations' play out in reality. Developer claims form a central part of project marketing and sales strategies, yet the capacity to challenge these claims in the early stages of a development is limited. A limitation of this research is our focus on material collected, and processes observed during the planning and early stages of development. As Gleeson (2012) notes, there is an important distinction between idealised and material urban realities, a disjuncture between the New Urbanist suburb and the actually existing constraints and realities which frame development. The utopian, New Urbanist development model promoted in Huntlee (and other new town centre developments on the fringes of metropolitan Australia) is likely to be tested and reconfigured as residents move in over the next two decades. Only time will tell whether Huntlee will live up to the developer's vision and marketing.

Notes

1 Research funded by an Australian Research Council Discovery Grant (DP130100582). This chapter draws on interviews with six key stakeholders.
2 Early 2017.
3 My Huntlee Facebook page accessed 2 August 2016.
4 http://www.environment.nsw.gov.au/threatenedspeciesapp/profile.aspx?id=10599

References

ABC News (2015a) 'Hunter Expressway opening up development opportunities', *ABC News*, 6 February, http://www.abc.net.au/news/2015-02-06/hunter-expressway-opening-up-development-opportunities/6074616 (accessed 22 August 2016).

ABC News (2015b) 'Huntlee to become Australia's first off-grid town', *ABC News*, 7 November, http://www.abc.net.au/news/2015-11-07/huntlee-to-become-australia27s-first-off-grid-town/6921088, (accessed 22 August 2016).

ABC News (2015c) 'Huntlee works with Royal Botanic Gardens to save endangered plant', *ABC News*, 14 July, http://www.abc.net.au/news/2015-07-14/huntlee-works-with-royal-botanic-gardens-to-save-endangered-pla/6617240, (accessed 22 August 2016).

Anderson, Kay (2006) 'Introduction: After sprawl: Post-suburban Sydney', in K. Anderson, R. Dobson, F. Allon and B. Neilson (eds), *Post-Suburban Sydney: The City in Transformation*, Conference Proceedings, 22–22 November, 2005, Parramatta: University of Western Sydney.

Atkinson, R. (1999) 'Discourses of partnership and empowerment in contemporary British urban regeneration', *Urban Studies*, 36(1): 59–72.

Beattie, L., and Haarhoff, E. (2014) 'Delivering quality urban consolidation on the urban fringe: A case study of University Hill, Melbourne, Australia'. *Journal of Urban Regeneration & Renewal*, 7(4): 329–342.

Boyd, R. (ed) (2010) *The Great Australian Ugliness*, Melbourne: Text Publishing.

Brunton, N. (2011) 'Minister for Planning again thwarted: Sweetwater Action Group Inc v Minister for Planning [2011] NSWLEC 106', *National Environmental Law Review*, 3: 29–33.

Bunker, R. (2012) 'Reviewing the path dependency in Australian metropolitan planning', *Urban Policy and Research*, 30(4): 443–452.

Burke, M. (2016) 'Problems and prospects for public transport in Australian cities', *Built Environment*, 42(1): 37–54.

Burton, P. (2015) 'The Australian good life: The fraying of a suburban template', *Built Environment*, 41(4): 504–518.

Buxton, M. and Scheurer, J. (2007) 'Density and outer urban development in Melbourne', *Urban Policy and Research*, 25(1): 91–112.

Cessnock Advertiser (2014) 'Ideas shine bright in Huntlee design competition', 25 June, http://www.cessnockadvertiser.com.au/story/2375184/ideas-shine-bright-in-huntlee-park-design-competition/ (accessed 22 August 2016).

Charmers, E., and Keil, R. (2015) 'The politics of post-suburban densification in Canada and France'. *International Journal of Urban and Regional Research*, 39(3): 581–602.

Cheshire, L., Wickes, R. and White, G. (2013) 'New suburbs in the making? Locating master planned estates in a comparative analysis of suburbs in South East Queensland', *Urban Policy and Research*, 31, 3, 281–299.

Cochrane, A., Colenutt, B. and Field, M. (2015) 'Living on the edge: Building a sub/urban region', *Built Environment*, 41, 4, 567–578.

Cook, N. and Ruming, K.J. (2008) 'On the fringe of neoliberalism: Residential development in outer suburban Sydney', *Australian Geographer*, 39(3): 211–228.

Davison, G. (2013)'The suburban idea and its enemies'. *Journal of Urban History*, 39(5): 829–847.

Dodson, J. (2010) 'In the wrong place at the wrong time? Assessing some planning, transport and housing market limits to urban consolidation policies', *Urban Policy and Research*, 28(4): 487–504.

Dodson, J. (2016) 'Suburbia in Australian urban policy', *Built Environment*, 42(1): 7–20.

Dodson, J. and Sipe, N. (2008) *Unsettling Suburbia: The New Landscape of Oil and Mortgage Vulnerability in Australian Cities*. Urban Research Program Research Paper No. 17, Brisbane: Griffith University.

Dowling, R. and Mee, K. (2000) 'Tales of the city: Western Sydney at the end of the millennium', in J. Connell (ed.) *Sydney: The Emergence of a World City*, Melbourne: Oxford University Press.

Foley, P. and Martin, S. (2000) 'A new deal for the community? Public participation in regeneration and local service delivery', *Policy and Politics*, 28(4): 479–491.

Forster, C. (2006) 'The challenge of change: Australian cities and urban planning in the new millennium', *Geographical Research*, 44(2): 173–182.

Foster, S., Hooper, P., Knuiman, M., Bull, F., and Giles-Corti, B. (2015) 'Are liveable neighbourhoods safer neighbourhoods? Testing the rhetoric on new urbanism and safety from crime in Perth, Western Australia'. *Social Science & Medicine*, 164: 150–157.

Gibson, M. (2012) 'Bildung in the 'burbs: Education for the suburban nation'. *International Journal of Cultural Studies*, 15(3): 247–257.

Gleeson, B. (2006) 'Towards a new Australian suburbanism'. *Australian Planner*, 43(1): 10–13.

Gleeson, B. (2012) 'The urban age: Paradox and prospect'. *Urban Studies*, 49(5): 931–943.

Gleeson, B. (2016) *Australian Heartlands: Making Space for Hope in the Suburbs*, Sydney: Allen and Unwin.

Gleeson, B. and Low, N. (2000), *Australian Urban Planning: New Challenges, New Agendas*, Sydney: Allen & Unwin.

Gordon, D. L. A., Maginn, P. J., Biermann, S., Sisson, A., Huston, I., and Moniruzzaman, M. (2015) *Estimating the Size of Australia's Suburban Population*. PATREC Perspectives October 2015, Perth: The Planning and Transport Research Centre, University of Western Australia.

Grant, J (2006) *Planning the Good Community: New Urbanism in Theory and Practice*, New York: Routledge.

Johnson, L. C. (2010) 'Master planned estates: Pariah or panacea?' *Urban Policy and Research*, 28(4): 375–390.

Kane, A. (2016) 'Australian town Huntlee could be first off-grid, but what about everyone else?' *The Guardian*, 3 February, https://www.theguardian.com/sustainable-business/2016/feb/03/australian-town-huntlee-could-be-first-off-grid-but-what-about-everyone-else, (accessed 22 August 2016).

Keil, R. (2015) 'Towers in the park, bungalows in the garden: Peripheral densities, metropolitan scales and political cultures of post-suburbia', *Built Environment*, 41(4): 579–596.

Kirkwood, I. (2013) 'Housing approval has subsidence concerns', *Newcastle Herald*, 29 April, http://www.theherald.com.au/story/1466047/ian-kirkwood-housing-approval-has-subsidence-concerns/ (accessed 26 October 2016).

Kirkwood, I. (2014) 'Huntlee development opened', *Newcastle Herald*, 24 February, http://www.theherald.com.au/story/2111692/huntlee-development-opened/, (accessed 22 August 2016).

LWP (2016) 'About us', https://www.lwppropertygroup.com.au/about-lwp/lwp-who-we-are (accessed October 26, 2016).
LWP (n.d.a) *Huntlee Project Brochure*, LWP: Branxton.
LWP (n.d.b) *Planning for a Community*, LWP: Branxton.
LWP (n.d.c.) *Huntlee + Water*, LWP: Branxton.
MacKillop, F. (2013) 'Sustainable as a basis of affordable? Understanding the affordability "crisis" in Australian housing'. *Australian Planner*, 50(1): 2–12.
Maginn, P. (2007) 'Towards more effective community participation in urban regeneration: The potential of collaborative planning and applied ethnography', *Qualitative Research*, 7(1): 25–43.
May, A. (2009) 'Ideas from Australian cities: Relocating urban and suburban history'. *Australian Economic History Review*, 49(1): 70–86.
McDougall, A. and Maharaj, V. (2011) 'Closing gaps on the urban fringe of Australian capital cities: An investment worth making'. *Australian Planner*, 48(3): 131–140.
McGuirk, P. and Dowling, R. (2009) 'Neoliberal privatisation? Remapping the public and the private in Sydney's masterplanned residential estates', *Political Geography*, 28(3): 174–185.
Mee, K. (2002) 'Prosperity and the suburban dream: Quality of life and affordability in western Sydney'. *Australian Geographer*, 33(3): 337–351.
Morison, I. (2000) 'The corridor city: Planning for growth in the 1960s', in S. Hamnett, and R. Freestone (eds) *The Australian Metropolis: A Planning History*, Sydney: Allen & Unwin.
Newman, P. (2014) 'Density, the sustainability multiplier: Some myths and truths with application to Perth', *Australia Sustainability*, 6(9): 6467–6487.
Newton, P., Newman, P., Glackin, S., and Trubka, R. (2012) 'Greening the greyfields: Unlocking the redevelopment potential of the middle suburbs in Australian cities'. *International Journal of Social, Behavioral, Educational, Economic and Management Engineering*, 6(11): 2870–2889.
NSWDoP (New South Wales Department of Planning) (2005) *City of Cities: A Plan for Sydney's Future*, Sydney: NSW Department of Planning.
NSWDPE (New South Wales Department of Planning and Environment) (2014) *A Plan for Growing Sydney*, Sydney: NSW Government.
NSWDPI (New South Wales Department of Planning and Infrastructure) (2013) *Huntlee Development Control Plan 2013*, Sydney: New South Wales Department of Planning and Infrastructure.
Nichols, L. (2013) 'Huntlee approved while council waits', *Singleton Argus*, 30 April, http://www.singletonargus.com.au/story/1464961/huntlee-approved-while-council-waits/ (accessed 26 October 2016).
Page, D. (2013) 'Old mine site slated for Huntlee, video', *Newcastle Herald*, 15 November, http://www.theherald.com.au/story/1911657/old-mine-site-slated-for-huntlee-video/ (accessed 22 August 2016).
RPS (2014) *Huntlee Conservation Management Plan*, Broadmeadow: RPS.
Ruming, K.J. (2005) 'Partnership, master planning and state provision: A case study of "actually existing neoliberalism" on the Central Coast, *Geographical Research*, 43(1): 82–92.
Ruming, K.J. (2009) 'Following the actors: Mobilizing an actor–network theory methodology in geography', *Australian Geographer*, 40(4): 451–469.
Sander, T.H. (2002) 'Social capital and New Urbanism: Leading a civic horse to water?' *National Civic Review*, 91: 213–235.

Spearritt, P. and DeMarco, C. (1988) *Planning Sydney's Future*, Sydney: Allen & Unwin.

Thompson, S. (2014) 'Huntlee: A new town and what it means', LWP Blog, https://www.lwppropertygroup.com.au/about-lwp/blog/our-blog/2014/02/24/huntlee-a-new-town-and-what-it-means (accessed 13 October 2016).

Touati-Morel, A. (2015) 'Hard and soft densification policies in the Paris city-region'. *International Journal of Urban and Regional Research*, 39(3): 603–612.

Troy, P. (1996) *The Perils of Urban Consolidation*, Sydney: Federation Press.

Walsh, N. (2009) 'Decision invalid, Hunter planning strategy in limbo', *ABC News*, 19 October, http://www.abc.net.au/news/2009-10-19/decision-invalid-hunter-planning-strategy-in-limbo/1108174 (accessed 24 October 2016).

Wonnarua Nation Aboriginal Corporation (2015a) Community Newsletter July 2015, Singleton: Wonnarua Nation Aboriginal Corporation.

Wonnarua Nation Aboriginal Corporation (2015b) Community Newsletter September 2015, Singleton: Wonnarua Nation Aboriginal Corporation.

Index

Printed in the United States
by Baker & Taylor Publisher Services